Mechanics and Control of Soft-fingered Manipulation

Takahiro Inoue • Shinichi Hirai

Mechanics and Control of Soft-fingered Manipulation

Springer

Takahiro Inoue, Dr.Eng.
Department of Systems Engineering for Sports
Okayama Prefectural University
111 Kuboki, Soja
Okayama 719-1197
Japan

Shinichi Hirai, Dr.Eng.
Department of Robotics
Ritsumeikan University
Noji-higashi 1-1-1, Kusatsu
Shiga 525-8577
Japan

ISBN 978-1-84996-808-9 e-ISBN 978-1-84800-981-3

DOI 10.1007/978-1-84800-981-3

A catalogue record for this book is available from the British Library

Cover design: eStudio Calamar S.L., Girona, Spain

Printed on acid-free paper

9 8 7 6 5 4 3 2 1

springer.com

Foreword

Dexterity in manipulation differentiates the human from all other animals. Indeed, civilizations are built with hands that manipulate objects and produce artifacts. Dexterous manipulation is made possible with high degree-of-freedom fingers, effective motor control, and a high level of perception and task planning capabilities. Despite profound scientific interest and practical importance, manipulation was poorly understood until the robotics community addressed it as a general science and engineering problem. The robotics community elaborated upon two unique modeling methodologies. One is a mechanical hand as a physical model, and the other is mathematical and computational modeling. Roboticists have studied not only the human hand, but also general multi-fingered hand systems. Addressing manipulation as a generic issue, and by validating new findings with physical hand systems, robotics has established manipulation science as its own field, clearly departing from biomechanics and other relevant disciplines. Over the last 30 years, a number of significant discoveries have been made, and effective principles and techniques have been developed in the field of manipulation science and engineering. Yet, there are many challenging issues to address. This book co-authored by Takahiro Inoue and Shinichi Hirai describes their outstanding contributions to the field as well as a broad survey of manipulation science and engineering. Inspired by the human fingers, they have studied the elasticity of fingertips and found effective control strategies and system architecture. The static and dynamic analyses are rigorous and insightful. The control methodology based on non-Jacobian, distributed control is novel and effective. The mechanics and control of their soft fingers is a landmark study of dexterous manipulation. They have addressed many issues that are central to the field of robotics and beyond. I celebrate their contributions and hope that the reader will find their book a useful reference in manipulation science and engineering.

Cambridge, Massachusetts, September 2008 *H. Harry Asada*

Preface

This book elaborates on the contact mechanics and new control strategies of a two-fingered robotic hand having a pair of hemispherical soft fingertips on the distal position of each finger. The onset of this research dates back about 6 years; at that time I had matriculated at Ritsumeikan University as a first-year master's student intending to specialize in robotics and its controls. At the first meeting of the laboratory at the time, my supervisor, Prof. Hirai, told us why the traditional control schemes of multifingered robotic hands are extremely complicated, whereas human pinching movements are very simple tasks. In that meeting, he immediately demonstrated the precise control of a randomly placed object that is grasped by two finger-like LEGO blocks. We by necessity expected the soft fingertips on the toy to have a simple but important unknown mechanics that directly affected the stable and dexterous manipulation of the object.

This book starts from similar observations about the dynamic behavior of an object using a pair of soft-fingered robotic hands, implying that the formulation of a new mathematical model of soft fingertips is needed to sufficiently describe the oscillatory convergence of a grasped object, which could be found in experimental observations. In the process of modeling, we have analytically discovered the minimal energy induced by the elastic deformation of a fingertip. The change of the elastic energy during manipulation is consistently and constantly exerted to stable grasping and dexterous manipulation all the time, based on which new control methods for two-fingered pinching motions are proposed. In particular, in the latter half of the book, we focus on the design concept of the control method and its performance within soft-fingered manipulation and demonstrate in experiments that the robotic hand can be modeled on a straightforward sensor-based control without using the fingertip model.

The index finger is commonly used with the opposable thumb to create a pinching motion in order to pick up small or thin objects such as needles, pins, or pens. Knitting, crocheting, weaving, general needlework, and other such skills all require considerable effort from the index finger and thumb.

However, none of the conventional robotic hands have attained a high level of dexterity in the performance of given tasks. Considering that reason, it is completely natural to consider that the functional roles of each finger are probably distinct from each other during skilled tasks because the individual physical parameter and musculoskeletal structure of the index finger and thumb are inherently different each other. Based on this point of view, this book reveals that the control loops of both fingers can be completely separated to that of each finger and the objective of each finger movement for control can be determined independently for, say the index finger and the thumb.

We believe that this proposed control scheme for a soft-fingered hand can be implemented in other robotic systems within which one or more other closed-link structures or redundant mechanisms exist. Thus, we expect this book to be given a close and careful reading by researchers who specialize in the field of robotics and who are interested in multivariable feedback control associated with the anthropomorphic robotic hand.

This monograph is based mainly on my doctoral thesis; therefore, I would like to sincerely thank my principal adviser, Prof. Shinichi Hirai. Through 5 years of master's and doctoral courses at Ritsumeikan University, he immediately suggested a cutting-edge topic in completely unexplored territory. He consistently provided me with stimulating discussion about our topic wherever and whenever I desired. As a result, many papers had fortunately been accepted for IEEE conferences and journals.

Finally, I would like to thank my family, who encourages me every day and understands what is involved in writing a book.

Okayama, Japan, October 2008 *Takahiro Inoue*

This book focuses on grasping and manipulation performed by soft fingertips. Most of the chapters are based on the Ph.D. thesis of Dr. Inoue, the coauthor of this book. He joined my laboratory in 2002 as a graduate student. I have been interested in human dexterity in object manipulation, and at that time, I often asked myself why humans can exhibit dexterity in object manipulation despite the delay in the human brain-nerve system. Signal transmission in biological systems is slower than that in computers but current robotics cannot realize human dexterous manipulation. Furthermore, the key to human dexterity has not yet been revealed. At that time, Professor Cutkosky's group and Professor Kao's group had been investigating modeling and mechanics of soft fingertips. Professor Arimoto and his colleagues had been investigating soft-fingered grasping and manipulation extensively with mathematical formulation. Inspired by these researches, Dr. Inoue and I discussed the human grasping and manipulation and stuck upon the idea that human soft fingertips might contribute to dexterity. Dr. Inoue then began to analyze the mechanics of soft fingertips in contact and developed the *parallel-distributed model*, which is the main concept of this book. From 2002 to 2005, he studied extensively the mechanics and control of manipulation by soft fingertips. This was a very fruitful period. He was nominated two successive years as a finalist for the Best Manipulation Paper Award at the IEEE International Conference on Robotics and Automation, 2005 and 2006. In addition, he received the IEEE Kansai Section Student Paper Award in 2006.

We believe that human anatomy contributes to the outstanding human ability to manipulate objects. Our parallel-distributed model reflects the human structure of a soft fingertip and a hard fingernail. Manipulation has been one main topic in robotics research. Many excellent studies have contributed to the area of manipulation in robotics, and I hope that our research will be a valuable contribution.

The authors would like to thank Mr. Anthony Doyle, editor of Springer, for his invitation to write this book, Mr. Simon Rees, editorial assistant, for his assistance in the publication of this book, and Ms. Sorina Moosdorf, production editor of le–tex publishing services, for her editing of this book. In addition, I would like to thank my wife and sons for their cheerful encouragement.

Kusatsu, October 2008 *Shinichi Hirai*

Contents

Acronyms

CMC	Carpometacarpal
CNS	Central nervous system
CSM	Constraint stabilization method
DOFs	Degrees of freedom
DMA	Direct memory access
GIE	Generalized inertia ellipsoid
IP	Interphalangeal
LMEE	Local minimum of elastic energy
LMEEwC	LMEE with constraints
LSM	Least-squares method
MP	Metacarpophalangeal
NLSM	Nonlinear least-squares method
RP	Revolute and prismatic joints
RR	Revolute and revolute joints
TM	Trapeziometacarpal

Notation

Fingertip model

a	Radius of fingertip
E	Young's modulus of fingertip material
ν	Poisson's ratio of fingertip material
k	Spring constant of infinitesimal component of soft fingertip
K	Total stiffness of soft fingertip
c	Viscous modulus of fingertip material
d	Maximum displacement in fingertip deformation
d_{t}	Tangential displacement in fingertip deformation
d_{n}	Normal displacement in fingertip deformation
θ_{p}	Relative angle between fingertip and object surface

Statics and dynamics of soft-fingered manipulation

$\mathcal{L}$	Lagrangian
K	Kinetic energy
P	Potential energy
I	Internal energy
$M_{\mathrm{f}i}$	Mass of ith finger
$I_{\mathrm{f}i}$	Inertia of ith finger
g	Acceleration of gravity
W_{obj}	Width of grasped object
M_{obj}	Mass of grasped object
I_{obj}	Inertia of grasped object
L	Length of robotic finger
$2W_{\mathrm{B}}$	Base distance between both fingers
$2d_{\mathrm{f}}$	Thickness of robotic finger
x_{obj}	x-coordinate of object
y_{obj}	y-coordinate of object
θ_{obj}	Object orientation
$\theta_{\mathrm{f}i}$	Rotational angle of ith finger

O_i	Center of ith fingertip
C_{n}	Geometric constraint for normal direction
C_{t}	Rolling constraint for tangential direction
λ_{n}	Constraint force for normal direction
λ_{t}	Constraint force for tangential direction
$\boldsymbol{\Phi}_{\mathrm{n}}$	Constraint matrix for C_{n}
$\boldsymbol{\Phi}_{\mathrm{t}}$	Constraint matrix for C_{t}
α	Critical damping parameter for C_{n}
β	Critical damping parameter for C_{t}

Control of soft-fingered manipulation

$\boldsymbol{J}$	Jacobian matrix
K_{P}	Proportional gain
K_{D}	Differential gain
K_{I}	Integral gain
τ_{b}	Biased torque
$\boldsymbol{f}_{\mathrm{p}}$	Generalized force vector due to potential energy
$\boldsymbol{f}_{\mathrm{ext}}$	External force vector applied to target object
$\boldsymbol{u}$	Control input vector

Three-dimensional manipulation

$\boldsymbol{x}_{\mathrm{obj}}$	Position vector of object
R_{obj}	Rotation matrix of object
$\boldsymbol{q}_{\mathrm{obj}}$	Quaternion corresponding to object rotation
$d_{\mathrm{n}k}$	Normal displacement of kth fingertip
$d_{\mathrm{u}k}$	Tangential displacement of kth fingertip along $\boldsymbol{u}_k$
$d_{\mathrm{v}k}$	Tangential displacement of kth fingertip along $\boldsymbol{v}_k$
$C_{\mathrm{n}k}$	Normal constraint of kth fingertip
$\dot{C}_{\mathrm{u}k}$	Rolling velocity constraint of kth fingertip along $\boldsymbol{u}_k$
$\dot{C}_{\mathrm{v}k}$	Rolling velocity constraint of kth fingertip along $\boldsymbol{v}_k$
$\theta_{\mathrm{p}k}$	Relative angle between kth fingertip and its contacting surface
$\phi_{\mathrm{t}k}$	Rotation of kth fingertip on its contacting surface

Chapter 1
Introduction

1.1 Goal

In this chapter, we look back on the history of anthropomorphic robotic hands and robotic manipulators that evolved rapidly during the early 1970s. Such an overview is useful because it will allow us to clarify which technologies have already been explored and the theoretical and practical challenges that remain to be overcome.

Approaches to the development of anthropomorphic robotic hands can be classified broadly into two parts. The first is based on conventional robot technology that contains kinematic and dynamic problems of manipulators and theoretical investigations. The second approach is predicated on a humanlike and anatomically based approach that has emerged recently (in the late 1990s). As a result, the latest research associated with the robotic hand has spread to several scientific and engineering societies or institutes, *e.g.*, robotics, biomechanical, hand surgery, and orthopedic research. It is therefore important for researchers investigating anthropomorphic robotic hands to become familiar with multiple scientific fields.

Refined humanlike robotic hands having multiple fingers have been produced by several universities and research institutes. These robotic hands look conspicuously like the human hand structure in a mechanical sense but rarely resemble the human hand with respect to biomechanical or anatomical structure.

A great deal of theoretical research has been done on dexterous multifingered robotic hands and the development of practical applications for anthropomorphic mechanical hands. For that time, academic scientists, leading companies, and members of the general public have also come to believe that humanlike robots with hands that have a sort of *artificial intelligence* would be useful in various fields, such as healthcare, public welfare, and security systems. Such challenges are being met, in part, and it appears that specialized robots for specific tasks and services will become widely used in homes and businesses.

As mentioned above, the field of robotic hands has seen excellent progress since the 1970s.

1.2 A Brief History of Articulated Robot Hands

1.2.1 The 1970s

Numerous scientists and researchers have devoted considerable attention to the feasibility of humanlike articulated robotic hands having complex characteristics and extreme versatility. In academic institutes and in industry, numerous prototypes have been developed based on traditional anatomical observations and the biomechanical characteristics of the human hand. For example, Taylor *et al.* proposed six basic prehensile patterns [TS55], some of which were originally classified by Schlesinger [Sch19].

Based on the study by Taylor *et al.*, Skinner developed a three-fingered robotic hand adopting a triangular (120° opposed) mechanism on its base [Ski75]. At the same time, the optimal number of degrees of freedom (DOFs) for stable grasping of basic geometric object shapes was considered. These basic shapes are categorized into rectangular and triangular prisms, spheres, and cylinders. The study by Skinner enabled various experimental grasps to be achieved using only four motors and the corresponding control inputs. Skinner's report may be the first attempt to design an anthropomorphic robotic hand suitable for practical use that can achieve multiple prehensions.

From the early 1970s, Inoue [Ino71] implemented peg-in-hole tasks by means of a computer-aided redundant 7-DOF manipulator made by Mitsubishi Electric [Tak74]. To our knowledge, this was the first task-oriented study using a simple tactile mechanism composed of 28 switching contacts and is therefore the first implementation of *hybrid control*. Although the manipulator does not make use of a torque sensor, it can perform the peg-in-hole assembly task because the position error of end effectors in every steady state during the movement of the robot enables the load torque generated on the end effector to be estimated when a grasped object comes into contact with the surrounding environment. At the time, this study was cutting edge in that a hybrid control was implemented based on a computer-controlled manipulator.

Around that time, the path planning of manipulators and the well-positioning control method, in which the tip of the manipulator was free in 3D space and not constrained to any particular environment, were studied extensively. In general, it was difficult to accomplish manipulation tasks by robotic hands when the grasped object was in contact with the external environment and constraints were present, especially in assembly operations such as mating parts.

In order to overcome these difficulties, researchers at the Charles Stark Draper Laboratory presented a conceptual explanation and discussion on the amount of sensory information that mechanical systems receive from their working environment and the sophistication of control strategies that can be applied to force feedback to accomplish assembly tasks [NW73, NW75, Whi76, Whi77]. In the late 1970s, Paul introduced the ideas of *compliance* and *compliant motion* and indicated for the first time that compliance could be controlled in a Cartesian coordinate system [PS76, Pau79a, Pau79b]. He described translational and rotational manipulator motions using 4 x 4 transformation matrices by contriving methods by which to set up an arbitrary coordinate system that does not involve manipulator link frames, whether the additional coordinate system remains fixed or tracks a moving system [Pau72]. In addition, he presented a new approach whereby "pick-and-place" motions performed by manipulators on a constraint workspace could be achieved by modifying the transformation matrix effectively in accordance with dynamic or unexpected changes in the environment. Note that the robot implementation based on the new approach can no longer be successfully performed unless some type of force sensor system is mounted on the robots.

Originally, a qualitative understanding that the position control and force/torque control of manipulators should be played simultaneously and independently had already been suggested in a few reports [Ino71, Tak74]. This concept was called *bilateral control* but was later referred to in the field of robotics as *hybrid control*, which was first defined by Raibert and Craig [RC81].

In 1974, an improved model of the manipulator designed by Takase [Tak74] was able to accommodate direct control signals related to each joint torque. In addition, Takase generalized the kinematics and dynamics of a 6-DOF manipulator in an arbitrary task space to implement the *computed torque method*, which is effective for analysis when a target object is constrained along a predetermined coordinate system or when the task coordinate system is moving in an arbitrary direction [Tak77].

In 1976, an industrial robot capable of conducting precision insertion tasks was created by Takeyasu *et al.* [TGI76]. This robot, which had a flexible mechanical structure with 2D suspension springs in its end effector, could perform a given peg-in-hole task in only 3 s. The end-effector mechanism provided a sort of mechanical compliance during the inserting operation.

An energy-based analysis in connection with robotic manipulation tasks, which is induced by the elastic deformation of three elastic fingers equipped on a robotic hand, was performed by Hanafusa and Asada [HA77a, HA77b]. They proposed that the most suitable grasping position using three fingers could be determined by computing a stable prehension condition from visual information and the stable condition could be obtained by calculating one or multiple points that satisfy the local minimum of the sum of the elastic and gravitational energies. Their studies were a significant contribution in the sense that an energy-based approach using not only geometry but also mechanics had been formulated.

In addition, they defined the forces exerted on a grasped object as a *handling force* and proposed that the handling force and the rigidity of a prehension system were adjustable with finger control [HA77b], which is now called *stiffness control* [Sal80] or *mechanical impedance control* [Hog80]. In particular, they indicated for the first time that the handling force could be divided into a *grasping force* and a *manipulating force* when the peripheral contour of the grasped object is known mathematically in advance.

In 1979, Craig and Raibert attempted to experimentally verify hybrid position/force control using a 2-DOF manipulator in contact with a moving environment, where a force sensor (strain gage) was mounted on the wrist of the manipulator [CR79]. While the use of a strain gage, rather than an on-off tactile sensor [Ino71, TGI76], had not been used before, the measurement of the sensor response while in contact with the environment revealed the presence of a small mechanical oscillation induced by a wrist sensor placed between the environment and the manipulator. They summarized that the advancement of computer hardware performance would be strongly expected to solve aliasing problems caused by low sampling rate and continuous matrix operations for multilinked manipulators. A sort of defect of the noise problems, however, still exists even today. To date, well-designed and sophisticated anthropomorphic robotic hands and manipulators with refined force sensors capable of interacting directly with humans have not yet been developed.

As stated above, new concepts such as compliance/impedance control and hybrid control were proposed for precise task operations. These theoretical concepts would be systematized compactly by Mason [Mas81, BHJ+82] and Craig and Raibert [CR79] during the 1980s.

1.2.2 The 1980s

1.2.2.1 Manipulators

The hybrid control scheme [RC81, BHJ+82] was systematically formulated by Raibert and Craig, and this scheme had been defined conclusively by the end of the 1980s. The conceptual definition of the hybrid control, shown in Fig. 1.1, is to independently control both the position of a manipulator (above the dotted line) and the contact forces generated at the end effector (below the dotted line). In addition to their hybrid control design, a set of force feedforward inputs is added in order to achieve stable grasping in robotic hand operations. The corresponding control input vector can be represented as follows:

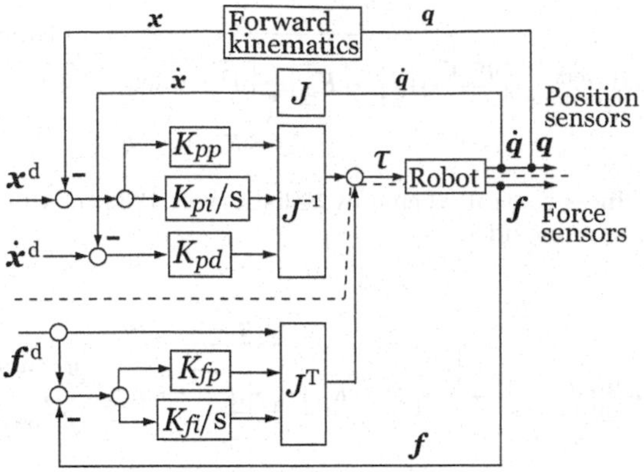

Fig. 1.1 Hybrid control scheme by Raibert and Craig in 1981, in which a force feedforward input is added in order to achieve simultaneous position and force control of mechanical manipulators. The symbols $\boldsymbol{q}$, $\boldsymbol{x}$, $\boldsymbol{J}$, and $\boldsymbol{f}$ denote the joint angles, the position of the end effector in Cartesian coordinates, the Jacobian matrix, and the contact forces, respectively

$$\begin{aligned}\boldsymbol{\tau} = \boldsymbol{J}^{-1}&\left\{K_{pp}(\boldsymbol{x}^{\mathrm{d}} - \boldsymbol{x}) + K_{pd}(\dot{\boldsymbol{x}}^{\mathrm{d}} - \dot{\boldsymbol{x}}) + K_{pi}\int_0^t (\boldsymbol{x}^{\mathrm{d}} - \boldsymbol{x})\mathrm{d}\tau\right\}\\ &+ \boldsymbol{J}^{\mathrm{T}}\left\{K_{fp}(\boldsymbol{f}^{\mathrm{d}} - \boldsymbol{f}) + K_{fi}\int_0^t (\boldsymbol{f}^{\mathrm{d}} - \boldsymbol{f})\mathrm{d}\tau + \boldsymbol{f}^{\mathrm{d}}\right\}. \qquad (1.1)\end{aligned}$$

The force control is applied along the normals of constraint surfaces, and the position control complies with the tangents under certain frictional circumstances. To put it another way, both the position and force spaces are geometrically orthogonal. This orthogonality can be defined rigorously when the contact mechanism between the manipulators and environments is formulated by utilizing a point-contact theory. Since the hybrid control is based on orthogonality, if the orthogonal relationship is not satisfied or theoretical formulations in terms of the contact for environments cannot be described well, the hybrid control scheme would collapse from the beginning. In particular, in the case of soft-fingered manipulation, since the precise modeling of a soft fingertip is difficult, the hybrid control technique cannot be applied to soft-fingered robotic hand systems. Therefore, other control designs for manipulation tasks with soft fingers should be proposed.

In addition, stiffness control [Sal80] by Salisbury, resolved-acceleration control [LWP80] by Luh *et al.*, and impedance control [Hog85a, Hog85b, Hog85c, KHS86a, KSH86, KHS86b] were newly proposed in the early 1980s. Stiffness control, shown in Fig. 1.2, does not require the sensing of force signals, at least in principle, and controls the force measurement implicitly at an inner loop of the system. The reason for this is that the contact force, $\boldsymbol{f} = K_p\boldsymbol{J}(\boldsymbol{q}^{\mathrm{d}} - \boldsymbol{q})$, obtained from joint angles is subsumed before multiplying the transpose of

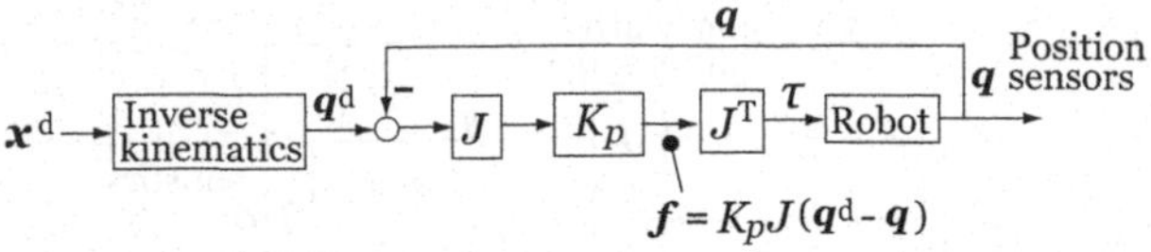

Fig. 1.2 Basic stiffness control scheme by Salisbury at 1980, in which no force feedback control inputs are included

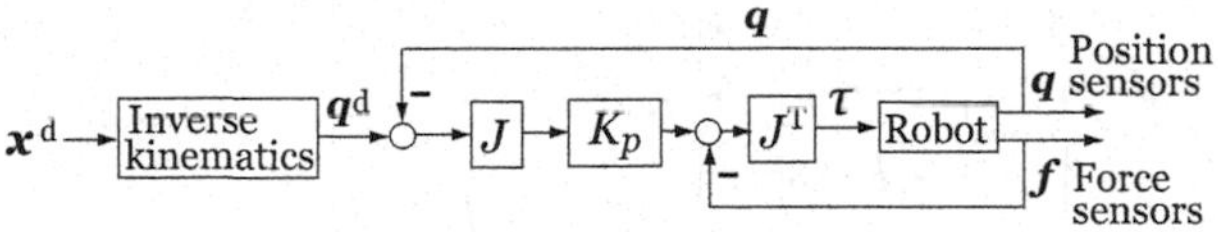

Fig. 1.3 An extended stiffness control scheme by Salisbury, in which force feedback control inputs are included

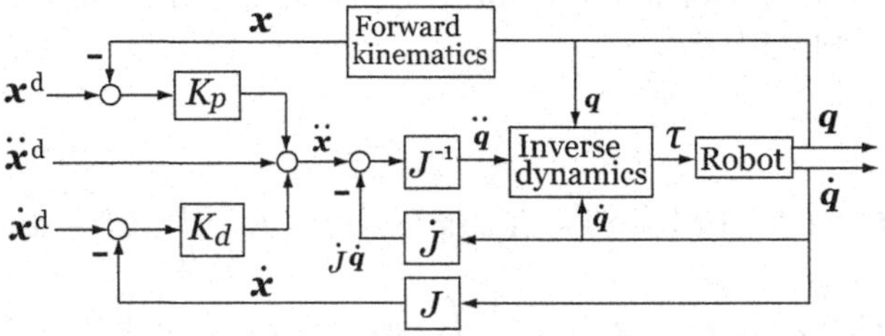

Fig. 1.4 Resolved-acceleration-control method for manipulator position and orientation control by Luh *et al.* [LWP80] in 1980

the Jacobian matrix, as shown in Fig. 1.2. However, this control design may allow the sensitivity of force feedback to decrease according to the magnitude of each element of the stiffness matrix K_p described in diagonal form, resulting in a reduction in performance for force resolution. Hence, if strict force control is needed, it is preferable to add a force feedback input to the stiffness control, as shown in Fig. 1.3. An analogy between the hybrid control and stiffness control is that no inverse-dynamics computation of manipulators is necessary, while the inverse transformation of the Jacobian matrix appears only on the hybrid control.

The resolved-acceleration control shown in Fig. 1.4 newly introduced a set of desired accelerations that is essential for fine trajectory control of manipulators, which has, to date, obtained widespread recognition for manipulator control designers without using a particular nomenclature. This method is restricted to the position and orientation control of manipulators, and its control law is expressed according to the task-space description:

$$\ddot{\boldsymbol{x}} = \ddot{\boldsymbol{x}}^{\mathrm{d}}(t) + K_p(\dot{\boldsymbol{x}}^{\mathrm{d}}(t) - \dot{\boldsymbol{x}}(t)) + K_d(\boldsymbol{x}^{\mathrm{d}}(t) - \boldsymbol{x}(t)). \tag{1.2}$$

Using the fundamentals of velocity kinematics, the Jacobian matrix, a couple of kinematic relationships are obtained:

$$\dot{\boldsymbol{x}} = \boldsymbol{J}\dot{\boldsymbol{q}}, \tag{1.3}$$

$$\ddot{\boldsymbol{x}} = \dot{\boldsymbol{J}}\dot{\boldsymbol{q}} + \boldsymbol{J}\ddot{\boldsymbol{q}}. \tag{1.4}$$

Substituting Eq. 1.4 into Eq. 1.2 and eliminating $\ddot{\boldsymbol{x}}$, the angular acceleration of joints $\ddot{\boldsymbol{q}}$ is represented as follows:

$$\ddot{\boldsymbol{q}} = \boldsymbol{J}^{-1}\left\{\ddot{\boldsymbol{x}}^{\mathrm{d}} + K_p(\dot{\boldsymbol{x}}^{\mathrm{d}} - \dot{\boldsymbol{x}}) + K_d(\boldsymbol{x}^{\mathrm{d}} - \boldsymbol{x}) - \dot{\boldsymbol{J}}\dot{\boldsymbol{q}}\right\}. \tag{1.5}$$

From Eq. 1.5 $\ddot{\boldsymbol{q}}$ can readily be calculated if $\boldsymbol{q}$ and $\dot{\boldsymbol{q}}$ are measured by means of encoders mounted on DC motors. If the dynamics of the manipulator is already known as $F(\boldsymbol{q}, \dot{\boldsymbol{q}}, \ddot{\boldsymbol{q}}) = \boldsymbol{\tau}$, appropriate torque quantities for each joint motor can be computed, although this depends on the accuracy of the manipulator dynamics. This systematic control scheme can smoothly be applied to the *computed torque method*, which was investigated by Paul [Pau72] and Raibert and Horn [RH78] for the first time. This control method was reported to exhibit the best performance for position and orientation control of the manipulator in comparison with any other control method [AAH88]. Furthermore, an extended resolved-acceleration-control method, in which a force feedback is superimposed for force control of manipulators in contact with environments was proposed by Shin *et al.* [SL85].

The impedance control proposed by Hogan was based on the anatomical muscular structure of the human/animal forearm, with the elbow joints consisting of a pair of agonist and antagonist muscles, and its mechanical model is depicted in Fig. 1.5 [Hog80]. This simple mechanical model has a couple of important implications common to soft-finger manipulations in that displacement of the limb from the equilibrium position results in the generation of a restoring torque, which is a function of the mechanical properties of the muscle. In addition, Hogan postulated that if the activities of the antagonist muscles are increased simultaneously, the equilibrium condition of the joint can remain unchanged while the stiffness for the equilibrium condition is increased. A mechanical structure based on the parallel arrangement of multiple linear springs corresponds to a soft-fingertip model that will be discussed below. Hogan concluded that the coactivation of antagonist muscles provides an important vehicle for adaptive tuning of the system parameters, which is independent of feedback, that is, open-loop control [Hog80]. This prospective observation will be verified for soft-fingered manipulation.

Furthermore, it has been said that a muscle is mechanically analogous to a spring, and a muscle's force is a function of its length. The position at which the length-dependent forces due to opposing muscles are equal is the equilibrium position of the limb. Consequently, the central nervous system (CNS) may maintain a desired joint position by simultaneous activation of agonist and antagonist muscles [Fel66, NH76, Kel77, Coo79, KH80, BAC+84].

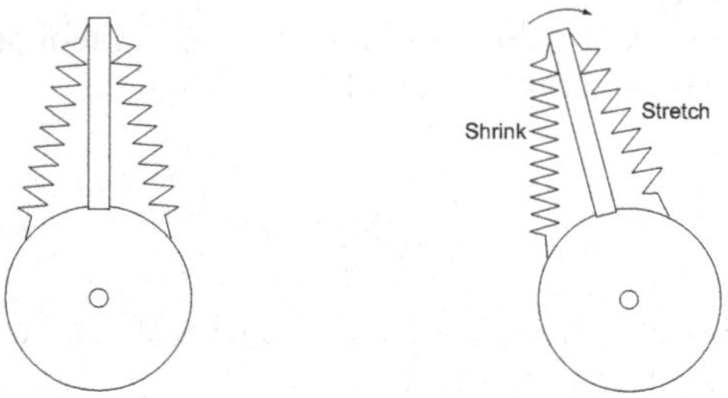

(**a**) Equilibrium configuration (**b**) Energy storage of springs

Fig. 1.5 A simple mechanical model of the muscular structure for human/animal forearms with their elbow joints in which agonist and antagonist muscles are located in a symmetric configuration

In addition, experimental studies of visually triggered head and arm movements in trained monkeys [BPM76, PB79, BAC+84] have shown that a final head and/or forearm position is indeed an equilibrium point between opposing forces. As a result, Bizzi concluded that the CNS controls simple large movements by specifying only the final equilibrium point, and the details of the movement trajectory are determined by the inherent inertial and viscoelastic properties of the limb and the muscles. In addition, he postulated that the CNS could define a final limb position by setting the spring constants of agonist and antagonist muscles, even in the absence of peripheral feedback [BAC+84]. This experimental hypothesis, which is referred to as *final position control* and also be renamed as *virtual trajectory hypothesis* [HH00], is largely analogous to our soft-fingered manipulation scheme, in which potential energy due to elastic deformation of soft fingertips inevitably satisfies an equilibrium point during manipulating and pinching movements. This fact enables the control strategy for the soft-fingered manipulation to be simplified and permits large temporal delay of the control cycle up to approximately 66 ms, which will be discussed in Chap. 8. Since Hogan's claim in terms of impedance control was somewhat conceptual for most roboticians at the time, theoretical extension and practical application of impedance control was performed for the first time by Kazerooni, among others, [Kaz88, AA88, Gol88, AS88], in the late 1980s. Hogan published thought-provoking and interesting themes related to impedance control throughout the 1980s [Hog84a, Hog87, Hog84b, Hog88].

Although a number of theoretical explorations and contributions had been published, great difficulties associated with force sensors arose in the practical application of manipulators and robot hands, including a problem in which the force reading from the sensors is unsuitable for force feedback control. Wu proposed a joint torque sensor system for conventional wrist/hand force sensor systems to avoid sensor uncertainties caused by high-frequency contact vibrations with the environment, although this is the oldest sensing technique [WP80, LFP83]. This technique had advantages in that the joint sensor system was closed around each corresponding joint, gear reduction, and actuator.

In additon to these findings, new frameworks and architectures of practical control methods, control theory, and stability problems for manipulators had been investigated starting in the early 1980s. Takegaki and Arimoto introduced a desirable artificial potential function added to control inputs based on certain sensory feedback so that the manipulator inevitably transfers to a stable equilibrium configuration [TA81, MA85]. Arimoto and Miyazaki proved the asymptotic stability of the feedback control method for positioning in joint space under the full dynamics of a robotic system in the case of a control scheme that contains linear feedback loops relating to position and velocity of a robot with a simple nonlinear compensator for the gravity term [AM85]. During the same period, a useful evaluation index for manipulator design based on system dynamics, which is referred to as the *generalized inertia ellipsoid* (GIE) and *generalized inertia moment*, was introduced by Asada [Asa83, Asa84]. In practical manipulator design, by monitoring the shape of the GIE and attempting to maintain the shape as a multidimensional perfect sphere, the dynamic performance of the manipulator is improved dramatically. Since the geometric representation of the GIE can exhibit the dynamic characteristics of multilinked manipulators in terms of the mechanical inertia of the manipulators, this visible representation method facilitates the design of manipulators, even in the presence of nonlinear factors such as Coriolis force and centrifugal force. Yoshikawa introduced a scalar value with a manipulator Jacobian capable of investigating optimal design and control of manipulators, which is referred to as the *measure of manipulability*, and specialized in the evaluation of manipulator statics [Yos84, Yos85]. This observation method is an analogous concept to the GIE in the sense that the degree of manipulator performance and design evaluation could be presented using the visible multidimensional ellipsoid. Due to their convenience, the GIE and Yoshikawa's manipulability have been been used in numerous studies.

Up until the mid-1980s, manipulator control problems were well known worldwide through robotic researchers and industrial companies, and numerous control laws based on the above new control strategies had been proposed. As a result, since the late 1980s, coordinated control architectures of multiple manipulators have displaced the single-manipulator design as a subsequent new generation design, that is, *robotic fingers and hands.*

1.2.2.2 Robotic Fingers and Hands

Robotic fingers and hands are usually confronted with a situation in which multiple contact points of each finger occur simultaneously and must be maintained so that a grasped object does not fall due to gravitational force. To this end, a number of control methodologies for achieving stable and dexterous manipulation tasks was proposed, and a number of related studies were published in the late 1980s. Salisbury and Roth for the first time defined *internal forces* that allow complete immobilization of a gripped object relative to the fingers and indicated that the internal forces are arbitrarily determined while maintaining a secure grasp. They also proposed a geometric approach for determining these forces [SR83]. Their study noted that the concepts of *force closure* [RK76] and *form closure* [Lak78] are important to the complete restraint of the grasped object. Later, force closure was systematically formulated by Nguyen [Ngu86, Ngu87, Ngu88]. Hence, strategies for computing the optimal internal force were thereafter actively discussed.

Ji and Roth presented a direct-computation method for grasping force and provided an optimal internal force so that the dependence of the internal forces on contact friction remains minimized [JR88]. Kerr and Roth reported that the determination of internal grasp forces required for stable grasping could be reduced to a linear programming problem that considers friction and joint torque limit constraints [KR86]. These studies, however, were based not on dynamics but on statics with kinematic and geometric description. Cutkosky *et al.* used a *shearing model* to describe the contact friction based on material mechanics and raised for the first time a soft-fingered manipulation problem. They also compared the individual capability of different types of fingers in terms of their contribution to the stiffness and stability of a simple grasp [CW86]. When the material of the finger becomes soft and extremely flexible, the static analysis or geometric approach to grasping alone is not sufficient to reveal the type of mechanical phenomenon occurring around the grasped object or why the stable grasping can be maintained. Furthermore, Cole *et al.* formulated for the first time the kinematics of rolling between the object and fingertip surfaces using velocity and normal constraints and systematically derived the dynamics of grasps containing the rolling task [CHS89]. In addition, Li and Sastry discussed the optimal grasp of an object using differential geometry so that a task could be executed efficiently and the robot motion planning could be found under nonholonomic rolling constraints [LS88, SL89]. Kobayashi for the first time explicitly defined grasping and manipulating forces that could be used independently for the control input applied to the robotic hand [Kob85]. Moreover, Yoshikawa *et al.* [YN87, YN88] and Nakamura *et al.* [NNY89] systematized the control architecture based on the decomposition scheme of grasping and manipulating forces in the presence of coulomb friction between the target object and the robotic fingers. This type of fundamental control scheme, such as independent grasping and manipulating forces, has been clarified and is often employed by robotics and automation researchers. Therefore, a tremendous number of theoretical analyses of manipulators and robotic hands have been proposed, and their successful application in control strategy became a well-known way to move manipulators and robotic hands. After the 1990s, these manipulation problems were further developed, and these control techniques were applied to a number of robots. Research and studies conducted during the 1990s and 2000s are referred to throughout the body of this book, which provides a further exploration of robotic hands and related applications.

1.3 Overview

Here we present an overview of this book. Chapter 2 describes observations of grasping and manipulation performed by a pair of 1-DOF fingers with hemispherical soft tips. Based on these observations, in Chap. 3, we propose a new fingertip model called the parallel-distributed model. The model is extended to a model with tangential deformation in Chap. 4. Chapter 5 summarizes the variational principles in statics and dynamics, which are fundamental in the following chapters. Chapter 6 describes the statics of soft-fingered grasping and manipulation, and Chap. 7 describes the dynamics of soft-fingered grasping and manipulation. Using static and dynamic formulations, Chap. 8 focuses on the control of soft-fingered grasping and manipulation. Chapter 9 describes nonlinear fingertip models, which are extensions of the proposed model. Chapter 10 describes grasping and manipulation by robotic hands of various configurations. Chapter 11 describes the formulation of 3D grasping and manipulation. The final chapter summarizes the contributions of this book and discusses future areas for study.

Chapter 2
Observation of Soft-fingered Grasping and Manipulation

2.1 Introduction

Through historical observation of the cutting-edge technology of robotic hands we know that compliant motions or compliance in mechanical systems is important for controlling robotic systems that are in contact with external environments. Impedance control and stiffness control have been investigated extensively, especially from the perspective of control strategy, in order to achieve difficult robotic tasks. The characteristics of mechanical systems that are in contact with the environment are the physical rigidity and inflexibility of these systems. On the other hand, if the contact is flexible, what control scheme is adequate to the robotic system?

In daily life, we are surrounded by soft objects and materials such as paper, cloth, string, meat, and pizza dough. Humans can be considered flexible mechanical systems, which is a crucial difference between the human and traditional robotic mechanical systems. The flexibility and deformability of the human mechanical system enable dexterous motions and versatile abilities in human movement.

Based on the above, we may have to consider that robots possess soft regions in their bodies, regardless of whether the region is in contact with the external environment, because soft materials are able to readily conform to the shape of the target object along with the contacting environment. On the other hand, when dealing with *flexible robots*, in contrast to conventional rigid systems, the challenge of developing an appropriate control strategy will be difficult.

This study focuses on a soft-tipped robotic hand designed to mimic the flexibility of human fingers, which are capable of dexterous manipulation. In order to focus on how flexibility affects the controllability of dynamic systems, we assume that the number of degrees of freedom (DOFs) of the robotic hand system remains minimal throughout this study, *i.e.,* a pair of 1-DOF rotational joints with soft fingertips is mounted on a robotic hand.

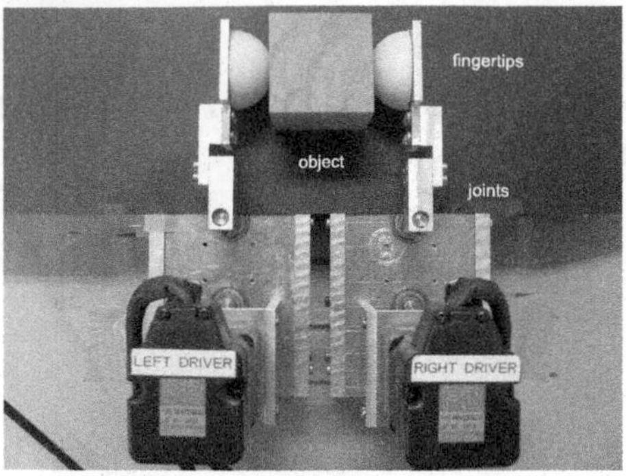

Fig. 2.1 A pair of 1-DOF fingers with soft fingertips

Fig. 2.2 Deformation of fingertips when two fingers rotate in opposite directions

Fig. 2.3 Motion of object when two fingers rotate in the same direction

Furthermore, the target object is assumed to be a rigid, cubic shape. In what follows, we introduce related research on soft-fingered manipulation.

2.2 Object Pinching by a Pair of 1-DOF Fingers

Let us observe the grasping and manipulation of a rigid object by a pair of soft-tipped fingers, each with a 1-DOF rotational joint driven by an actuator. Figure 2.1 shows such a pair of fingers, driven by stepping motors, grasping a rectangular rigid object. When the two fingertips move inward, *i.e.*, toward each other, they become more deformed, suggesting that they are applying a larger grasping force, as shown in Fig. 2.2. This observation suggests that the

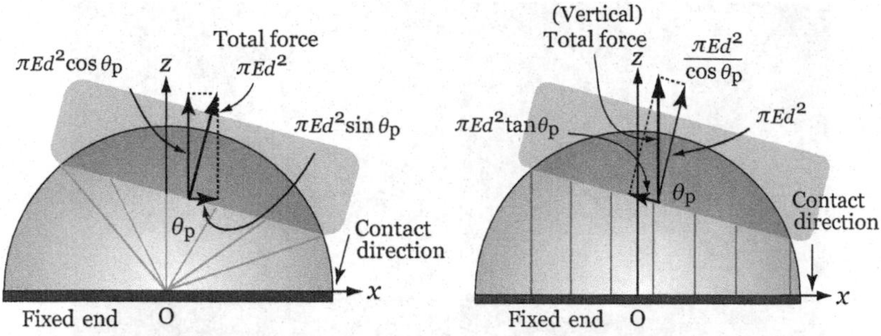

(**a**) Radially-distributed model (**b**) Parallel-distributed model

Fig. 2.4 Fingertip models

grasping force can be regulated by the two joint angles, *i.e.*, secure grasping can be achieved by a pair of 1-DOF fingers with soft fingertips. When one fingertip moves inward and the other moves outward, *i.e.*, both fingertips rotate in the same direction, the object rotates in the opposite direction, implying that the orientation of the object can be regulated by the two joint angles, as shown in Fig. 2.3. This observation suggests that orientation can be controlled by a pair of 1-DOF fingers with soft fingertips.

These findings indicate that a pair of 1-DOF fingers with soft fingertips can control both grasping force and object orientation independently, in contrast to an Arimoto's conclusion [HAK+01]. In their study on the dynamics of soft-fingered grasping and manipulation, they reported that a pair of 1-DOF rotational fingers with soft fingertips can control grasping force but not object orientation. They asserted that both a 1-DOF finger and a 2-DOF finger are required to control both grasping force and object orientation. Thus, there is a discrepancy between our observations and their claim, which suggests that their model may not be valid in this case. In their *radially-distributed model*, Fig. 2.4a, the contact force passes through the center of a soft hemispherical fingertip, and its magnitude is dependent on the maximum displacement of the fingertip but not on the relative orientation between the fingertip and the object. Since opposing contact forces by both fingers would apply a non-zero moment to the object, an additional DOF is needed to cancel out the unexpected moment and stabilize object rotation. Therefore, we need another fingertip model that properly describes the maximum displacement and the relative orientation between the fingertips and the object. We introduce a new model, called a *parallel-distributed* model (Fig. 2.4b), in Chap. 3.

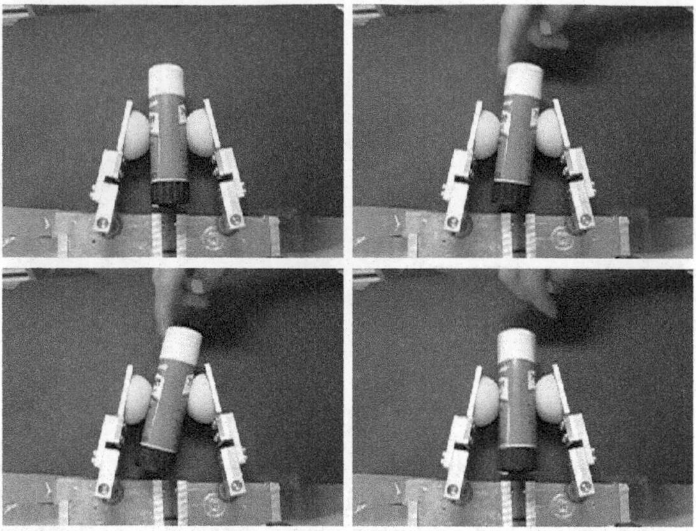

Fig. 2.5 Rotation of a pinched object by an external force

2.3 Rotation of a Pinched Object by External Force

When an external force is applied to a rigid object pinched by two fingers fixed at given orientations, the object may rotate, as shown in Fig. 2.5. When the applied force is relaxed, the object returns to its initial orientation. The object, therefore, does not slip, and so there are geometric constraints, which are referred to as *rolling constraints.* Assigning constant values to both of the joint angles and solving the two rolling constraints of the total of four constraints results in the orientation of the grasped object being uniquely determined, without solving the other two constraints. This is another discrepancy with respect to the report by Arimoto *et al.*, and it is due to the lack of tangential deformation in their model. In their radially-distributed model, any point on the hemispherical surface of a soft fingertip moves along a line normal to the surface, which determines the shrinkage of an elastic element inside the fingertip. That is, each elastic element deforms normally but not tangentially. Therefore, it is necessary to introduce tangential deformation into the fingertip model in order to describe the rotation of a pinched object by an external force. We extend the parallel-distributed model without tangential deformation (Fig. 2.6a) to a parallel-distributed model with tangential deformation (Fig. 2.6b) in Chap. 4.

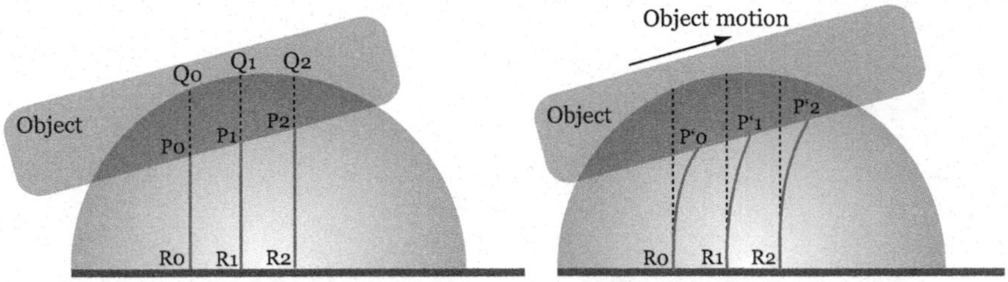

(**a**) Without tangential deformation (**b**) With tangential deformation

Fig. 2.6 Tangential deformation in the parallel-distributed model

2.4 Concluding Remarks

We observed rigid body manipulation performed by a pair of 1-DOF fingers with soft fingertips. The observations call for a new model, which is described in the following chapters.

Chapter 3
Elastic Model of a Deformable Fingertip

3.1 Introduction

To date, much research has been done on the manipulation of objects by soft-fingered robotic hands. Most of these studies, particularly the earlier ones, focused only on contact mechanisms on various soft fingers. More recently, there has been an increase in the number of studies investigating the sensing mechanisms of the human hand and designing control systems in robotic applications to emulate the human capabilities that are applicable to robotic hands. Conventional studies, however, have not explicitly provided any analytical exploration of the simplicity in grasping and manipulating motions in terms of soft-fingered handling. It has therefore been very difficult to derive a fine elastic model of soft materials used in fingertips.

Yokokohji *et al.* proposed a control scheme with visual sensors that can cancel the frictional twist/spin moment at the contact point of soft fingertips and achieved stable grasping by spherical soft fingertips [YSY99, YSY00]. Maekawa *et al.* developed a finger-shaped tactile sensor covered with a soft, thin material and proposed a control algorithm based on tactile feedback using a sensor that requires no information about the geometry of the grasped object [MKT92, MTK+92]. They managed to control the position of an object along a desired trajectory. In these papers, point-contacts were used to represent the constraints of rolling contact in their theoretical models, although the fingertips were made from a soft material such as rubber. Arimoto *et al.* verified the passivity of equations of motion for a total handling system using a Lagrangian function incorporating the elastic potential energy due to the deformation of soft fingertips [ANH+00] and compensated for the gravity effect in 3D space [ADN+02]. An elastic force model was also derived for the elastic potential energy of a system in which, for simplicity, virtual linear springs were arranged in a radial pattern inside hemispherical soft fingertips. Doulgeri *et al.* discussed the problem of stable grasping with deformable fingertips on which rolling constraints were described as nonholonomic because of the change in the effective rolling radius of the soft fingertip [DFA02, DF03]. The above-mentioned studies, however, focused mainly on deriving a control law to realize stable grasping and attitude control of the grasped object rather than on revealing a physically appropriate deformation model, which also contains the nonlinear characteristics of a hemispherical soft fingertip.

On the other hand, Xydas *et al.* proposed an exact deformation model based on the mechanics of materials having nonlinear characteristics and performed finite element analysis (FEA) for a hemispherical soft fingertip [XK99, XBK00]. Kao *et al.* experimentally demonstrated that the elastic force due to deformation satisfied a power law with respect to the displacement of the fingertip and insisted that their theory subsumed Hertzian contact [KY04]. These studies, however, did not distinguish between the material nonlinearity of the soft fingertip and the geometric nonlinearity caused by the hemispherical shape of the fingertip; they also defined a parameter that accounts for the effects of both nonlinearities. Consequently, the cause of the discrepancy between the results of the simulation based on their model and the results of actual experiments was not apparent. In addition, as a result of the complexity of these models, these studies do not generally lend themselves to the analysis of equations of motion for a soft-fingered manipulation system. While FEA may enable us to derive a stress distribution and an elastic force on the soft fingertip, these simulation results depend on the selected mesh pattern. Although FEA based on a certain arbitrary mesh pattern may prove the stability for equations of motion of the handling system, it does not always provide proof of stability for equations derived from other mesh patterns.

We herein propose a static elastic model of a hemispherical soft fingertip in a physically reasonable and straightforward form suitable for theoretical analysis of robotic handling motions. This model is assumed to be composed of 1D linear springs placed within a hemispherical soft fingertip that stand perpendicularly on the bottom and is called a *parallel-distributed model*. We distinguish between geometric nonlinearity due to the hemispherical shape and material nonlinearity of soft materials, *i.e.*, the nonlinearity of the Young's modulus of the soft material, allowing us to focus only on the geometric nonlinearity of the soft fingertip and analytically formulate the elastic force and elastic potential energy equations for the deformation of the fingertip. We show that each equation is a function of two variables: the maximum displacement of the fingertip and the orientation angle of the contacted object. We also show that when the object is positioned normal to the fingertips, the elastic potential energy is minimal. Finally, we validate the static elastic model by conducting a compression test of the hemispherical soft fingertip and evaluating the results.

3.2 Static Elastic Model of a Hemispherical Soft Fingertip

3.2.1 Fingertip Stiffness

We treat the fingertips as if they were composed of an infinite number of virtual linear springs that are standing vertically. Figure 3.1 shows one such spring. We formulate elastic force and elastic potential energy equations for the deformation of the fingertip. In order to simplify the derivation process of both equations, two assumptions associated with material characteristics are given as follows: (1) The incompressibility of elastomer materials is not considered and (2) Young's modulus is assumed to be constant. Note that the contact condition being discussed in the present study is restricted to the case in which a force applied to the fingertip is assumed to be along the z-axis of the fingertip. In addition, we assume that the object never comes into contact with the bottom plane of the fingertip.

Let us apply an infinitesimal virtual spring QR with sectional area $\mathrm{d}S$ inside the soft fingertip, as shown in Fig. 3.1. Let $\mathrm{d}F$ be the infinitesimal elastic force due to the shrinkage PQ of the virtual spring. Let θ_p be the orientation angle of the contacting object, and let a be the fingertip radius. Let d be the maximum displacement of the fingertip, and let $a_\mathrm{c} = \sqrt{a^2 - (a-d)^2}$ be the radius of the contacting circle. Let P be the point at which the spring is in contact with the object. Finally, let θ be the angle subtended between line PQ and the z-axis, and let ϕ be the azimuthal angle on the xy-plane. Using the contact surface equation, $x \sin\theta_\mathrm{p} + z\cos\theta_\mathrm{p} = a - d$ (Appendix A.1), the length of PR is then expressed as

$$\mathrm{PR} = \frac{a - d - x\sin\theta_p}{\cos\theta_p}. \tag{3.1}$$

Since the length of QR becomes $\sqrt{a^2 - (x^2 + y^2)}$ due to the hemispherical feature, the infinitesimal elastic force $\mathrm{d}F$ on a single virtual spring QR is given by

$$\mathrm{d}F = k \cdot \mathrm{PQ} = k\left\{\sqrt{a^2 - (x^2+y^2)} - \frac{a - d - x\sin\theta_\mathrm{p}}{\cos\theta_\mathrm{p}}\right\}, \tag{3.2}$$

where k is the linear spring constant of the spring QR. Note that k is proportional to the sectional area $\mathrm{d}S$ and inversely proportional to the natural length $\sqrt{a^2 - (x^2 + y^2)}$. Letting E be Young's modulus of soft-finger materials, k is described as (Appendix A.2)

$$k = \frac{E\,\mathrm{d}S}{\sqrt{a^2 - (x^2+y^2)}}. \tag{3.3}$$

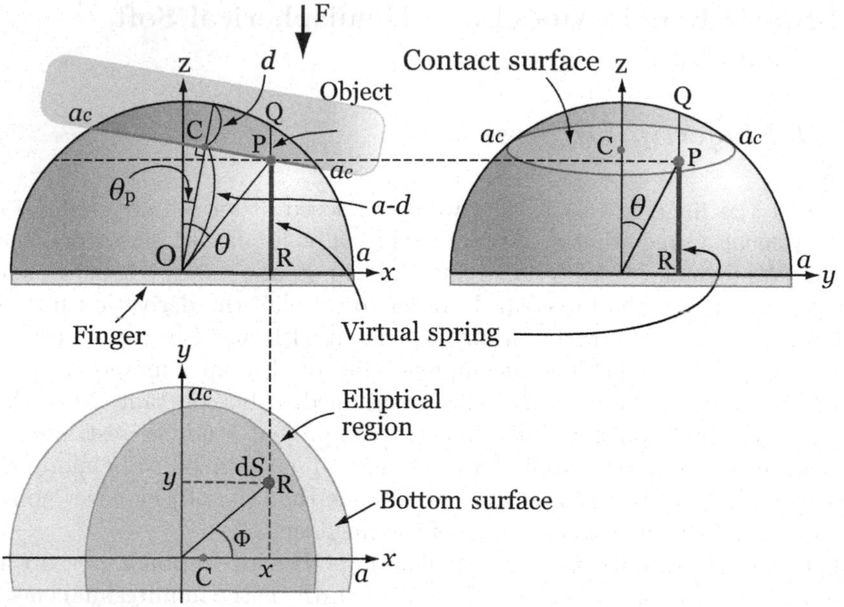

Fig. 3.1 Contact mechanism

Letting K be the fingertip stiffness on the entire deformed part illustrated in Fig. 3.1, based on Eq. 3.3, K can be expressed as

$$K = \int_{ell} k = E \int_{-a_c}^{a_c} \int_{b_1(y)}^{b_2(y)} \frac{\mathrm{d}x\,\mathrm{d}y}{\sqrt{a^2 - (x^2 + y^2)}}, \tag{3.4}$$

where

$$b_1(y) = (a - d)\sin\theta_p - \cos\theta_p \sqrt{a_c^2 - y^2}, \tag{3.5}$$

$$b_2(y) = (a - d)\sin\theta_p + \cos\theta_p \sqrt{a_c^2 - y^2}, \tag{3.6}$$

and *ell* denotes the elliptical region shown in Fig. 3.2. Applying a numerical integration to Eq. 3.4, we find that the fingertip stiffness is almost constant with respect to the object orientation, which is depicted as continuous lines, as shown in Fig. 3.3. This occurs even when the maximum displacement d changes several values. This indicates that the fingertip stiffness K is independent of the object orientation θ_p. Hence, in the present study we need a third assumption: (3) The fingertip stiffness is independent of the object orientation as long as the maximum displacement remains constant. Using the above numerical assumption, we formulate the fingertip stiffness K as an analytical formula.

Next, performing the substitution whereby $x = r\cos\phi\cos\theta_p + (a-d)\sin\theta_p$ and $y = r\sin\phi$, Eq. 3.4 is then transformed into (Appendix A.3)

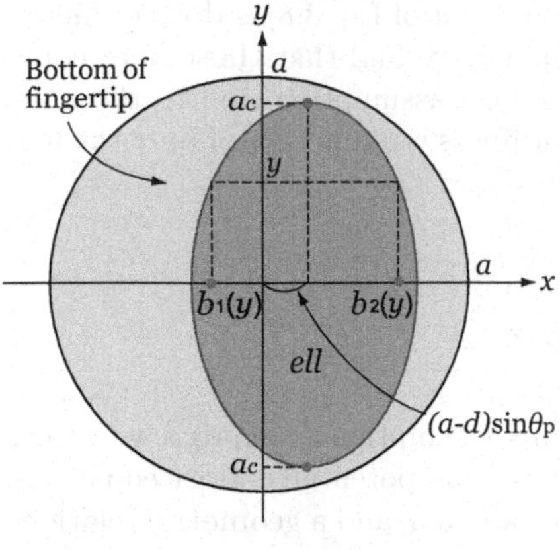

Fig. 3.2 Integration area obtained by projecting the contact plane onto the xy-coordinate

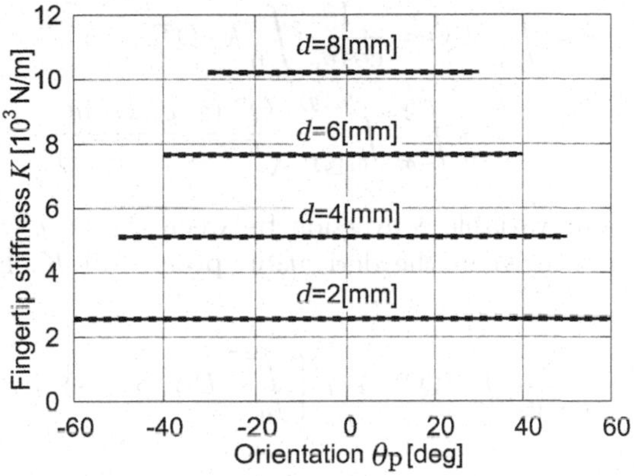

Fig. 3.3 Comparison between the numerical results of Eq. 3.4 (*continuous lines*) and analytical simulations of Eq. 3.8 (*dotted lines*) for $E = 0.2032$ MPa

$$\begin{aligned} K &= E \int_0^{a_c} r \left\{ \int_0^{2\pi} \frac{\cos\theta_p \, d\phi}{\sqrt{a^2 - \{x^2(r,\phi) + y^2(r,\phi)\}}} \right\} dr \\ &= E \int_0^{a_c} r \int_0^{2\pi} B(r,\phi) d\phi \, dr. \end{aligned} \tag{3.7}$$

Since assumption (3) requires that K is independent of θ_p, we can substitute $\theta_p = 0$ into Eq. 3.7 and obtain

$$K = E \int_0^{a_c} r \left\{ \int_0^{2\pi} \frac{d\phi}{\sqrt{a^2 - r^2}} \right\} dr = 2\pi E d. \tag{3.8}$$

Plotting the simulation result of Eq. 3.8 as dotted lines onto Fig. 3.3 together with the results of Eq. 3.4, we find that these lines coincide with each other. This implies that the third assumption due to the numerical observation is appropriate, and the stiffness is a function of only the maximum displacement d.

3.2.2 Elastic Force

Likewise, using the third assumption associated with the fingertip stiffness, we formulate the elastic force and potential energy equations in a straightforward manner. Using Eqs. 3.2 and 3.3, and a geometric relationship $\mathrm{QT} = \mathrm{PQ}\cos\theta_\mathrm{p}$ (Fig. A.2 in Appendix A.3), the elastic force F on the total deformed region can be written as

$$\begin{aligned} F &= \int_{ell} k\mathrm{PQ} = \frac{1}{\cos\theta_\mathrm{p}} \int_{ell} k \cdot \mathrm{QT} \\ &= \frac{E}{\cos\theta_\mathrm{p}} \int_{-a_\mathrm{c}}^{a_\mathrm{c}} \int_{b_1(y)}^{b_2(y)} \frac{\mathrm{QT}(x,y)\,\mathrm{d}x\,\mathrm{d}y}{\sqrt{a^2-(x^2+y^2)}}. \end{aligned} \tag{3.9}$$

Performing the same variable conversion between the (x, y)-coordinate and the (r, ϕ)-coordinate used in the derivation process of K, Eq. 3.9 is then transformed as

$$F = \frac{E}{\cos\theta_\mathrm{p}} \int_0^{a_\mathrm{c}} \mathrm{QT}(r) \cdot r \left\{ \int_0^{2\pi} B(r,\phi)\,\mathrm{d}\phi \right\} \mathrm{d}r, \tag{3.10}$$

where (Fig. A.2)

$$\mathrm{QT}(r) = \sqrt{a^2 - r^2} - (a - d). \tag{3.11}$$

In Eq. 3.10, $B(r, \phi)$ corresponds to the integrand within the braces in Eq. 3.7. Here, applying assumption (3) to $B(r, \phi)$ as well as Eq. 3.8, F is finally calculated as

$$F = \frac{E}{\cos\theta_\mathrm{p}} \int_0^{a_\mathrm{c}} \mathrm{QT}(r) \cdot r \left\{ \int_0^{2\pi} \frac{\mathrm{d}\phi}{\sqrt{a^2-r^2}} \right\} \mathrm{d}r = \frac{\pi E d^2}{\cos\theta_\mathrm{p}}. \tag{3.12}$$

Thus, we can obtain a straightforward equation that will be applicable to the analytical validation for manipulation due to the simplicity.

3.2.3 Elastic Potential Energy

Note that Eqs. 3.12 and 3.15 indicate that the elastic force and elastic potential energy on the entire deformed part of a hemispherical soft fingertip are functions of two variables, namely, the maximum displacement d and the object orientation angle θ_p. Furthermore, we find that both formulae have a local minimum when the orientation angle remains zero. In particular, we describe the minimum value of elastic energy as the *local minimum of elastic potential energy* (*LMEE*).

In addition to Eq. 3.9, the elastic potential energy P on the entire deformed region is expressed as

$$\begin{aligned} P &= \frac{1}{2}\int_{ell} k\mathrm{PQ}^2 = \frac{1}{2\cos^2\theta_\mathrm{p}}\int_{ell} k\cdot\{\mathrm{QT}(x,y)\}^2 \\ &= \frac{E}{2\cos^2\theta_\mathrm{p}}\int_{-a_\mathrm{c}}^{a_\mathrm{c}}\int_{b_1(y)}^{b_2(y)} \frac{\{\mathrm{QT}(x,y)\}^2\mathrm{d}x\,\mathrm{d}y}{\sqrt{a^2-(x^2+y^2)}}. \end{aligned} \tag{3.13}$$

Performing the same variable conversion between the (x, y)-coordinate and the (r, ϕ)-coordinate used in the derivation process of F, Eq. 3.13 is then transformed as

$$P = \frac{E}{2\cos^2\theta_\mathrm{p}}\int_0^{a_\mathrm{c}} \{\mathrm{QT}(r)\}^2\cdot r\left\{\int_0^{2\pi} B(r,\phi)\,\mathrm{d}\phi\right\}\mathrm{d}r. \tag{3.14}$$

Again, applying assumption (3) to $B(r, \phi)$ in Eq. 3.14, P is finally calculated as

$$\begin{aligned} P &= \frac{E}{2\cos^2\theta_\mathrm{p}}\int_0^{a_\mathrm{c}} \mathrm{QT}^2(r)\cdot r\left\{\int_0^{2\pi}\frac{\mathrm{d}\phi}{\sqrt{a^2-r^2}}\right\}\mathrm{d}r \\ &= \frac{\pi E d^3}{3\cos^2\theta_\mathrm{p}}. \end{aligned} \tag{3.15}$$

Finally, in order to confirm the transformations of formulae from Eq. 3.9 to Eq. 3.12 and from Eq. 3.13 to Eq. 3.15, we compare the numerical analysis of Eqs. 3.9 and 3.13 with the simulation results of Eqs. 3.12 and 3.15. Figure 3.4 shows a good result, and concludes that both Eqs. 3.12 and 3.15 are mathematically reasonable formulae in the present study.

3.2.4 Relationship Between Elastic Force and Elastic Energy

While the individual virtual spring used in the present study is based on a linear elasticity, the entire fingertip model obtained by completing the double

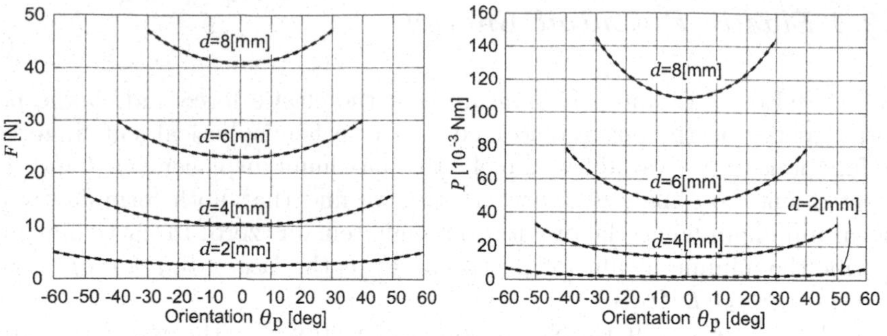

(**a**) Elastic force (**b**) Elastic energy

Fig. 3.4 Comparison between the numerical integration and the analytical simulation of F and P, respectively. The *continuous lines* correspond to the numerical results of Eqs. 3.9 and 3.13, and the *dotted lines* correspond to the analytical results of Eqs. 3.12 and 3.15

integration on an elliptical region exhibits a geometric nonlinearity caused by the hemispherical shape of the fingertip. In other words, the completed fingertip model has a variable fingertip stiffness with respect to d, which is expressed as Eq. 3.8. Hence, when we compute the total force Eq. 3.12 from the energy Eq. 3.15, we must define an *equivalent displacement* to be used for the differential calculation.

In the case of normal contact corresponding to $\theta_\mathrm{p} = 0$ in Eq. 3.15, elastic models are given as follows:

$$P = \frac{\pi E d^3}{3}, \tag{3.16}$$

$$\frac{\partial P}{\partial d} = \pi E d^2 = F, \tag{3.17}$$

$$\frac{\partial^2 P}{\partial d^2} = 2\pi E d = K, \tag{3.18}$$

where d corresponds to the equivalent displacement. Next, let us consider the case of diagonal contact when $\theta_\mathrm{p} \neq 0$. We define Δz_{eq} as an equivalent displacement, which must satisfy

$$\frac{\partial P}{\partial \Delta z_{eq}} = \frac{1}{3}\frac{\partial\left(\pi E \Delta z_{eq}^3 \cos\theta_\mathrm{p}\right)}{\partial \Delta z_{eq}} = \pi E \Delta z_{eq}^2 \cos\theta_\mathrm{p} = \frac{\pi E d^2}{\cos\theta_\mathrm{p}} = F, \tag{3.19}$$

$$\frac{\partial^2 P}{\partial \Delta z_{eq}^2} = \frac{\partial\left(\pi E \Delta z_{eq}^2 \cos\theta_\mathrm{p}\right)}{\partial \Delta z_{eq}} = 2\pi E \Delta z_{eq} \cos\theta_\mathrm{p} = 2\pi E d = K. \tag{3.20}$$

The displacement Δz_{eq} to fulfill Eqs. 3.19 and 3.20 can be found such that a geometric relationship $d = \Delta z_{eq} \cos\theta_\mathrm{p}$ is maintained as shown in Fig. A.2.

Δz_{eq} is the true maximum displacement among all of the virtual springs in any case that includes $\theta_p = 0$ and $\theta_p \neq 0$.

3.3 Comparison with Hertzian Contact

In 1881, Hertz proposed a contact theory for two elastic objects having arbitrary curved surfaces [Joh85]. He showed that a normal contact force generated between an elastic sphere and a plane, whose Young modulus is infinity, can be expressed as

$$F = \frac{4\sqrt{R}}{3}\left(\frac{E}{1-\nu^2}\right)d^{\frac{3}{2}}, \tag{3.21}$$

where R is the radius, E is the Young modulus of the object, ν is the Poisson ratio, and d is the maximum displacement of the sphere. Since the above equation is useful from a practical viewpoint, it has been widely used for computing the contact stress between, for example, a wheel and a rail, a roll and material, or a retainer and a ball in a bearing. However, in Hertzian contact, it is assumed that both elastic objects are open elliptic paraboloids with an arbitrary radius of curvature. Consequently, no boundary conditions are used in the Hertzian contact model.

Kao *et al.* defined the parameter c_d corresponding to a material and geometric nonlinearity [KY04] and transformed Eq. 3.21 into

$$F = c_d d^{\zeta}. \tag{3.22}$$

They conducted a vertical compression test using a hemispherical soft fingertip and estimated the parameter c_d empirically using a weighted least-squares method (LSM). It has been shown that ζ is approx. 2.3 or 1.75 when the rate of deformation of the finger is above or below 20%, respectively. In other words, the parameter ζ is not identical to 3/2 in the contact model of soft fingertips. Thus, the Hertzian contact theory cannot be adopted for deriving the elastic model of the hemispherical soft fingertip.

Figure 3.5 shows a comparison result in which the elastic force value with respect to the displacement d is plotted when a hemispherical soft fingertip of radius 20 mm is compressed vertically. The vertically oriented spring model is more suitable for deriving an elastic force up to the midrange displacement of the fingertip. This is because our model contains a geometric nonlinearity due to the hemispherical shape of the fingertip, that is, the present model indicates that ζ becomes not 3/2, but 2, which appears within 1.75 and 2.3, only by adopting an appropriate natural length for the individual springs.

Soft materials exhibit nonlinear characteristics, even for infinitesimal deformations. Tatara newly derived a nonlinear Young's modulus with respect to compressive strain [Tat91]. Furthermore, the concept of the contact angle

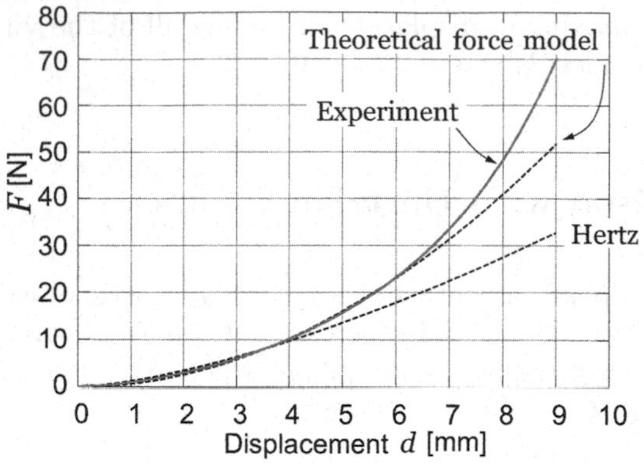

Fig. 3.5 Comparison between the Hertzian contact model and the present elastic force model (Eq. 3.12) when $\theta_p = 0$ and $E = 0.2032$ MPa obtained by a corresponding identification test mentioned in the next section

of the object is not incorporated in the Hertzian contact theory. While the Hertzian contact theory can be used for a simple contact pattern corresponding to the normal contact, no contact at any other arbitrary angle or rolling contact can be defined. On the other hand, the elastic models proposed in this chapter cover any contact angle of the object, and therefore, these models can be used to analyze grasping and manipulating motions for various possible types of contact by a soft-fingered robotic hand.

3.4 Measurement of Young's Modulus

In the present study, the Young modulus of the soft fingertip was measured by conducting a compression test on six cylinders of polyurethane gel. Three cylinders were 20 mm in diameter and 15, 20, and 25 mm in height, and three were 30 mm in diameter and 15, 20, and 25 mm in height, as shown in Fig. 3.6a.

Figure 3.7a shows an overall view of a measured stress-strain diagram, and an enlarged view of part of the diagram is shown in Fig. 3.7b. The numerical values shown in both graphs denote the specimen height on the left side and the specimen diameter on the right side. The data were averaged and smoothed using the LSM, as shown in Fig. 3.8. We assumed the maximum deformation of the soft fingertip to be 50% of the radius. Furthermore, in order to focus predominantly on the geometric nonlinearity due to the hemispherical shape, we did not consider the material's nonlinearity, which, for soft materials, is directly related to the Young's modulus. Consequently, we performed a linear approximation for a 50% strain, as in Fig. 3.8 and estimated the Young modulus as 0.2032 MPa.

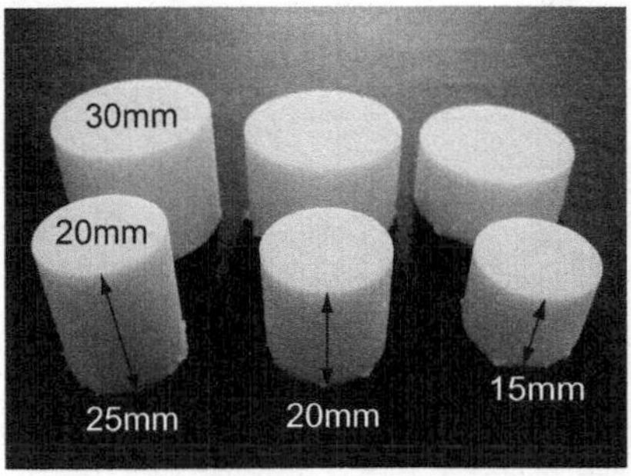

(**a**) Several specimens

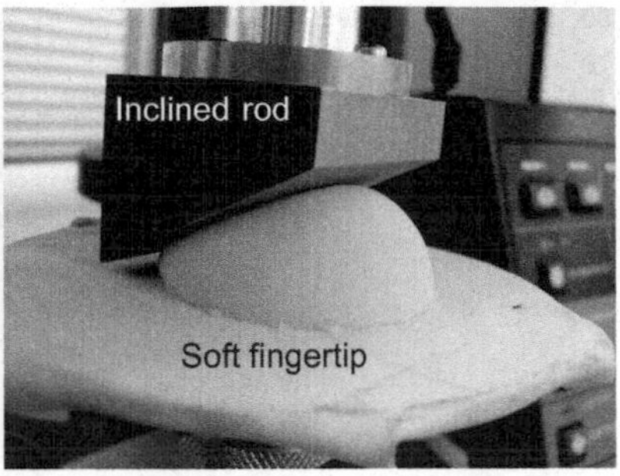

(**b**) Compression test

Fig. 3.6 Compression test of a hemispherical soft fingertip

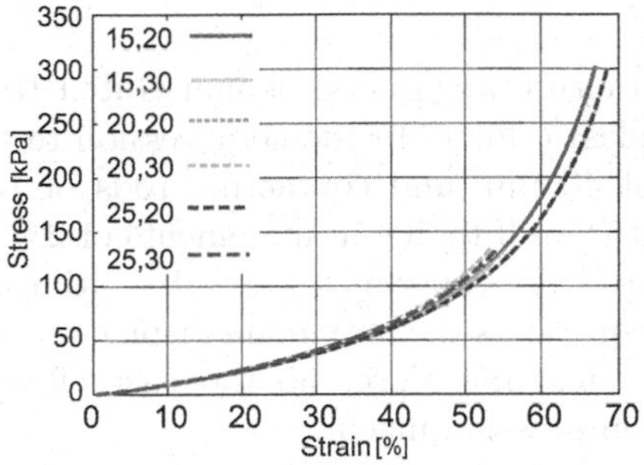

(**a**) Stress-strain diagram

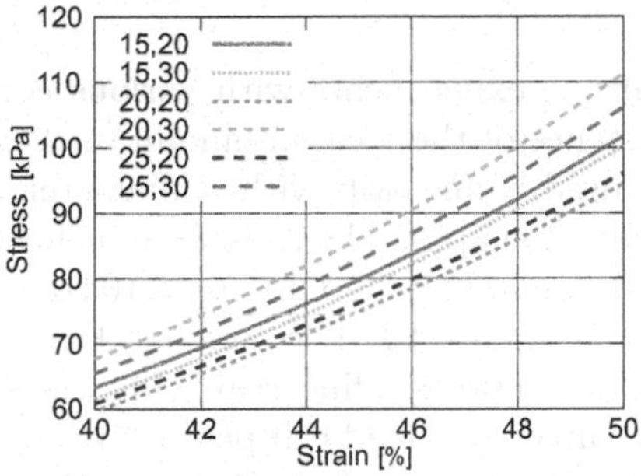

(**b**) Enlarged view

Fig. 3.7 Stress-strain diagram of polyurethane rubber

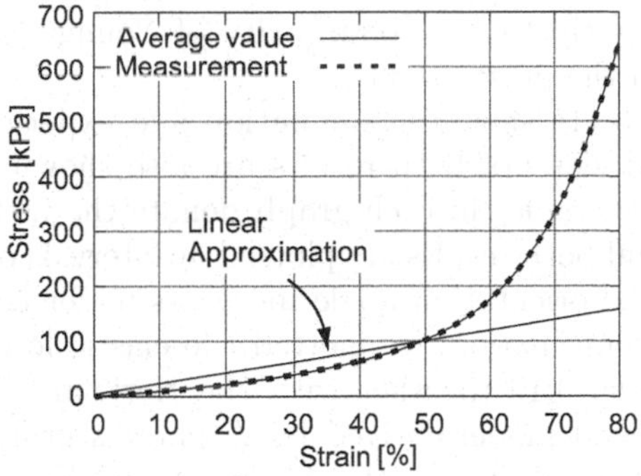

Fig. 3.8 Average value of stress-strain diagram

3.5 Compression Test

By compressing a hemispherical soft fingertip made of polyurethane gel along the normal direction, as shown in Figs. 3.1 and 3.6b, we verified the validity of the elastic force model represented in Eq. 3.12. Furthermore, by conducting

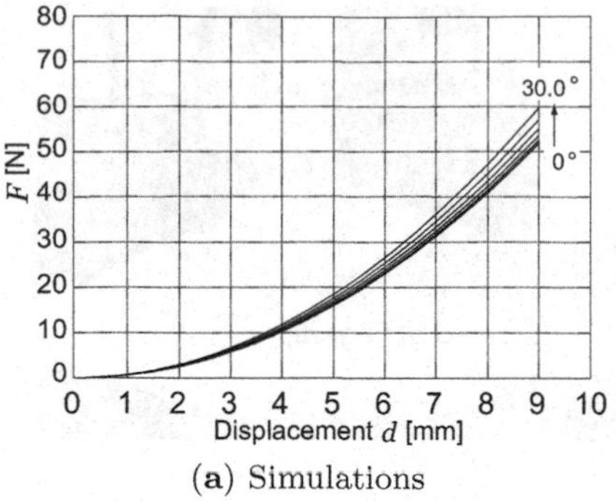

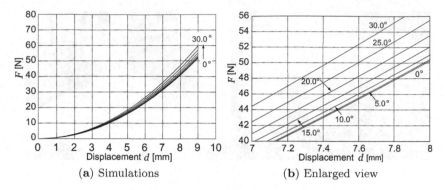

Fig. 3.9 Simulation results of elastic force

multiple experiments with various contacting angles, we demonstrated the existence of the local minimum of the elastic force. In the compression test, we used a fingertip with a diameter of 40 mm and contacting rods of 13 different shapes. The rods were inclined from 0 to 30° in increments of 2.5°, as shown in Fig. 3.6b. Figure 3.10 compares the experimental results with the simulation results. The horizontal axis represents the maximum displacement of the compressed fingertip, while the vertical axis represents the elastic force measured by a load cell placed in the compression machine.

In all of the graphs in Fig. 3.10, the simulation and experimental results are almost identical up to $d = 6.0$ mm, after which the discrepancies increase with the magnitude of the displacement. The discrepancies stem from the linear approximation of the experimental stress-strain diagram shown in Fig. 3.8. The effect leads directly to the nonlinearity of Young's modulus, which is outside the scope of the present study.

Figure 3.9a and 3.11a show the simulation and experimental results, respectively. Enlarged views of both results are also shown in Figs. 3.9b and 3.11b. The numerical values in each graph denote the inclined angle of the contacted object, and both results are plotted at intervals of 5.0°. The elastic force increases as the orientation angle increases under constant maximum displacement. For confirmation, we plotted the elastic force with respect to $\theta_{\rm p}$ of Eq. 3.12 in Fig. 3.12 together with the Hertzian contact model and the radially-distributed model derived by Arimoto's group. The numerical values shown in the graph denote the maximum displacement d. Note that the Hertzian force is depicted at a point because the model does not define the object orientation. At approx. 0°, there is a clear local minimum of the elastic force, and the change in elastic force with respect to $\theta_{\rm p}$ is greatest when the displacement is maximum, that is, 8.0 mm. The same tendency can also be seen in the simulation results. The results therefore indicate that the proposed elastic model is able to represent a distinctive phenomenon, *i.e.*, a local minimum elastic force, even when the derivation process is represented simply using linear virtual springs aligned in the normal direction. On the

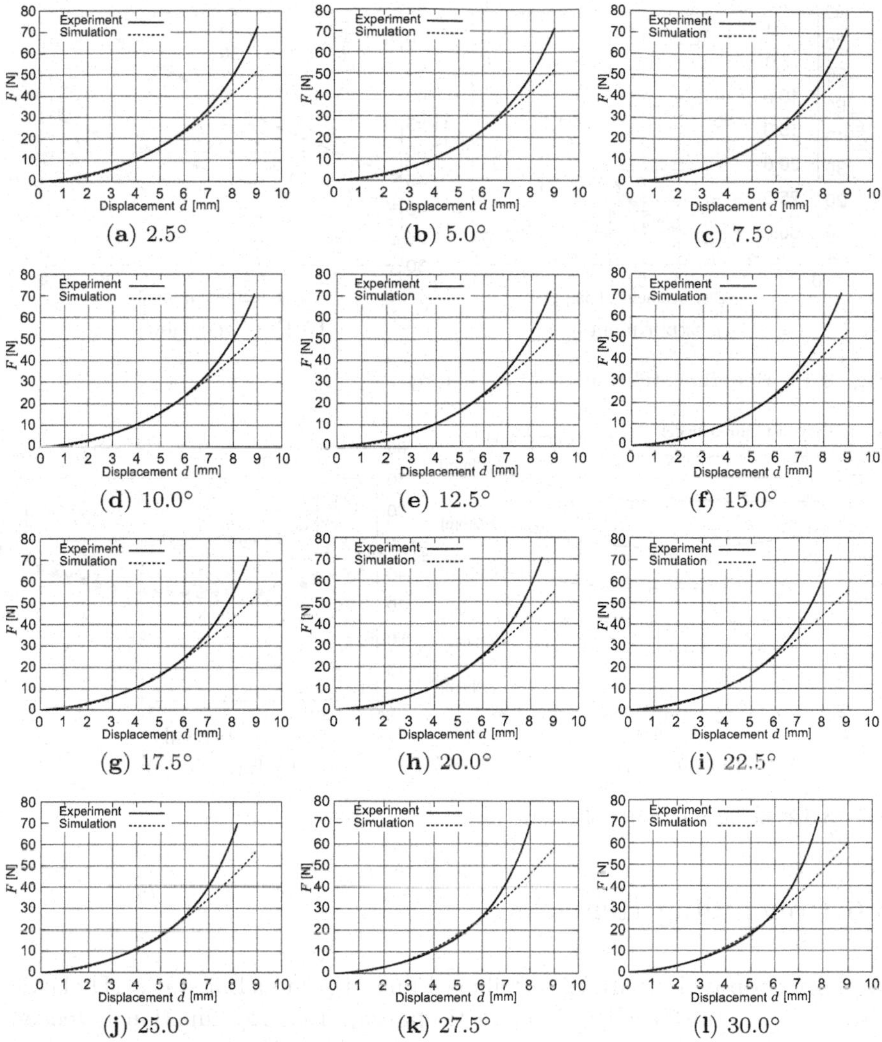

Fig. 3.10 Elastic forces in experiments

other hand, the discrepancy in the large displacement shown in Fig. 3.12 would be reduced if the Young modulus could be defined as a nonlinear function of compression strain and be used to adopt the model to accommodate the nonlinearity of the material. However, this elastic force model focuses on the geometric nonlinearity, and the derivation process including both nonlinearities will be addressed in Chap. 9.

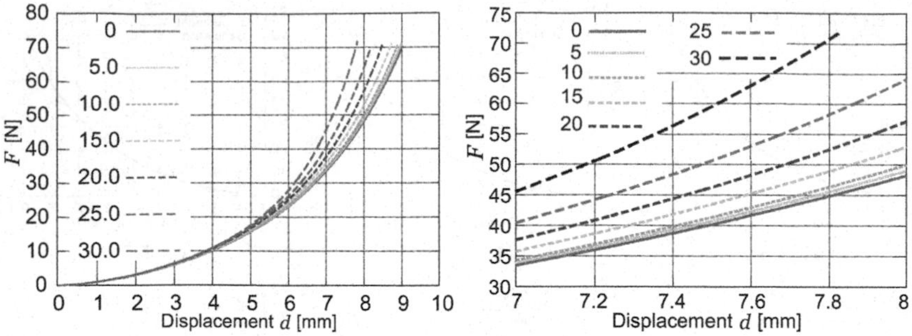

(a) Experiments (b) Enlarged view

Fig. 3.11 Experimental results of elastic force

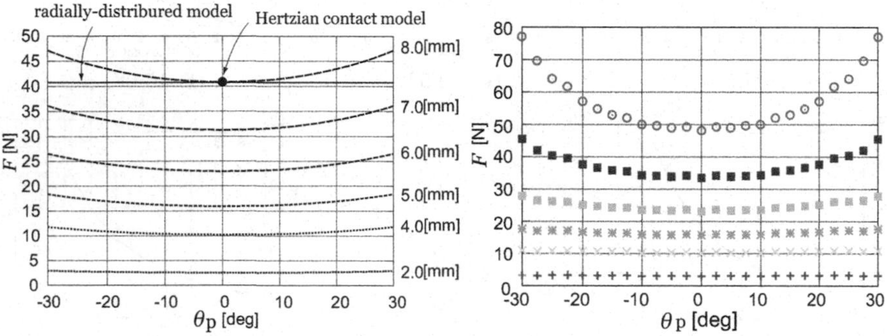

(a) Simulations (b) Experiments

Fig. 3.12 Local minimum of elastic force

3.6 Concluding Remarks

We have proposed a parallel-distributed fingertip model, a static 1D elastic force model, and formulated an elastic potential energy function based on virtual springs inside a hemispherical soft fingertip. We have also proven the existence of an LMEE and experimentally demonstrated that the elastic force due to the deformation has a local minimum. The proposed model requires only the measurement of the Young's modulus of a corresponding material to be used in robotic fingertips. In future studies, we will consider the constant-volume deformation of incompressible elastomer materials and derive elastic models incorporating a nonlinear elasticity.

By expanding the new concept of LMEE in the development of grasping and manipulation theory using a soft-fingered robotic hand, it is expected that the stable grasping and the pose control of a grasped object by a minimal-DOF two-fingered hand may be achieved and a succinct control system will be designed.

Chapter 4
Fingertip Model with Tangential Deformation

4.1 Introduction

The only difference between hard-contact handling and soft-fingered handling is the hardness/softness of the contact surface, but this has a significant impact on manipulation ability. A model of a soft finger that is applicable to theoretical analysis is important in order to describe and understand the static and dynamic behavior of an object being grasped by a multifingered robotic hand with soft fingertips.

Recently, a number of studies have investigated soft-fingered manipulation [XK99, KY04, XBK00, NA01]. Most of these studies used soft fingers/covers on a robotic hand merely as an intermediate object in order to increase the coefficient of friction or the frictional force, and therefore the soft materials themselves were not recognized as important for secure grasping and fine manipulation. Applying a reasonable and relatively rigorous model to the soft-fingered system enables us to reveal physical mechanisms related to stable and robust manipulation.

This chapter proposes a new parallel-distributed fingertip model (*2D elastic model*) of a hemispherical soft fingertip, which is an extension of the 1D model of virtual springs described in Sect. 3.2. We show the existence of a *local minimum of elastic potential energy (LMEE)* even in the case of the 2D model developed in this chapter. The physical mechanism of the LMEE is explained and the LMEE with constraints (LMEEwC) caused by the closed-link mechanism during the manipulating motion is newly introduced. Using the elastic energy model derived herein, we define an internal energy function with geometric constraints and numerically clarify the LMEEwC during manipulation with/without the gravitational force. Finally, we validate the proposed 2D model by evaluating the locus of the object grasped by two rotational fingers with soft fingertips.

4.2 Two-dimensional Elastic Energy Model

4.2.1 Derivation of the Energy Equation

Figure 4.1 shows an extended 2D model of a fingertip, in which a virtual spring having spring constant k is placed vertically inside the fingertip. The spring can be compressed and also deforms in the lateral direction, which corresponds to its bending motion. We assume that each constant related to compression and bending is equal in this model.

We first consider the $\mathrm{O}-xyz$ coordinate system, as shown in Fig. 4.1. Let a be the radius, and let d_{n} be the normal displacement with respect to the contacted surface. Let d_{t} be the tangential displacement on the fingertip, and let θ_{p} be the attitude of the object with respect to the z-axis. From the above assumption, the common spring constant, k, along the vertical and lateral direction is identical to Eq. 3.3 and can thus be rewritten as

$$k = \frac{E\,\mathrm{d}S}{\sqrt{a^2-(x^2+y^2)}}, \tag{4.1}$$

where E is the Young's modulus of the fingertip material and $\mathrm{d}S$ corresponds to the infinitesimal sectional area of the spring. The elastic energy P induced by vertical and lateral deformation can then be expressed as

$$P = \frac{1}{2}\int_{ell} k\left\{(\mathrm{PQ}+d_{\mathrm{t}}\sin\theta_{\mathrm{p}})^2 + d_{\mathrm{t}}^2\cos^2\theta_{\mathrm{p}}\right\}. \tag{4.2}$$

Equation 4.2 is transformed into

$$\begin{aligned} P &= \frac{1}{2}\int_{ell} k\left\{\mathrm{PQ}^2 + d_{\mathrm{t}}^2\sin^2\theta_{\mathrm{p}} + 2\mathrm{PQ}d_{\mathrm{t}}\sin\theta_{\mathrm{p}} + d_{\mathrm{t}}^2\cos^2\theta_{\mathrm{p}}\right\} \\ &= \frac{1}{2}\int_{ell} k\left\{\mathrm{PQ}^2 + 2\mathrm{PQ}d_{\mathrm{t}}\sin\theta_{\mathrm{p}} + d_{\mathrm{t}}^2\right\}. \end{aligned} \tag{4.3}$$

Since k, $k\,\mathrm{PQ}$, and $(1/2)k\,\mathrm{PQ}^2$ in Eq. 4.3 have already been computed using a numerical assumption described in Sect. 3.2, Eq. 4.3 can be written as

$$P(d_{\mathrm{n}}, d_{\mathrm{t}}, \theta_{\mathrm{p}}) = \pi E\left\{\frac{d_{\mathrm{n}}^3}{3\cos^2\theta_{\mathrm{p}}} + d_{\mathrm{n}}^2 d_{\mathrm{t}}\tan\theta_{\mathrm{p}} + d_{\mathrm{n}}d_{\mathrm{t}}^2\right\}. \tag{4.4}$$

Note that Eq. 4.4 is an equation having three independent variables, d_{n}, d_{t}, and θ_{p}, while the previous 1D model had only two variables, d_{n} and θ_{p}.

In addition, when we consider manipulating motions by the two-fingered hand shown in Fig. 4.2, the sum of the elastic energy induced on both fingers can be given by

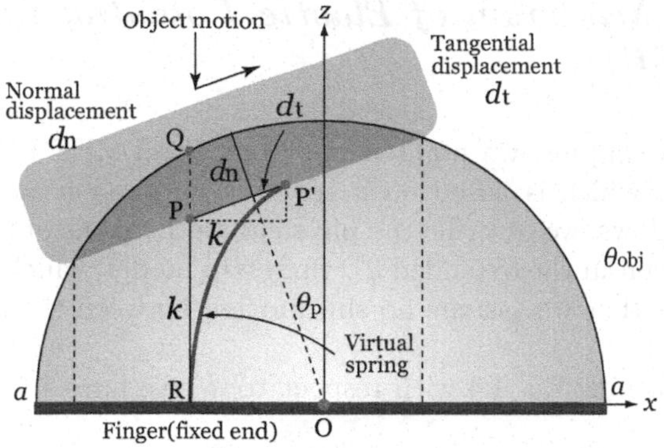

Fig. 4.1 Two-dimensional model of the soft fingertip

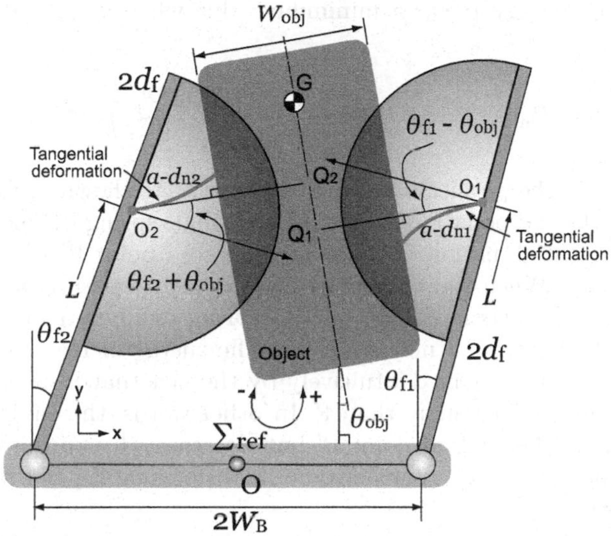

Fig. 4.2 Two-fingered handling

$$P = \pi E \sum_{i=1}^{2} \left\{ \frac{d_{ni}^3}{3\cos^2\theta_{pi}} + d_{ni}^2 d_{ti} \tan\theta_{pi} + d_{ni} d_{ti}^2 \right\}, \tag{4.5}$$

where i denotes the ith finger/fingertip of the robotic hand. Therefore, θ_{pi} is equivalent to $\theta_{fi} + (-1)^i \theta_{obj}$ in Fig. 4.2.

4.2.2 Local Minimum of Elastic Potential Energy (LMEE)

In the previous chapter, we pointed out that the LMEE exists in the 1D fingertip model, which is based on a spring compression along its vertical axis. In what follows, we describe the physical mechanisms of the LMEE that are generated even in the extended 2D fingertip model, which is represented as Eq. 4.4. Note that we assume no slip motion between the grasped object and soft fingers.

First, transforming Eq. 4.4 with respect to d_t, we have

$$P = \pi E d_\mathrm{n}\left\{\left(d_\mathrm{t} + \frac{d_\mathrm{n}}{2}\tan\theta_\mathrm{p}\right)^2 + \frac{d_\mathrm{n}^2}{3\cos^2\theta_\mathrm{p}} - \frac{d_\mathrm{n}^2}{4}\tan^2\theta_\mathrm{p}\right\}. \tag{4.6}$$

Thus, potential energy P has a minimum value when $d_\mathrm{t} = -(d_\mathrm{n}/2)\tan\theta_\mathrm{p}$, and it yields

$$P_{\mathrm{min}}|_{d_\mathrm{t}} = \pi E d_\mathrm{n}^3\left(\frac{1}{3\cos^2\theta_\mathrm{p}} - \frac{\tan^2\theta_\mathrm{p}}{4}\right). \tag{4.7}$$

Next, we explain the physical meaning of Eq. 4.7. As shown in Fig. 4.3a, we consider that the end point P of a compressed single spring PR moves according to the tangential motion of the object up to a point P′ without changing the angle θ_p. When the tangential displacement satisfies $d_\mathrm{t} = (\mathrm{PP'}) = -(d_\mathrm{n}/2)\tan\theta_\mathrm{p}$, the elastic energy on the spring exhibits a minimal value. Although, at first glance, it may seem that the energy increases slightly, this phenomenon can be explained intuitively by the fact that P′ is biased to the vertical direction slightly more than P. In other words, the energy increases due to the lateral bending at point P′ but decreases due to the vertical decompression of the spring. Furthermore, transforming Eq. 4.4 with respect to θ_p, we have

$$P = \pi E d_\mathrm{n}\left\{\frac{d_\mathrm{n}^2}{3}\left(\tan\theta_\mathrm{p} + \frac{3d_\mathrm{t}}{2d_\mathrm{n}}\right)^2 + \frac{d_\mathrm{n}^2}{3} + \frac{d_\mathrm{t}^2}{4}\right\}. \tag{4.8}$$

Thus, the potential energy has a minimum value when $\tan\theta_\mathrm{p} = -3d_\mathrm{t}/(2d_\mathrm{n})$, which yields

$$P_{\mathrm{min}}|_{\theta_\mathrm{p}} = \pi E d_\mathrm{n}\left(\frac{d_\mathrm{n}^2}{3} + \frac{d_\mathrm{t}^2}{4}\right). \tag{4.9}$$

Likewise, this result means that the energy on the single spring also has a minimal value when PP′ is constant, as shown in Fig. 4.3b.

On the other hand, since d_n is positive during successful manipulation, as shown in Fig. 4.2, Eq. 4.4 is a monotonically increasing function with respect to d_n. Therefore, the analysis regarding the LMEE with d_n should

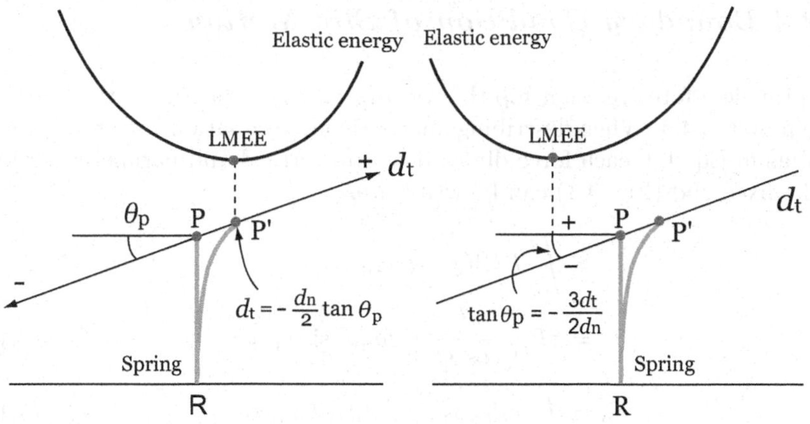

(a) With respect to d_t (Eq. 4.7) (b) With respect to θ_p (Eq. 4.9)

Fig. 4.3 LMEE

be performed in the case of a two-fingered grasping motion. Hence, it is obvious that a minimum value of elastic energy exists in the d_n-direction. The individual minimal energy with respect to d_n, d_t, and θ_p plays an important role in stable grasping and robust manipulation. The actual effectiveness of the LMEE in manipulating operations is mentioned in the subsequent section.

4.2.3 Restoring Moment for a Contacted Object

Next, we explain the most important physical phenomenon that directly affects stable grasping and robust manipulation in the soft-fingered handling.

By partially differentiating the energy equation P written as Eq. 4.4 with respect to θ_p, we obtain

$$\frac{\partial P}{\partial \theta_p} = \frac{\pi E d_n^2}{\cos^2 \theta_p} \left(d_t + \frac{2}{3} d_n \tan \theta_p \right). \tag{4.10}$$

In Eq. 4.10, the first term denotes a *restoring moment* related to the lateral bending motion of the fingertip, and the second term is another moment associated with the vertical compression of the fingertip. This implies that the vertical and lateral elastic moments induced by soft deformation always contribute to secure grasping and manipulation. This theoretical perspective can be verified by a compression test of a soft fingertip, as shown in Fig. 3.12, in which the elastic force measured by a load cell increases gradually as θ_p becomes larger.

4.2.4 Boundary Condition of Slip Motion

We provide a force relationship that occurs on the right side of the hand, as shown in Fig. 4.4. When describing an elastic force equation with the integral sign, as in Eq. 4.2, each force directed to the vertical and parallel directions to the fixed end (Fig. 4.1) can be written as

$$F_{\mathrm{v}} = \int_{ell} k\left(\mathrm{PQ} + d_{\mathrm{t}} \sin\theta_{\mathrm{p}}\right) = \pi E\left(\frac{d_{\mathrm{n}}^2}{\cos\theta_{\mathrm{p}}} + 2d_{\mathrm{n}}d_{\mathrm{t}}\sin\theta_{\mathrm{p}}\right), \tag{4.11}$$

$$F_{\mathrm{p}} = \int_{ell} kd_{\mathrm{t}}\cos\theta_{\mathrm{p}} = 2\pi Ed_{\mathrm{n}}d_{\mathrm{t}}\cos\theta_{\mathrm{p}}. \tag{4.12}$$

Letting μ be the static friction coefficient between an object and soft fingers, the local boundary condition of slip motion on one side of the hand satisfies

$$\frac{\pi Ed_{\mathrm{n}}^2\tan\theta_{\mathrm{p}} + 2\pi Ed_{\mathrm{n}}d_{\mathrm{t}} - \lambda_{\mathrm{t}1}}{\pi Ed_{\mathrm{n}}^2 + \lambda_{\mathrm{n}1}} < \mu. \tag{4.13}$$

Note that an accurate boundary condition can be obtained by also treating the force relationship on another side in two-fingered manipulation.

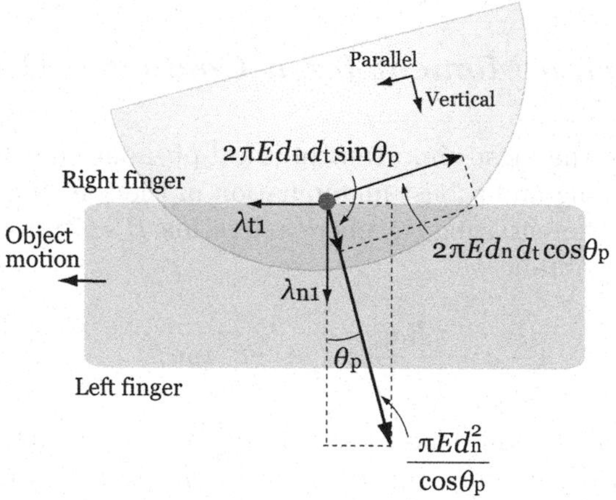

Fig. 4.4 Force relationship around the contact region for the right side of the finger

4.3 Formulation of Geometric Constraints

4.3.1 Normal Constraints

Not only in the case of hard contact manipulation but also in actual soft-fingered manipulation there exist geometric constraints between the object and soft fingers as long as the contact continues. These constraints can be found in the normal and tangential directions to the object surface. In what follows, we formulate two sets of geometric constraints, normal and tangential, for a soft-fingered robotic hand with two rotational joints while grasping and manipulating a planar rigid object, as illustrated in Fig. 4.2. Subsequent processes of the formulation are based on the point-contact modeling between two bodies [MLS94] and the contact modeling between a soft fingertip and a rigid body formulated by Arimoto *et al.* [ANH+00, ATY+01, ADN+02, HAK+01].

These geometric constraints act in the normal and tangential directions with respect to the object surface, which is grasped by the two-fingered hand. Let a be the radius of the fingertip, and let $d_{\mathrm{n}i}$ be the maximum displacement of the ith fingertip to the normal direction. Let L be the length of the finger, and let W_{obj} be the width of the grasped object. Let $2W_{\mathrm{B}}$ be the width of the fingers at their base, and let $2d_{\mathrm{f}}$ be the thickness of the finger. Furthermore, let $\theta_{\mathrm{f}i}$ be the joint angle of the ith finger, and let O_i be the ith fingertip origin. Let O be the origin of the reference coordinate system Σ_{ref} with respect to the midpoint of $2W_{\mathrm{B}}$. Letting θ_{obj} be the orientation angle of the object with respect to Σ_{ref}, which is positive counterclockwise, the coordinate of point O_i is described with respect to Σ_{ref} as follows:

$$\mathrm{O}_{ix} = -(-1)^i W_{\mathrm{B}} + (-1)^i L \sin\theta_{\mathrm{f}i} + (-1)^i d_{\mathrm{f}} \cos\theta_{\mathrm{f}i}, \tag{4.14}$$

$$\mathrm{O}_{iy} = L\cos\theta_{\mathrm{f}i} - d_{\mathrm{f}} \sin\theta_{\mathrm{f}i}, \tag{4.15}$$

where $i = 1, 2$ denotes the right and left fingers, respectively.

As shown in Fig. 4.5, for the left fingertip, let C be the geometric center position, and let $\mathrm{G}(x_{\mathrm{obj}}, y_{\mathrm{obj}})$ be the center of gravity of the grasped object, which is placed at (w, h) far from point C. Letting A be the center of contact of the circular area and Q_i the foot of a perpendicular to line $\mathrm{O}_i\mathrm{A}$, the length of $\mathrm{O}_i\mathrm{Q}_i$ can be represented as follows:

$$\mathrm{O}_i\mathrm{Q}_i = (-1)^i(x_{\mathrm{obj}} - \mathrm{O}_{ix})\cos\theta_{\mathrm{obj}} + (-1)^i(y_{\mathrm{obj}} - \mathrm{O}_{iy})\sin\theta_{\mathrm{obj}}. \tag{4.16}$$

On the other hand, considering the soft-fingertip deformation, a geometric constraint, $C_{\mathrm{n}i}$, between the ith fingertip and the object, which is normal to the object surface, is then expressed by

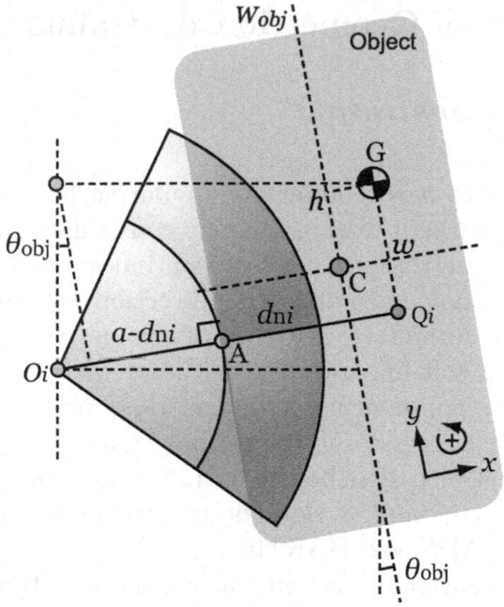

Fig. 4.5 Geometric relationship in the normal direction between the ith fingertip and a grasped object

$$C_{ni} = (-1)^i(x_{\text{obj}} - \text{O}_{ix})\cos\theta_{\text{obj}} + (-1)^i(y_{\text{obj}} - \text{O}_{iy})\sin\theta_{\text{obj}} \\ -(a - d_{\text{n}i}) - \frac{W_{\text{obj}}}{2} - (-1)^i w = 0. \quad (4.17)$$

Note that the geometric normal constraints for rigid-fingered manipulation can be represented by substituting $d_{\text{n}i} = 0$ into Eq. 4.17.

4.3.2 Tangential Constraints

To formulate the tangential constraints appearing on both fingertips during rolling motion, *i.e.*, *rolling constraints*, we assume that there is no slip between the fingertips and the grasped object. As well as the previous formulation of the normal constraints, we refer to the rolling contact modeling between rigid bodies [MLS94] and the rolling contact modeling between a soft fingertip and a rigid body [ANH+00, ATY+01, ADN+02, HAK+01]. Here, we incorporate the tangential deformation into the rolling contact model.

As shown in Fig. 4.6, for the left fingertip, when the object rolls on the soft fingertip of radius $a - d_{\text{n}i}$, the rolling distance AA′ can be expressed by

$$\text{AA}' = (a - d_{\text{n}i})\left\{\theta_{\text{f}i} + (-1)^i\theta_{\text{obj}}\right\}. \quad (4.18)$$

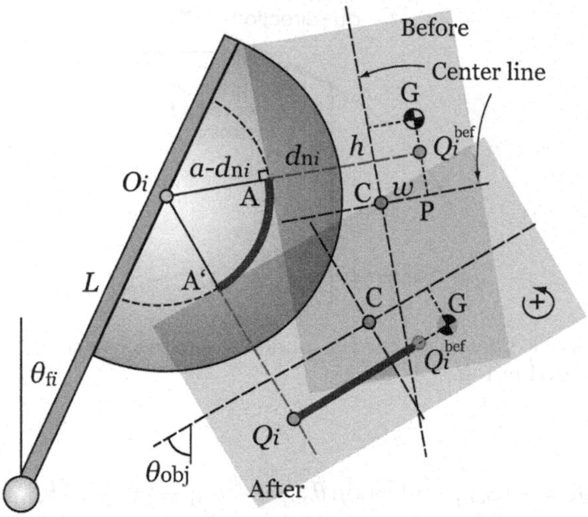

Fig. 4.6 Geometric relationship along the tangential direction between the ith fingertip and a grasped object

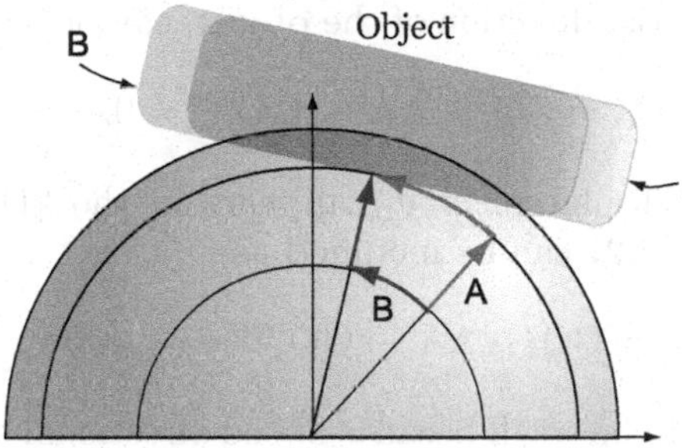

Fig. 4.7 Discrepancy of an object due to nonholonomic rolling

This type of rolling motion corresponds to a nonholonomic constraint because d_{ni} is a variable number during the rolling. The nonholonomic equation causes a discrepancy in the rolling path (Fig. 4.7) such that each different selection for the path between A and B leads to the deviation of the final position of the object when iterating the rolling motion back and forth. Hence, in this chapter, we assume the maximum displacement d_{ni} to be small, which results in Eq. 4.18 being transformed into [CC95]

$$\mathrm{AA'} = a\left\{\theta_{\mathrm{f}i} + (-1)^i\theta_{\mathrm{obj}}\right\}. \tag{4.19}$$

On the other hand, the length GQ_i is expressed by the geometric relationship (Fig. 4.5) as follows:

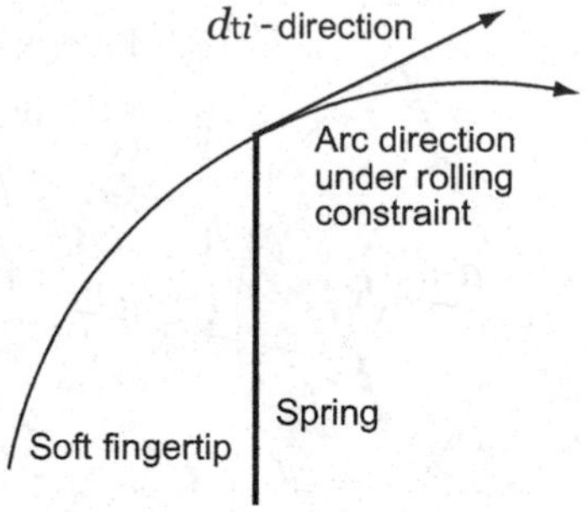

Fig. 4.8 Directional difference between tangential displacement $d_{\mathrm{t}i}$ and rolling distance AA′ illustrated in Fig. 4.6

$$\mathrm{GQ}_i = -(x_{\mathrm{obj}} - \mathrm{O}_{ix}) \sin \theta_{\mathrm{obj}} + (y_{\mathrm{obj}} - \mathrm{O}_{iy}) \cos \theta_{\mathrm{obj}}, \tag{4.20}$$

where θ_{obj} is a positive number during counterclockwise rolling. Since the rolling distance should generally be the sum of all past paths on the fingertip from an initial state, the tangential constraint, $C_{\mathrm{t}i}$, on the ith finger, which is directed in the tangential direction of the object, can be written as (Fig. 4.6)

$$C_{\mathrm{t}i} = \mathrm{GQ}_i - \mathrm{AA}' - \mathrm{GQ}_i^{\mathrm{bef}} = 0. \tag{4.21}$$

Since the tangential displacement $d_{\mathrm{t}i}$ appears in the 2D model shown in Figs. 4.1 and 4.8, Eq. 4.21 can be modified as

$$C_{\mathrm{t}i} = \mathrm{GQ}_i - \mathrm{AA}' - \mathrm{GQ}_i^{\mathrm{bef}} - d_{\mathrm{t}i} = 0, \tag{4.22}$$

where $d_{\mathrm{t}i}$ is assumed to be small because of the directional difference between $d_{\mathrm{t}i}$ and AA′, as shown in Fig. 4.8. Equation 4.22 is then rewritten as

$$\begin{aligned} C_{\mathrm{t}i} = & -(x_{\mathrm{obj}} - \mathrm{O}_{ix}) \sin \theta_{\mathrm{obj}} + (y_{\mathrm{obj}} - \mathrm{O}_{iy}) \cos \theta_{\mathrm{obj}} \\ & - a \left\{ \theta_{\mathrm{f}i} + (-1)^i \theta_{\mathrm{obj}} \right\} - \mathrm{GQ}_i^{\mathrm{bef}} - d_{\mathrm{t}i} = 0, \end{aligned} \tag{4.23}$$

where $\mathrm{GQ}_i^{\mathrm{bef}}$ is previously determined by the corresponding rolling on the previous movement and is a constant value in Eq. 4.23. Utilizing Eq. 4.23, particularly all of the paths $\mathrm{GQ}_i^{\mathrm{bef}}$ on the rolling motion, results in a cumbersome procedure, especially in the analysis of dynamic systems. In this chapter, however, we deal with manipulating tasks by soft fingers from the viewpoint of static analysis based on force and moment equilibrium conditions. Hence, in the subsequent section, we make use of the normal and tangential constraints described as a pair of positional equations throughout this chapter.

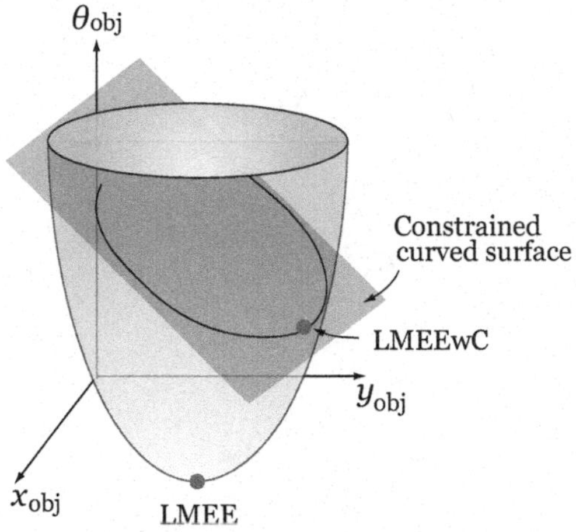

Fig. 4.9 Image diagram of LMEEwC

4.3.3 LMEE with Constraints (LMEEwC)

Now, in actual object handling, the object grasped by soft-fingered hands does not come to rest at the LMEE discussed in the previous chapter. In fact, it goes to an LMEE by satisfying the above four constraints, as illustrated in Fig. 4.9. Since the analytical approach for the LMEEwC is difficult due to the nonlinearity of the elastic energy equation represented in Eq. 4.5, we show the existence of the LMEEwC in Chap. 6 through a procedure by which we define an internal energy function and treat it as a static equilibrium problem.

4.4 Concluding Remarks

An enhanced 2D elastic model of a hemispherical soft fingertip has been introduced, and the mechanisms of stable grasping and robust manipulation in the soft-fingered hand have been clarified. In addition, we have explained the physical meaning of LMEE and LMEEwC that always exists in the handling motion of soft fingers.

Chapter 5
Variational Formulations in Mechanics

5.1 Introduction

Variational principles in mechanics [CKK+68, GPS02] provide a straightforward method to formulate a complex mechanical system under constraints. The principles are directly related to numerical optimization and numerical integration of ordinary differential equations under constraints. This chapter reviews the variational principles in mechanics, which are applied to formulate statics and dynamics in grasping and manipulation, as well as numerical methods related to these principles.

5.2 Variational Principles

5.2.1 Variational Principle in Statics

Variational principle in statics requires that a system be in equilibrium if and only if the variation of internal energy vanishes for any geometrically admissible variations. This is equivalent to the assertion that the internal energy of a system reaches its minimum at the stable state of the system. Let us explain this principle in terms of the motion of a simple pendulum.

Let us investigate the motion of a simple pendulum of length l and mass m suspended from point C, as illustrated in Fig. 5.1. An external torque τ is applied to the pendulum around point C. Let θ be the angle of the pendulum. Let us select a coordinate system $\mathrm{O}-xy$, as shown in the figure. Let P be the potential energy of the system, and let W be the work done by the external torque. Energy P and work W are described as follows:

$$P = mgy = mgl(1 - \cos\theta), \qquad W = \tau\theta.$$

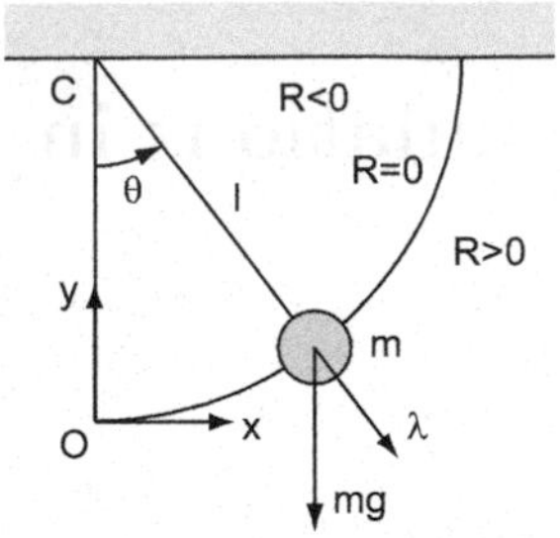

Fig. 5.1 Simple pendulum

According to the variational principle in statics, *internal energy* $I = P - W$ must reach its minimum at the stable state. Thus, we have the following minimization problem:

$$\min\ I(\theta) = mgl(1 - \cos\theta) - \tau\theta. \tag{5.1}$$

We can minimize function $I(\theta)$ by solving the following equation:

$$\frac{\mathrm{d}I}{\mathrm{d}\theta} = mgl\sin\theta - \tau = 0.$$

Consequently, we have

$$\tau - mgl\sin\theta = 0. \tag{5.2}$$

This equation implies that external torque τ and the moment of the gravitational force around point C are balanced.

The variation of the internal energy can be computed as follows:

$$\begin{aligned}\delta I &= \delta P - \delta W \\ &= mgl\sin\theta\delta\theta - \tau\delta\theta \\ &= (mgl\sin\theta - \tau)\,\delta\theta.\end{aligned}$$

According to the variational principle of statics, the variation δI must vanish for any $\delta\theta$. This also yields Eq. 5.2.

Let us formulate the motion of the pendulum in the Cartesian coordinate system $\mathrm{O} - xy$. Let $\boldsymbol{x} = [x, y]^{\mathrm{T}}$ be the position vector of mass m. An external force $\boldsymbol{f} = [f_x, f_y]^{\mathrm{T}}$ is applied to the mass of the pendulum. The potential energy P and work done by the external force W are described as follows:

$$P = mgy, \qquad W = \boldsymbol{f}^{\mathrm{T}}\boldsymbol{x} = f_x x + f_y y.$$

The following geometric constraint must be satisfied:

$$R(x, y) = \left\{x^2 + (y - l)^2\right\}^{\frac{1}{2}} - l = 0. \tag{5.3}$$

The internal energy $I = P - W$ must reach its minimum at the stable state under the above geometric constraint. Thus, we have the following conditional minimization problem:

$$\begin{aligned} \min \ & I(x,y) = mgy - f_x x - f_y y \\ \text{subject to } \ & R(x,y) = 0. \end{aligned}$$

Introducing a Lagrange multiplier λ, the above conditional minimization becomes an unconditional minimization:

$$\min \ J(x,y,\lambda) = I(x,y) - \lambda R(x,y). \tag{5.4}$$

We can minimize the function $J(x,y,\lambda)$ by solving the following equations:

$$\begin{aligned} \frac{\partial J}{\partial x} &= -f_x - \lambda \frac{x}{\{x^2 + (y-l)^2\}^{\frac{1}{2}}} = 0, \\ \frac{\partial J}{\partial y} &= mg - f_y - \lambda \frac{y-l}{\{x^2 + (y-l)^2\}^{\frac{1}{2}}} = 0, \\ \frac{\partial J}{\partial \lambda} &= \{x^2 + (y-l)^2\}^{\frac{1}{2}} - l = 0. \end{aligned} \tag{5.5}$$

From the above equations we have

$$f_x(l-y) + f_y x - mgx = 0. \tag{5.6}$$

This equation implies that the moment of the external force $\boldsymbol{f}$ and the moment of the gravitational force around point C are balanced. Consequently, the above equation is equivalent to Eq. 5.2.

The function $J(x,y,\lambda)$ can be rewritten as follows:

$$J = P - \{\tau\theta + \lambda R(x,y)\}.$$

The Lagrange multiplier λ has the dimension of force because the quantity R has the dimension of length. The above equation suggests that λ characterizes a force to constrain the mass m on the circle $R(x,y) = 0$. This force is referred to as a *constraint force.* A constraint force acts in the normal direction of a constraint. Since R is positive outside the circle and negative inside the circle, the force λ represents the magnitude of the outward constraint force corresponding to $R(x,y) = 0$, as illustrated in the figure.

The variation of the internal energy can be computed in $\mathrm{O}-xy$ as follows:

$$\begin{aligned} \delta I &= \delta P - \delta W \\ &= mg\delta y - f_x \delta x - f_y \delta y. \end{aligned}$$

The variation of geometric constraint $R(x,y) = 0$ yields a condition for geometrically admissible displacement. The variation of the constraint can be computed as follows:

$$\delta R(x,y) = \frac{\partial R}{\partial x}\delta x + \frac{\partial R}{\partial y}\delta y$$
$$= \frac{x}{\{x^2+(y-l)^2\}^{\frac{1}{2}}}\delta x + \frac{y-l}{\{x^2+(y-l)^2\}^{\frac{1}{2}}}\delta y.$$

According to the variational principle in statics, variation δI must vanish for any δx and δy that satisfy $\delta R(x,y) = 0$. Eliminating δy yields

$$\delta I = \left\{ -f_x - \frac{x}{y-l}(mg - f_y) \right\} \delta x.$$

The variation δI must vanish for any δx. This also yields Eq. 5.6. Alternatively, $\delta I = 0$ under $\delta R(x,y) = 0$ can be converted into the requirement that the variation $\delta I - \lambda \delta R$ vanishes for any δx and δy, where λ denotes a Lagrange multiplier. Computing the variation, we have

$$\delta I - \lambda \delta R$$
$$= \left[-f_x - \lambda \frac{x}{\{x^2+(y-l)^2\}^{\frac{1}{2}}} \right] \delta x + \left[mg - f_y - \lambda \frac{y-l}{\{x^2+(y-l)^2\}^{\frac{1}{2}}} \right] \delta y.$$

The variation $\delta I - \lambda \delta R$ must vanish for any δx and δy. This also yields Eq. 5.5, which is followed by Eq. 5.6.

5.2.2 Variational Principle in Dynamics

The *variational principle in dynamics* requires that a geometrically admissible motion of a holonomic system between two configurations at specified times is natural if and only if the variation of action integral vanishes for any variations. This is equivalent to the Lagrange equations of motions. Let us explain the principle using the motion of a simple pendulum illustrated in Fig. 5.1.

Assume that an external torque $\tau(t)$ is applied to the pendulum around point C at time t. Let $\theta(t)$ be the angle of the pendulum at time t. Let K be the kinetic energy of the system, and let P be its potential energy. Let W be the work done by the external torque. Recall that the inertial moment of the pendulum around point C is given by ml^2. Energies K and P and work W are described as follows:

$$K = \frac{1}{2}(ml^2)\dot{\theta}^2, \qquad P = mgl(l - \cos\theta), \qquad W = \tau\theta.$$

The *Lagrangian* of a system is defined as $\mathcal{L} = K - P + W$. Thus, the Lagrangian of the pendulum is described as

$$\mathcal{L}(\theta, \dot{\theta}) = \frac{1}{2}(ml^2)\dot{\theta}^2 - mgl(l - \cos\theta) + \tau\theta.$$

Note that

$$\frac{\partial \mathcal{L}}{\partial \theta} = -mgl \sin\theta + \tau, \qquad \frac{\partial \mathcal{L}}{\partial \dot{\theta}} = (ml^2)\dot{\theta}.$$

The Lagrange motion equation of the pendulum is given as follows:

$$\frac{\partial \mathcal{L}}{\partial \theta} - \frac{\mathrm{d}}{\mathrm{dt}}\frac{\partial \mathcal{L}}{\partial \dot{\theta}} = -mgl \sin\theta + \tau - (ml^2)\ddot{\theta} = 0. \tag{5.7}$$

From this equation we have

$$(ml^2)\ddot{\theta} = \tau - mgl \sin\theta. \tag{5.8}$$

Note that $\tau - mgl \sin\theta$ denotes the resultant torque around point C. Consequently, the above equation is equivalent to the equation of rotational motion.

According to the variational principle in dynamics, the variation of the action integral vanishes for any geometrically admissible variations. Let us compute the variation of action integral

$$\text{V.I.} = \int_{t_1}^{t_2} \delta\mathcal{L} \, \mathrm{dt}. \tag{5.9}$$

Computing the variation of the Lagrangian, we have

$$\begin{aligned}
\delta\mathcal{L} &= \frac{1}{2} ml^2 \, \delta(\dot{\theta}^2) - mgl \, \delta(l - \cos\theta) + \tau \, \delta\theta \\
&= ml^2\dot{\theta} \, \delta(\dot{\theta}) - mgl \sin\theta \, \delta\theta + \tau \, \delta\theta \\
&= ml^2\dot{\theta} \, \frac{\mathrm{d}}{\mathrm{dt}}\delta\theta + (-mgl \sin\theta + \tau) \, \delta\theta.
\end{aligned}$$

Since variation $\delta\theta$ must be equal to zero at $t = t_1$ and $t = t_2$, we have

$$\begin{aligned}
\int_{t_1}^{t_2} ml^2\dot{\theta} \, \frac{\mathrm{d}}{dt}\delta\theta \, \mathrm{dt} &= \Big[ml^2\dot{\theta} \, \delta\theta\Big]_{t=t_0}^{t=t_1} - \int_{t_1}^{t_2} ml^2\ddot{\theta} \, \delta\theta \, \mathrm{dt} \\
&= -\int_{t_1}^{t_2} ml^2\ddot{\theta} \, \delta\theta \, \mathrm{dt}.
\end{aligned}$$

Consequently,

$$\text{V.I.} = \int_{t_1}^{t_2} \left\{-ml^2\ddot{\theta} - mgl \sin\theta + \tau\right\} \delta\theta \, \mathrm{dt}. \tag{5.10}$$

The variation of the action integral must vanish for any $\delta\theta$. This also yields Eq. 5.8.

Let us formulate the motion of the pendulum in a Cartesian coordinate system O $-$ xy. Let $\boldsymbol{x} = [x, y]^{\mathrm{T}}$ be the position vector of mass m. An external force $\boldsymbol{f} = [f_x, f_y]^{\mathrm{T}}$ is applied to the mass of the pendulum. Energies K and P and work W are described as follows:

$$K = \frac{1}{2}m(\dot{x}^2 + \dot{y}^2), \qquad P = mgy, \qquad W = f_x x + f_y y.$$

The following geometric constraint must be satisfied:

$$R(x, y) = \{x^2 + (y - l)^2\}^{\frac{1}{2}} - l = 0.$$

The Lagrangian of a system under a geometric constraint is described as $\mathcal{L} = K - P + W + \lambda R$, where λ is a Lagrange multiplier. Thus, the Lagrangian of the pendulum is described as

$$\mathcal{L}(x, y, \dot{x}, \dot{y}) = \frac{1}{2}m(\dot{x}^2 + \dot{y}^2) - mgy + (f_x x + f_y y) + \lambda R(x, y).$$

Note that

$$\frac{\partial \mathcal{L}}{\partial x} = f_x + \lambda \frac{x}{\{x^2 + (y - l)^2\}^{\frac{1}{2}}},$$
$$\frac{\partial \mathcal{L}}{\partial y} = -mg + f_y + \lambda \frac{y - l}{\{x^2 + (y - l)^2\}^{\frac{1}{2}}},$$
$$\frac{\partial \mathcal{L}}{\partial \dot{x}} = m\dot{x}, \qquad \frac{\partial \mathcal{L}}{\partial \dot{y}} = m\dot{y}.$$

The Lagrange motion equation of the pendulum is given as follows:

$$\frac{\partial \mathcal{L}}{\partial x} - \frac{\mathrm{d}}{\mathrm{d}t}\frac{\partial \mathcal{L}}{\partial \dot{x}} = f_x + \lambda \frac{x}{\{x^2 + (y - l)^2\}^{\frac{1}{2}}} - m\ddot{x} = 0, \tag{5.11}$$
$$\frac{\partial \mathcal{L}}{\partial y} - \frac{\mathrm{d}}{\mathrm{d}t}\frac{\partial \mathcal{L}}{\partial \dot{y}} = -mg + f_y + \lambda \frac{y - l}{\{x^2 + (y - l)^2\}^{\frac{1}{2}}} - m\ddot{y} = 0. \tag{5.12}$$

Eliminating the Lagrange multiplier λ yields

$$m\{(l - y)\ddot{x} + x\ddot{y}\} = f_x(l - y) + f_y x - mgx. \tag{5.13}$$

Note that $f_x(l - y) + f_y x - mgx$ denotes the resultant torque around point C. Since $x = l\sin\theta$ and $y = l(1 - \cos\theta)$, we have

$$\ddot{x} = l\{\ddot{\theta}\cos\theta - \dot{\theta}^2\sin\theta\}, \qquad \ddot{y} = l\{\ddot{\theta}\sin\theta + \dot{\theta}^2\cos\theta\},$$

which follows $m\{(l-y)\ddot{x}+x\ddot{y}\} = ml^2\ddot{\theta}$. This implies that Eq. 5.13 is equivalent to Eq. 5.8.

Let us compute the variation of the action integral. Computing the variation of the Lagrangian, we have

$$\delta\mathcal{L} = m\dot{x}\,\frac{\mathrm{d}t}{\mathrm{d}\delta}x + m\dot{y}\frac{\mathrm{d}t}{\mathrm{d}\delta}x - mg\,\delta y + (f_x\,\delta x + f_y\,\delta y) + \lambda\left\{\frac{\partial R}{\partial x}\delta x + \frac{\partial R}{\partial y}\delta y\right\}.$$

Since δx and δy are equal to zero at $t = t_1$ and $t = t_2$, we have

$$\int_{t_1}^{t_2} m\dot{x}\,\frac{\mathrm{d}t}{\mathrm{d}\delta}x\,\mathrm{d}t = -\int_{t_1}^{t_2} m\ddot{x}\,\delta x\,\mathrm{d}t,$$

$$\int_{t_1}^{t_2} m\dot{y}\,\frac{\mathrm{d}t}{\mathrm{d}\delta}y\,\mathrm{d}t = -\int_{t_1}^{t_2} m\ddot{y}\,\delta y\,\mathrm{d}t.$$

Consequently,

$$\begin{aligned}\text{V.I.} &= \int_{t_1}^{t_2}\left\{-m\ddot{x} + f_x + \lambda\frac{\partial R}{\partial x}\right\}\delta x\,\mathrm{d}t \\ &+ \int_{t_1}^{t_2}\left\{-m\ddot{y} - mg + f_y + \lambda\frac{\partial R}{\partial y}\right\}\delta y\,\mathrm{d}t. \qquad (5.14)\end{aligned}$$

The variation of the action integral must vanish for any δx and δy. This also yields Eqs. 5.11 and 5.12.

5.3 Numerical Optimization of Energy Functions

5.3.1 Nelder–Mead Method

This section introduces the *Nelder–Mead method* [NM65], which numerically optimizes an objective function without its derivatives. The steepest descent method and the quasi-Newton method are well-known algorithms used to minimize or maximize an objective function. These algorithms require not only an objective function, but also its derivatives. When the derivatives of an objective function are computable, we can apply the steepest descent method or the quasi-Newton method to compute the minimum value of the objective function. However, it is often the case that the objective function itself is computable but its derivatives are difficult to compute. One example is the case in which a computer simulation calculates the objective function value. Unless the derivatives of all of the functions in the simulation can be formulated, the derivatives of the object function are not computable. This suggests the requirement for numerical optimization of an objective function without its derivatives.

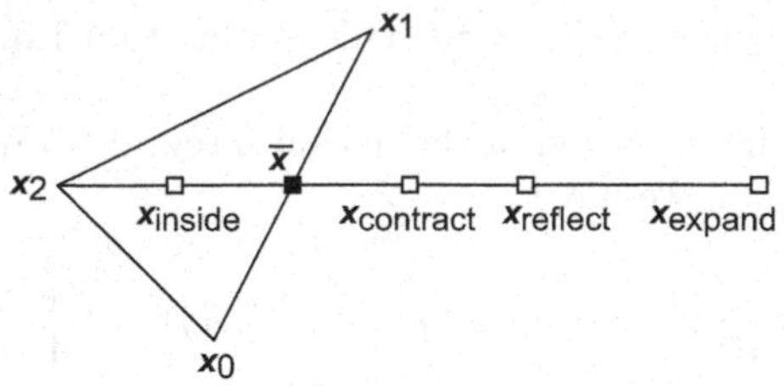

Fig. 5.2 Points to be evaluated in the 2D Nelder-Mead method ($\alpha = \gamma = 1$, $\beta = 1/2$)

The Nelder-Mead method evaluates the objective function at multiple points, compares the function values, and updates the set of points so that the function values decrease. Assume that the objective function depends on n variables, x_1 through x_n. Let $\boldsymbol{x} = [x_1, x_2, \cdots, x_n]^{\mathrm{T}}$ be a collective variable vector, and let $f(\boldsymbol{x})$ be an objective function that should be minimized with respect to $\boldsymbol{x}$. Consequently, the problem to be solved is described as follows:

$$\min\ f(\boldsymbol{x}). \tag{5.15}$$

Note that vector $\boldsymbol{x}$ is a point in n-dimensional space. Let us select $(n+1)$ points in the n-dimensional space. The closure consisting of these points is referred to as a *simplex*. A simplex in 2D space is a triangle and a simplex in 3D space is a tetrahedron. The Nelder-Mead method iteratively updates a simplex specified by $(n+1)$ points in n-dimensional space. An outline of this method is shown below:

Nelder-Mead method

Step 1 Generate the initial simplex.
Step 2 Stop if a convergence condition is satisfied.
Step 3 Update the simplex.
Step 4 Return to **Step 2**.

Let us explain the update of the simplex in the Nelder-Mead method. A simplex can be described by specifying its vertex points as follows:

$$S = \{\boldsymbol{x}_0, \boldsymbol{x}_1, \cdots, \boldsymbol{x}_{n-1}, \boldsymbol{x}_n\}. \tag{5.16}$$

Let us evaluate the objective function $f(\boldsymbol{x})$ at vertex points of the simplex. Let f_0 through f_n be function values at points $\boldsymbol{x}_0$ through $\boldsymbol{x}_n$, respectively. Without loss of generality, we assume that the following relationship is satisfied:

$$f_0 \leq f_1, \cdots, f_{n-2} \leq f_{n-1} \leq f_n. \tag{5.17}$$

That is, f_0 is the minimum value, f_n is the maximum value, and f_{n-1} is the second highest value among the $(n+1)$ function values. Point $\boldsymbol{x}_n$, which corresponds to the maximum value, is replaced by another candidate point, which may have a smaller function value. In order to compute a candidate point, let us compute centroid $\bar{\boldsymbol{x}}$ among vertex points other than $\boldsymbol{x}_n$ as follows:

$$\bar{\boldsymbol{x}} = \frac{1}{n}\sum_{k=0}^{n-1} \boldsymbol{x}_k.$$

Using the centroid, four points, reflection, contraction, expansion, and inner contraction, are defined as follows:

$$\begin{aligned} &\text{reflection} && \boldsymbol{x}_{\text{reflect}} = (1+\alpha)\bar{\boldsymbol{x}} - \alpha \boldsymbol{x}_n, && (5.18)\\ &\text{contraction} && \boldsymbol{x}_{\text{contract}} = (1-\beta)\bar{\boldsymbol{x}} + \beta \boldsymbol{x}_{\text{reflect}}, && (5.19)\\ &\text{expansion} && \boldsymbol{x}_{\text{expand}} = (1+\gamma)\boldsymbol{x}_{\text{reflect}} - \gamma\bar{\boldsymbol{x}}, && (5.20)\\ &\text{inner contraction} && \boldsymbol{x}_{\text{inner}} = (1-\beta)\bar{\boldsymbol{x}} + \beta \boldsymbol{x}_n. && (5.21) \end{aligned}$$

Figure 5.2 illustrates the above four points in the 2D case with $\alpha = \gamma = 1$ and $\beta = 1/2$. Let us evaluate the objective function upon reflection, contraction, expansion, and inner contraction, which are denoted as f_{reflect}, f_{contract}, f_{expand}, and f_{inner}, respectively. The current simplex is updated by comparing the function values at these points.

Update of a simplex

(A) If $f_{\text{reflect}} < f_0$, then

(A-1) If $f_{\text{expand}} \leq f_0$, then $S := \{\boldsymbol{x}_0, \cdots, \boldsymbol{x}_{n-1}, \boldsymbol{x}_{\text{expand}}\}$.
(A-2) If $f_{\text{expand}} > f_0$, then $S := \{\boldsymbol{x}_0, \cdots, \boldsymbol{x}_{n-1}, \boldsymbol{x}_{\text{reflect}}\}$.

(B) If $f_0 \leq f_{\text{reflect}} \leq f_{n-1}$, then $S := \{\boldsymbol{x}_0, \cdots, \boldsymbol{x}_{n-1}, \boldsymbol{x}_{\text{reflect}}\}$.

(C) If $f_{n-1} < f_{\text{reflect}} < f_n$, then

(C-1) If $f_{\text{contract}} \leq f_{\text{reflect}}$, then $S := \{\boldsymbol{x}_0, \cdots, \boldsymbol{x}_{n-1}, \boldsymbol{x}_{\text{contract}}\}$.
(C-2) If $f_{\text{contract}} > f_{\text{reflect}}$, then $S := \{\boldsymbol{x}_0, (\boldsymbol{x}_1 + \boldsymbol{x}_0)/2, \cdots, (\boldsymbol{x}_n + \boldsymbol{x}_0)/2\}$.

(D) If $f_n \leq f_{\text{reflect}}$, then

(D-1) If $f_{\text{inner}} \leq f_n$, then $S := \{\boldsymbol{x}_0, \cdots, \boldsymbol{x}_{n-1}, \boldsymbol{x}_{\text{inner}}\}$.
(D-2) If $f_{\text{inner}} > f_n$, then $S := \{\boldsymbol{x}_0, (\boldsymbol{x}_1 + \boldsymbol{x}_0)/2, \cdots, (\boldsymbol{x}_n + \boldsymbol{x}_0)/2\}$.

The update rules are summarized as Fig. 5.3.

Let us construct the initial simplex in n-dimensional space. Compute the following points

$$\begin{aligned} \boldsymbol{e}_0 &= [0, 0, 0, \cdots, 0, 0]^{\mathrm{T}},\\ \boldsymbol{e}_1 &= [\delta_1, \delta_2, \delta_2, \cdots, \delta_2, \delta_2]^{\mathrm{T}},\\ \boldsymbol{e}_2 &= [\delta_2, \delta_1, \delta_2, \cdots, \delta_2, \delta_2]^{\mathrm{T}}, \qquad (5.22)\\ &\;\;\vdots\\ \boldsymbol{e}_n &= [\delta_2, \delta_2, \delta_2, \cdots, \delta_2, \delta_1]^{\mathrm{T}}, \end{aligned}$$

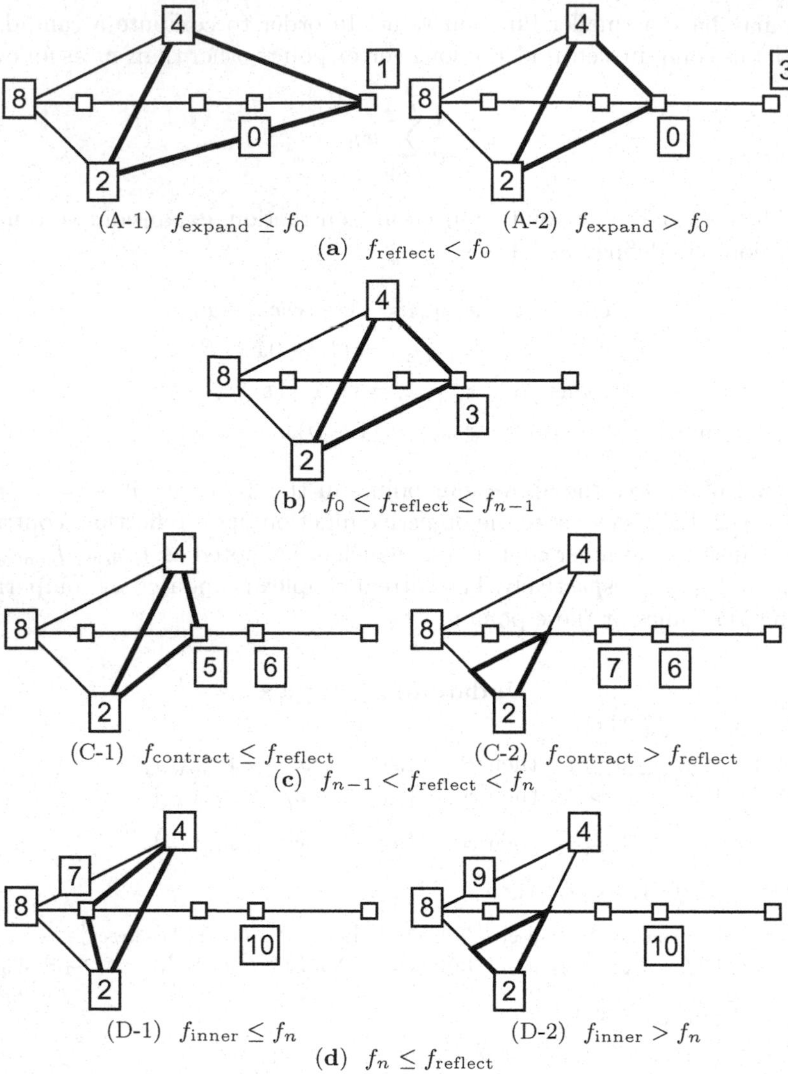

Fig. 5.3 Update of a simplex in the Nelder-Mead method

where

$$\delta_1 = \frac{\sqrt{n+1}+n-1}{\sqrt{2}n}, \qquad \delta_2 = \frac{\sqrt{n+1}-1}{\sqrt{2}n}.$$

The above points provide a simplex of edge length 1 with the coordinate origin. A simplex of edge length L with point $\boldsymbol{x}_{\text{init}}$ is given by

$$S_{\text{init}} = \{L\boldsymbol{e}_0 + \boldsymbol{x}_{\text{init}},\ L\boldsymbol{e}_1 + \boldsymbol{x}_{\text{init}}, \cdots,\ L\boldsymbol{e}_n + \boldsymbol{x}_{\text{init}}\}. \tag{5.23}$$

Given length L and point $\boldsymbol{x}_{\text{init}}$, the initial simplex is computed from the above equation. The iteration ends when function values at the vertex points of the simplex are sufficiently close to one another. Thus, we introduce the variance of function values f_0 through f_n. Let the iteration end when the variance is below a predetermined threshold ϵ, namely, the iteration ends when $\sigma^2 < \epsilon$ is satisfied where

$$\mu = \frac{1}{n+1}\sum_{k=0}^{N} f_k, \quad \sigma^2 = \frac{1}{n+1}\sum_{k=0}^{N}(f_k - \mu)^2.$$

We then find that the objective function takes its minimum value f_0 at $\boldsymbol{x}_0$.

5.3.2 Multiplier Method

This section introduces the *multiplier method*, which optimizes an objective function under equational and inequality constraints. The multiplier method [Hes69] converts an optimization problem with constraints into an unconstrained optimization problem. The converted problem can be optimized by descent methods or the Nelder–Mead method. Let us minimize an objective function $f(\boldsymbol{x})$ under an equational constraint $g(\boldsymbol{x}) = 0$:

$$\begin{aligned} &\min\ f(\boldsymbol{x}) \\ &\text{subject to}\ \ g(\boldsymbol{x}) = 0. \end{aligned} \tag{5.24}$$

Let $\boldsymbol{x}^*$ be the solution of the above minimization problem. Then there exists λ that satisfies the following condition:

$$\nabla f(\boldsymbol{x}^*) + \lambda \nabla g(\boldsymbol{x}^*) = \mathbf{0}. \tag{5.25}$$

That is, gradient vectors $\nabla f(\boldsymbol{x}^*)$ and $\nabla g(\boldsymbol{x}^*)$ are linearly dependent on each other, implying that they lie on a line. Then, we have $\nabla f(\boldsymbol{x}) \cdot \Delta\boldsymbol{x} = 0$, with the result that the value of $f(\boldsymbol{x})$ does not decrease for any $\Delta\boldsymbol{x}$ that satisfies this condition (Fig. 5.4a). On the other hand, when $\nabla f(\boldsymbol{x}^*)$ and $\nabla g(\boldsymbol{x}^*)$ are linearly independent, there exists $\Delta\boldsymbol{x}$ that satisfies $\nabla g(\boldsymbol{x}^*) \cdot \Delta\boldsymbol{x} = 0$ and $\nabla f(\boldsymbol{x}) \cdot \Delta\boldsymbol{x} < 0$. Thus, the value of $f(\boldsymbol{x})$ can be decreased under constraint $g(\boldsymbol{x}) = 0$, implying that $\boldsymbol{x}^*$ is not the minimum solution (Fig. 5.4b). Equation 5.25 is referred to as a *Kuhn–Tucker condition.*

Let us construct the *augmented Lagrangian function* for a minimization problem (Eq. 5.24):

$$L(\boldsymbol{x}; \lambda, r) = f(\boldsymbol{x}) + \lambda\, g(\boldsymbol{x}) + \frac{1}{2} r\, \{g(\boldsymbol{x})\}^2,$$

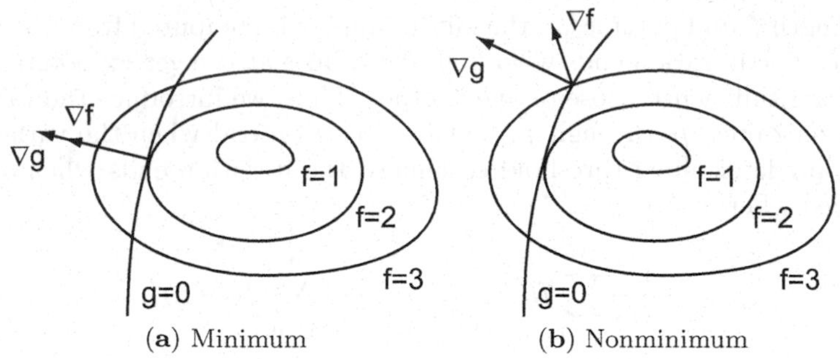

Fig. 5.4 Kuhn–Tucker condition

where r is a positive coefficient. When r is sufficiently large and λ satisfies the Kuhn–Tucker condition, the solution of minimization problem (Eq. 5.24) coincides with the local minimum solution of an augmented Lagrange function. Let us compute Lagrange multiplier λ iteratively. Let λ_k be the Lagrange multiplier in the kth iteration. Note that the local minimum solution of the above augmented Lagrange function satisfies

$$\frac{\partial L}{\partial \boldsymbol{x}}(\boldsymbol{x}^*) = \nabla f(\boldsymbol{x}^*) + \{\lambda + r\, g(\boldsymbol{x}^*)\}\, \nabla g(\boldsymbol{x}^*) = \mathbf{0}.$$

Thus, updating the Lagrange multiplier

$$\lambda_{k+1} = \lambda_k + r\, g(\boldsymbol{x}_k),$$

we can obtain the multiplier that satisfies the Kuhn–Tucker condition. Here it is assumed that coefficient r is sufficiently large; we need an update law of the coefficient. Let r_k be the coefficient value at the kth iteration. Selecting a constant $\alpha > 1$ and updating the coefficient by $r_{k+1} = \alpha r_k$ makes the coefficient sufficiently large. But when the iteration proceeds, the value of r diverges, making numerical computation unstable. Thus, let us modify the update laws of Lagrange multiplier λ and coefficient r. First, let us increase the value of coefficient r without updating the Lagrange multiplier until equational constraint $g(\boldsymbol{x}_k) = 0$ is satisfied so as to find a solution that satisfies the equational constraint. Namely, we introduce the following update law of a Lagrange multiplier:

$$\lambda_{k+1} = \begin{cases} \lambda_k + r\, g(\boldsymbol{x}_k) & \|\, g(\boldsymbol{x}_k)\, \| \le c_k, \\ \lambda_k & \|\, g(\boldsymbol{x}_k)\, \| \ge c_k, \end{cases}$$

where threshold c_k determines if a solution satisfies the constraint. Let us select a constant $\beta \in (0, 1)$ and introduce the following update law of coefficient r:

$$r_{k+1} = \begin{cases} r_k & \| g(\boldsymbol{x}_k) \| \leq \beta c_k \\ \alpha r_k & \| g(\boldsymbol{x}_k) \| \geq \beta c_k \end{cases} .$$

It has been proven that coefficient r can be computed stably by the above update law. Let us introduce an update law of threshold c_k. When the equational constraint is not satisfied at the kth iteration, that is, when $\| g(\boldsymbol{x}_k) \| \geq c_k$ is satisfied, let $c_{k+1} = c_k$ keep the condition for determining if the constraint is satisfied. By contrast, when the equational constraint is satisfied at the kth iteration, that is, when $\| g(\boldsymbol{x}_k) \| \leq c_k$ is satisfied, let $c_{k+1} = \| g(\boldsymbol{x}_k) \|$ make the condition more strict. The above update laws are summarized as follows:

$$\begin{array}{lll} \text{when } \| g(\boldsymbol{x}_k) \| \leq c_k & & \\ \quad \lambda_{k+1} = \lambda_k + r_k\, g(\boldsymbol{x}_k),\ c_{k+1} = \| g(\boldsymbol{x}_k) \| & & \\ \quad \text{when } \| g(\boldsymbol{x}_k) \| \leq \beta c_k & & r_{k+1} = r_k, \\ \quad \text{when } \| g(\boldsymbol{x}_k) \| \geq \beta c_k & & r_{k+1} = \alpha r_k, \\ \text{when } \| g(\boldsymbol{x}_k) \| \geq c_k & & \\ \quad \lambda_{k+1} = \lambda_k, & c_{k+1} = c_k, & r_{k+1} = \alpha r_k. \end{array}$$

Let us minimize an objective function $f(\boldsymbol{x})$ under an inequality constraint $h(\boldsymbol{x}) \leq 0$:

$$\begin{array}{l} \min \ f(\boldsymbol{x}) \\ \text{subject to } \ h(\boldsymbol{x}) \leq 0. \end{array} \tag{5.26}$$

Introduce a *slack variable* y to convert an inequality constraint into an equational constraint:

$$\begin{array}{l} \min \ f(\boldsymbol{x}) \\ \text{subject to } \ h(\boldsymbol{x}) + y^2 = 0. \end{array} \tag{5.27}$$

The augmented Lagrange function of the minimization problem (Eq. 5.27) is given by

$$L'(\boldsymbol{x}, y; \mu, s) = f(\boldsymbol{x}) + \mu\{h(\boldsymbol{x}) + y^2\} + \frac{1}{2}s\{h(\boldsymbol{x}) + y^2\}^2,$$

where s is a positive coefficient. The local minimum solution of the above augmented Lagrange function satisfies

$$\frac{\partial L'}{\partial y} = 2y\left[\mu + s\{h(\boldsymbol{x}) + y^2\}\right] = 0,$$

yielding the following equation:

$$y^2 = -\frac{\mu}{s} - h(\boldsymbol{x}).$$

When the right side of the above equation is positive or equal to zero, the inequality constraint $h(\boldsymbol{x}) \leq 0$ is satisfied. This implies that y^2 in the augmented Lagrange function L' can be replaced by the right side of the above equation. When the right side of the above equation is negative, the inequality constraint $h(\boldsymbol{x}) \leq 0$ is broken and the value of $h(\boldsymbol{x})$ is positive. Thus, let us impose an equational constraint $h(\boldsymbol{x}) = 0$ instead of the inequality constraint so that the value of $h(\boldsymbol{x})$ goes to zero. This suggests that y^2 in the augmented Lagrange function L' should be replaced by 0. Thus, we have the following update law:

$$y^2 = \begin{cases} -\mu/s - h(\boldsymbol{x}) & \mu + sh(\boldsymbol{x}) \leq 0, \\ 0 & \mu + sh(\boldsymbol{x}) \geq 0. \end{cases}$$

Consequently, minimization of the augmented Lagrange function $L'(\boldsymbol{x}, y; \mu, s)$ is equivalently converted into the minimization of an augmented Lagrange function without any slack variable given as follows:

$$L(\boldsymbol{x}; \mu, s) = \begin{cases} f(\boldsymbol{x}) - (1/2)\,\mu^2/s & \mu + sh(\boldsymbol{x}) \leq 0, \\ f(\boldsymbol{x}) + \mu\, h(\boldsymbol{x}) + (1/2)\, s\, \{h(\boldsymbol{x})\}^2 & \mu + sh(\boldsymbol{x}) \geq 0. \end{cases}$$

Update laws of coefficient r and multiplier s are described as follows:

$$\begin{array}{lll}
\text{when } \| h(\boldsymbol{x}_k) + y_k^2 \| \leq c_k & & \\
\quad \mu_{k+1} = \mu_k + s_k\{h(\boldsymbol{x}_k) + y_k^2\},\ c_{k+1} = \| h(\boldsymbol{x}_k) + y_k^2 \|, & & \\
\quad \text{when } \| h(\boldsymbol{x}_k) + y_k^2 \| \leq \beta c_k & & s_{k+1} = s_k, \\
\quad \text{when } \| h(\boldsymbol{x}_k) + y_k^2 \| \geq \beta c_k & & s_{k+1} = \alpha s_k, \\
\text{when } \| h(\boldsymbol{x}_k) + y_k^2 \| \geq c_k & & \\
\quad \mu_{k+1} = \mu_k, & c_{k+1} = c_k, & s_{k+1} = \alpha s_k.
\end{array}$$

Eliminating the slack variable yields

$$\begin{array}{ll}
H_k = \| \max\{h(\boldsymbol{x}_k), -\mu_k/s_k\} \| & \\
\text{when } H_k \leq c_k & \\
\quad \text{when } \mu_k + s_k h(\boldsymbol{x}_k) \leq 0\ \mu_{k+1} = 0, & \\
\quad \text{when } \mu_k + s_k h(\boldsymbol{x}_k) \geq 0\ \mu_{k+1} = \mu_k + s_k\, h(\boldsymbol{x}_k), & \\
\quad c_{k+1} = H_k, & \\
\quad \text{when } H_k \leq \beta c_k & s_{k+1} = s_k, \\
\quad \text{when } H_k \geq \beta c_k & s_{k+1} = \alpha s_k, \\
\text{when } H_k \geq c_k & \\
\quad \mu_{k+1} = \mu_k,\ c_{k+1} = c_k, & s_{k+1} = \alpha s_k.
\end{array}$$

Index H_k denotes how much the inequality constraint is satisfied at the kth iteration.

When we have multiple equational constraints, let us introduce λ and r for each constraint. When we have multiple inequality constraints, let us introduce μ and s for each constraint. Let us minimize an objective function

$f(\boldsymbol{x})$ under multiple equation constraints $g_i(\boldsymbol{x}) = 0,\ (i = 1, 2, \cdots)$ and multiple inequality constraints $h_j(\boldsymbol{x}) \leq 0,\ (j = 1, 2, \cdots)$. That is, we solve the following conditional minimization:

$$\begin{aligned} &\min\ f(\boldsymbol{x}) \\ &\text{subject to}\ \ g_i(\boldsymbol{x}) = 0,\ \ (i = 1, 2, \cdots), \\ &\qquad\qquad\quad h_j(\boldsymbol{x}) = 0,\ \ (j = 1, 2, \cdots). \end{aligned} \tag{5.28}$$

Introducing Lagrangian multipliers $\boldsymbol{\lambda} = [\lambda_1,\ \lambda_2,\ \cdots]^{\mathrm{T}}$ and parameters $\boldsymbol{r} = [r_1,\ r_2,\ \cdots]^{\mathrm{T}}$ for equation constraints as well as multipliers $\boldsymbol{\mu} = [\mu_1,\ \mu_2,\ \cdots]^{\mathrm{T}}$ and parameters $\boldsymbol{s} = [s_1,\ s_2,\ \cdots]^{\mathrm{T}}$ for inequality constraints, we have the following augmented Lagrangian function:

$$\begin{aligned} L(\boldsymbol{x}; \boldsymbol{\lambda}, \boldsymbol{\mu}, \boldsymbol{r}, \boldsymbol{s}) &= f(\boldsymbol{x}) + \sum_i \left\{ \lambda_i\, g_i(\boldsymbol{x}) + \frac{1}{2} r_i \{g_i(\boldsymbol{x})\}^2 \right\} \\ &+ \sum_j \begin{cases} -(1/2)\, (\mu_j)^2 / s_j & \mu_j + s_j h_j(\boldsymbol{x}) \leq 0, \\ \mu_j\, h_j(\boldsymbol{x}) + (1/2)\, s_j \{h_j(\boldsymbol{x})\}^2 & \mu_j + s_j h_j(\boldsymbol{x}) \geq 0. \end{cases} \end{aligned} \tag{5.29}$$

For the sake of simplicity, let us omit suffix k, which describes the number of iterations. The multiplier method is then described as follows:

Multiplier method

Step 1 $\alpha \approx 10$, $\beta \approx 1/4$. Set $\varepsilon > 0$ for convergence test.
Initialization: $\lambda_i = 0$, $\mu_j = 0$, $r_i = 10$, $s_j = 10$, $c = \infty$.
Step 2 Minimize the augmented Lagrangian function to obtain solution $\boldsymbol{x}$.
Step 3 Compute the following quantities:
$G_i = \|\ g_i(\boldsymbol{x})\ \|,\qquad (i = 1, 2, \cdots).$
$H_j = \|\ \max\{h_j(\boldsymbol{x}),\ -\mu_j/s_j\}\ \|,\qquad (j = 1, 2, \cdots).$
Compute $g_{\max} = \max\{G_1,\ G_2,\ \cdots\}$ and $h_{\max} = \max\{H_1,\ H_2,\ \cdots\}$.
Step 4 If $g_{\max} \geq c$ or $h_{\max} \geq c$, then go to **Step 7**.
Step 5 Set $c = \max\{g_{\max}, h_{\max}\}$. If $c \leq \varepsilon$, then stop.
Step 6 Update Lagrangian multipliers:
$\lambda_i := \lambda_i + r_i\, g_i(\boldsymbol{x}),\qquad (i = 1, 2, \cdots).$
$\mu_j := \max\{0,\ \mu_j + s_j h_j(\boldsymbol{x})\},\qquad (j = 1, 2, \cdots).$
Step 7 Update parameters:
If $G_i \leq \beta c$, then $r_i := \alpha r_i,\qquad (i = 1, 2, \cdots).$
If $H_j \leq \beta c$, then $s_j := \alpha s_j,\qquad (j = 1, 2, \cdots).$
Go back to **Step 2**.

Quantities G_i and H_j evaluate how much equation constraint g_i and inequality constraint h_j are satisfied. Unless all equation and inequality constraints are satisfied, that is, while $g_{\max} \geq c$ or $h_{\max} \geq c$ is satisfied, parameters r_i and s_j are increased so that all equation and inequality constraints are satisfied before Lagrangian multipliers are updated. Lagrangian multipliers are updated when all equation and inequality constraints are satisfied. We can apply a quasi-Newton method or the Nelder-Mead method to the minimization of the augmented Lagrangian function. In applying the quasi-Newton method, we need the gradient vectors of the objective function $f(\boldsymbol{x})$, equation constraints $g_i(\boldsymbol{x})$, and inequality constraints $h_j(\boldsymbol{x})$. In applying the Nelder-Mead method, we do not need these gradient vectors.

5.4 Numerical Integration of Equations of Motion

5.4.1 Runge–Kutta Method

Let us solve an ordinary differential equation

$$\dot{x} = f(x, t)$$

at discrete times $t_n = nT$, where T denotes a constant time interval. The following methods provide an iterative equation that computes $x_{n+1} = x(t_{n+1})$ from $x_n = x(t_n)$. This implies that, starting from the initial value $x_0 = x(0)$, we can obtain the value of x_n using the equation iteratively.

Euler method (one-stage method)

$$x_{n+1} = x_n + Tf(x_n, t_n) \tag{5.30}$$

Heun method (two-stage method)

$$\begin{aligned} x_{n+1} &= x_n + \frac{T}{2}(k_1 + k_2), \\ k_1 &= f(x_n, t_n), \\ k_2 &= f(x_n + Tk_1, t_n + T). \end{aligned} \tag{5.31}$$

Runge–Kutta method (four-stage method)

$$\begin{aligned} x_{n+1} &= x_n + \frac{T}{6}(k_1 + 2k_2 + 2k_3 + k_4), \\ k_1 &= f(x_n, t_n), \\ k_2 &= f(x_n + \frac{1}{2}Tk_1, t_n + \frac{1}{2}T), \\ k_3 &= f(x_n + \frac{1}{2}Tk_2, t_n + \frac{1}{2}T), \\ k_4 &= f(x_n + Tk_3, t_n + T). \end{aligned} \tag{5.32}$$

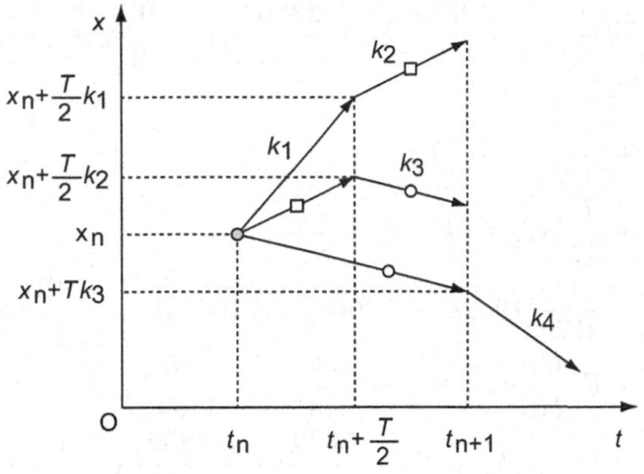

Fig. 5.5 Runge–Kutta method

The Runge–Kutta method is illustrated in Fig. 5.5. As shown in the figure, k_1 through k_4 are gradients at different pairs of x and t. The increment $x_{n+1} - x_n$ is given by a weighted sum of the four gradients. Since $\dot{x}(t_n) = f(x_n, t_n)$, both sides of the formula coincide with each other up to their first order with respect to time interval T in the Euler method. Since $k_2 = f + T(f_x f + f_t)$ and $\ddot{x}(t_n) = f_x f + f_t$, where $f = f(x_n, t_n)$, $f_x = \partial f/\partial x(x_n, t_n)$, and $f_t = \partial f/\partial t(x_n, t_n)$, both sides of the formula coincide with each other up to their second order in the Heun method. It can be shown that both sides of the formula coincide with each other up to their fourth order in the Runge–Kutta method. Consequently, the Euler method, the Heun method, and the Runge–Kutta method provide the first-order, second-order, and fourth-order solutions, respectively.

Time interval T should be constant in the above methods. In addition, an inappropriately large time interval may yield an incorrect solution. Thus, we have to select an appropriately small time interval to obtain a correct solution, which results in significant computation time. Adaptive selection of the time interval has been proposed to improve both solution correctness and time efficiency [Feh68, Feh69]. The following formula allows us to select an appropriate time interval during the computation.

Runge–Kutta–Fehlberg method (six-stage method)

$$\begin{aligned}
x_{n+1} &= x_n + T(\frac{16}{135}k_1 + \frac{6656}{12825}k_3 + \frac{28561}{56430}k_4 - \frac{9}{50}k_5 + \frac{2}{55}k_6),\\
k_1 &= f(x_n, t_n),\\
k_2 &= f(x_n + \frac{T}{4}k_1, t_n + \frac{1}{4}T),\\
k_3 &= f(x_n + \frac{T}{32}(3k_1 + 9k_2), t_n + \frac{3}{8}T),\\
k_4 &= f(x_n + \frac{T}{2179}(1932k_1 - 7200k_2 + 7296k_3), t_n + \frac{12}{13}T),\\
k_5 &= f(x_n + T(\frac{439}{216}k_1 - 8k_2 + \frac{3680}{513}k_3 - \frac{845}{4104}k_4), t_n + T),\\
k_6 &= f(x_n + T(-\frac{8}{27}k_1 + 2k_2 - \frac{3544}{2565}k_3 + \frac{1859}{4104}k_4 - \frac{11}{40}k_5), t_n + \frac{1}{2}T).
\end{aligned} \tag{5.33}$$

The Runge–Kutta–Fehlberg method provides a fifth-order solution. The following algorithm, which is referred to as the *Runge–Kutta–Fehlberg formula*, updates the time interval adaptively:

Step 1 Compute x_{n+1} using Eq. 5.33.
Step 2 Compute x^*_{n+1} given by

$$x^*_{n+1} = x_n + T(\frac{25}{216}k_1 + \frac{1408}{2565}k_3 + \frac{2197}{4104}k_4 - \frac{1}{5}k_5). \tag{5.34}$$

Step 3 Compute $\hat{T}$ given by

$$\hat{T} = \alpha T \left\{ \frac{\epsilon}{\|x^*_{n+1} - x_{n+1}\|} \right\}^{\frac{1}{5}}, \tag{5.35}$$

where ϵ is a small positive allowance and α denotes a safety ratio of approximately 0.8 to 0.9.
Step 4 Select time interval T equal to or smaller than $\hat{T}$.

It can be shown that x^*_{n+1} is a fourth-order solution of the ordinary differential equation. Using the difference of orders between x_{n+1} and x^*_{n+1}, an appropriate time interval T can be selected adaptively.

Any of the above methods can be applied to a set of differential equations

$$\dot{\boldsymbol{x}} = \boldsymbol{f}(\boldsymbol{x}, t),$$

where $\boldsymbol{x}$ consists of a set of state variables and $\boldsymbol{f}$ consists of a set of functions that compute the time derivatives of individual state variables. In one of the above methods, replacing the state variable x by the state variable vector $\boldsymbol{x}$, replacing the scalar function f by the vector function $\boldsymbol{f}$, and replacing the scalar k by the vector $\boldsymbol{k}$ yields a numerical method to integrate a set of ordinary differential equations.

5.4.2 Constraint Stabilization Method

The *constraint stabilization method*, abbreviated *CSM*, provides a numerical computation of a system of differential equations under geometric constraints [Bau72]. Let us explain this method using the motion equation of a simple pendulum.

Let us investigate the motion of a simple pendulum of length l and mass m suspended from point O, as illustrated in Fig. 5.1. Let $\theta(t)$ be the angle of the pendulum at time t. Then, angle θ satisfies the following differential equation:

$$(ml^2)\ddot{\theta} = -mgl\sin\theta,$$

Introducing $\omega = \dot{\theta}$, the above differential equation turns into a system of differential equations of the first order:

$$\dot{\theta} = \omega,$$
$$\dot{\omega} = -\frac{g}{l}\sin\theta.$$

Note that no geometric constraints are imposed in this example. Thus, we can simply apply the Euler method or the Runge–Kutta method to solve the above example numerically.

Let us formulate the motion of a pendulum in Cartesian coordinates. Let (x, y) be the position of the mass. The kinetic energy K of the pendulum and its potential energy P are described as

$$K = \frac{1}{2}m(\dot{x}^2 + \dot{y}^2), \qquad P = mgy.$$

The following geometric constraint must be satisfied:

$$R(x, y) = \left\{x^2 + (y - l)^2\right\}^{\frac{1}{2}} - l = 0. \tag{5.36}$$

The Lagrangian with a geometric constraint is then formulated as follows:

$$\begin{aligned} \mathcal{L} &= K - P + \lambda R \\ &= \frac{1}{2}m(\dot{x}^2 + \dot{y}^2) - mgy + \lambda\left[\left\{x^2 + (y - l)^2\right\}^{\frac{1}{2}} - l\right], \end{aligned} \tag{5.37}$$

where λ denotes a Lagrange multiplier, which corresponds to the magnitude of a constraint force because the quantity R has the dimension of length. Since R is positive outside the circular trajectory of the pendulum but negative inside the circular trajectory of the pendulum, the force λ represents the outward constraint force as illustrated in the figure. From Eq. 5.37 we can derive the Lagrange equations of motion:

$$\lambda R_x(x,y) - m\ddot{x} = 0, \tag{5.38}$$
$$-mg + \lambda R_y(x,y) - m\ddot{y} = 0, \tag{5.39}$$

where $R_x(x,y) = \partial R/\partial x$ and $R_y(x,y) = \partial R/\partial y$. These partial derivatives are given by $R_x(x,y) = xP(x,y)$ and $R_y(x,y) = (y-l)P(x,y)$, where $P(x,y) = (x^2+(y-l)^2)^{-1/2}$. Note that a geometric constraint Eq. 5.36 must be satisfied, but simple application of the Euler method or the Runge–Kutta method may violate the constraint. We have to incorporate the constraint into the numerical computation of differential equations.

Let us introduce a critical damping of the geometric constraint so that the constraint converges to zero during the computation:

$$\ddot{R} + 2\alpha\dot{R} + \alpha^2 R = 0, \tag{5.40}$$

where α is a predetermined angular frequency. Since the above equation provides a critical damping, the quantity R converges to zero quickly and the geometric constraint is again satisfied during the computation, even if the constraint is violated. Substituting Eq. 5.36 into the above equation, we have

$$\begin{aligned} &R_x(x,y)\ddot{x} + R_y(x,y)\ddot{y} + \{\dot{x}^2 + \dot{y}^2\}\, P(x,y) - \{x\dot{x} + (y-l)\dot{y}\}^2 P(x,y)^3 \\ &+2\alpha\{x\dot{x} + (y-l)\dot{y}\}\, P(x,y) + \alpha^2 R(x,y) = 0. \end{aligned} \tag{5.41}$$

Introducing $v_x = \dot{x}$ and $v_y = \dot{y}$, Eqs. 5.38, 5.39, and 5.41 are described as follows:

$$\begin{aligned} \dot{x} &= v_x, \\ \dot{y} &= v_y, \\ m\dot{v}_x - R_x(x,y)\,\lambda &= 0, \\ m\dot{v}_y - R_y(x,y)\,\lambda &= -mg, \\ -R_x(x,y)\,\dot{v}_x - R_y(x,y)\,\dot{v}_y &= C(x,y,v_x,v_y), \end{aligned}$$

where

$$\begin{aligned} C(x,y,v_x,v_y) = (v_x^2 + v_y^2)\, P(x,y) - \{xv_x + (y-l)v_y\}^2 P(x,y)^3 \\ +2\alpha\{xv_x + (y-l)v_y\}\, P(x,y) + \alpha^2 R(x,y). \end{aligned}$$

Consequently, we have the following linear equation on $\dot{x}$, $\dot{y}$, $\dot{v}_x$, $\dot{v}_y$, and λ:

$$\begin{bmatrix} 1 & 0 & 0 & 0 & 0 \\ 0 & 1 & 0 & 0 & 0 \\ 0 & 0 & m & 0 & -R_x(x,y) \\ 0 & 0 & 0 & m & -R_y(x,y) \\ 0 & 0 & -R_x(x,y) & -R_y(x,y) & 0 \end{bmatrix} \begin{bmatrix} \dot{x} \\ \dot{y} \\ \dot{v}_x \\ \dot{v}_y \\ \lambda \end{bmatrix} = \begin{bmatrix} v_x \\ v_y \\ 0 \\ -mg \\ C \end{bmatrix}.$$

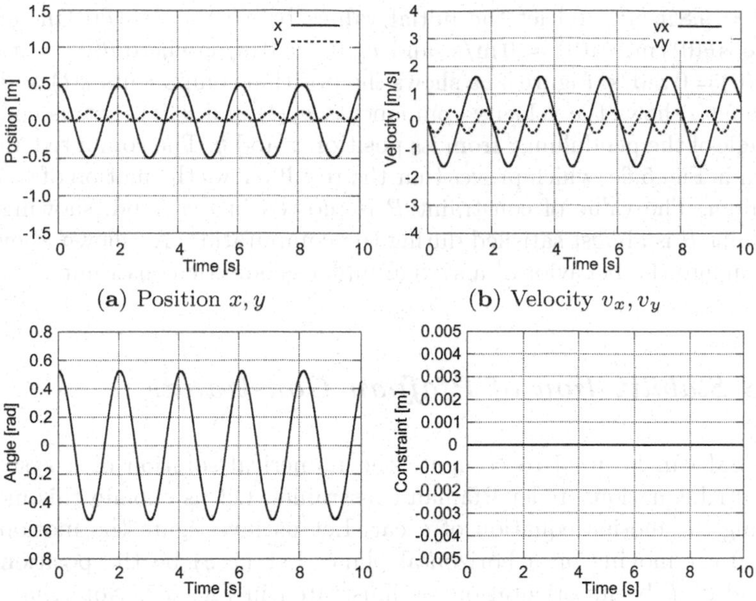

(a) Position x, y (b) Velocity v_x, v_y

(c) Angle θ (d) Constraint R

Fig. 5.6 Simulation of the simple pendulum motion described in Cartesian coordinates ($m = 0.01\,\text{kg}$, $l = 0.99\,\text{m}$, $g = 9.8\,\text{m/s}^2$, initial values: $x(0) = l\sin\theta(0)\,\text{m}$, $y(0) = l(1 - \cos\theta(0))\,\text{m}$, $v_x(0) = 0\,\text{m/s}$, $v_y(0) = 0\,\text{m/s}$, where $\theta(0) = \pi/6\,\text{rad}$ and $\omega(0) = 0\,\text{rad/s}$)

Note that the above linear equation is solvable because the matrix is regular, implying that we can compute $\dot{x}$, $\dot{y}$, $\dot{v}_x$, and $\dot{v}_y$. As a result, we can sketch x, y, v_x, and v_y using the Euler method or the Runge–Kutta method. Solving the above linear equation yields

$$\lambda = m\{gR_y(x, y) - C(x, y, v_x, v_y)\},$$

$$\dot{v}_x = \frac{R_x(x, y)}{m}\lambda, \qquad \dot{v}_x = \frac{R_y(x, y)}{m}\lambda - g.$$

From x, y, v_x, and v_y we can compute λ, then $\dot{v}_x$ and $\dot{v}_y$. Thus, we can apply the Euler method or the Runge–Kutta method to solve the above equation numerically. The above procedure to incorporate geometric constraints into the numerical solution of dynamical equations is referred to as the CSM.

Recall that the gradient vector $[\,R_x,\, R_y\,]^{\mathrm{T}}$ corresponds to the outward normal vector of constraint $R(x, y)$. This vector determines the direction of the constraint force, and the Lagrange multiplier λ describes its magnitude. Thus, the constraint force is given by $\lambda[\,R_x,\, R_y\,]^{\mathrm{T}}$.

Figure 5.6 shows the simulation result for the simple pendulum motion described in Cartesian coordinates. Assume that $m = 0.01\,\text{kg}$, $l = 0.99\,\text{m}$, and $g = 9.8\,\mathrm{m/s^2}$, and let the initial values be $x(0) = l \sin\theta(0)\,\mathrm{m}$, $y(0) = l(1 - \cos\theta(0))\,\mathrm{m}$, $v_x(0) = 0\,\mathrm{m/s}$, and $v_y(0) = 0\,\mathrm{m/s}$, where $\theta(0) = \pi/6\,\mathrm{rad}$ and $\omega(0) = 0\,\mathrm{rad/s}$. Figure 5.6a shows the position components $x(t)$ and $y(t)$. Figure 5.6b shows the velocity components $v_x(t)$ and $v_y(t)$. Let us compute the angle of the pendulum θ from its position x and y. The computed angle is plotted in Fig. 5.6c, which proves that the result shows the motion of a simple pendulum. The value of constraint R is plotted in Fig. 5.6d, showing that constraint R is almost satisfied during the computation. As shown above, we can compute the behavior of a system under geometric constraints.

5.4.3 Stabilization of Pfaffian Constraints

The CSM can be used to compute the numerical solution of a system of differential equations under Pfaffian constraints. Let us explain this method by using the motion equation of a car. Let us investigate the motion of a wheeled car moving on a horizontal plane. Let (x, y) be the position of a car, and let θ be its orientation, as illustrated in Fig. 5.7. Note that a car can reach an arbitrary position in the plane with an arbitrary orientation, implying that position components x and y and orientation θ are independent of each other. Let $[\dot{x}, \dot{y}]^{\mathrm{T}}$ be the velocity of the car, and let $\dot{\theta}$ be its angular velocity. A car can move forward or backward, but not sideways, implying that the velocity components $\dot{x}$ and $\dot{y}$ are not independent. The direction of velocity $[\dot{x}, \dot{y}]^{\mathrm{T}}$ should coincide with the orientation given by angle θ. Thus, velocity components $\dot{x}$ and $\dot{y}$ and orientation θ must satisfy the following constraint:

$$Q \triangleq \dot{x} S_\theta - \dot{y} C_\theta = 0. \tag{5.42}$$

The above constraint is not integrable with respect to time, that is, this is a *nonholonomic constraint.* This constraint is linear with respect to velocity components, as follows:

$$Q = \begin{bmatrix} S_\theta & C_\theta \end{bmatrix} \begin{bmatrix} \dot{x} \\ \dot{y} \end{bmatrix} = 0.$$

Such a constraint is referred to as a *Pfaffian constraint.* The above equation is a nonholonomic Pfaffian constraint. Any Pfaffian constraint is holonomic when it is integrable with respect to time.

Let us derive the equation of motion of a car. Let m be the mass of a car, and let I be its moment of inertia. Assume that a driving force $[f_x, f_y]^{\mathrm{T}}$ and torque τ are applied to the car due to the motion of wheels and steering. Assume that a viscous force is applied to the car and that the damping force is proportional to the velocity with coefficient b. Assume that a viscous torque is applied to the car and that the torque is proportional to the angular

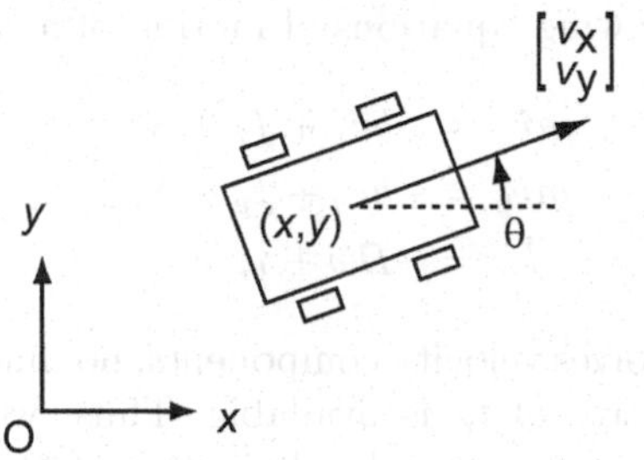

Fig. 5.7 Car moving on a plane

velocity of the car with coefficient B. Then, we have the following equations of motion of a car:

$$\begin{aligned} m\dot{v}_x &= -bv_x + f_x, \\ m\dot{v}_y &= -bv_y + f_y, \\ I\dot{\omega} &= -B\omega + \tau. \end{aligned} \tag{5.43}$$

Note that nonholonomic Pfaffian constraint Q must be satisfied.

In the stabilization of holonomic constraint $R(x, y) = 0$, we add the term λR_x to the equation of motion along the x-axis and the term λR_y to the equation of motion along the y-axis. Recall that a simple numerical integration of the equations of motion may violate the constraint, which would cause the position (x, y) of a mass to no longer satisfy the constraint. The stabilization terms denote the components of a constraint force, which modifies the motion of a mass so that position (x, y) satisfies the constraint again after the constraint has been violated. We can compute the Lagrange multiplier λ by adding a second-order differential equation $\ddot{R} + 2\alpha\dot{R} + \alpha^2 R = 0$ for the stabilization of the equations of motion. Let us introduce such stabilization terms for a Pfaffian constraint. Recall that the equations of motion of the car given in Eq. 5.44 are ordinary differential equations with respect to v_x, v_y, and ω. A simple numerical integration of equations of motion may violate the nonholonomic Pfaffian constraint Q, which would result in v_x, v_y, and ω no longer satisfying the constraint. In order to modify v_x, v_y, and ω such that constraint Q is satisfied, we introduce the following constraint stabilization terms:

$$\begin{aligned} \lambda \frac{\partial Q}{\partial v_x} &= \lambda S_\theta, \\ \lambda \frac{\partial Q}{\partial v_y} &= -\lambda C_\theta, \\ \lambda \frac{\partial Q}{\partial \omega} &= 0. \end{aligned}$$

Then we have the following equations of motion of a car:

$$\begin{aligned} m\dot{v}_x &= -bv_x + f_x + \lambda S_\theta, \\ m\dot{v}_y &= -bv_y + f_y - \lambda C_\theta, \\ I\dot{\omega} &= -B\omega + \tau. \end{aligned} \tag{5.44}$$

Since constraint Q involves velocity components, no differential equation with a second-order derivative of Q is available. Thus, we apply the following differential equation with a first-order derivative of Q:

$$\dot{Q} + \beta Q = 0, \tag{5.45}$$

where β is a predetermined damping ratio, which takes a large positive value. Then, the quantity Q converges quickly to zero, and the Pfaffian constraint is again satisfied during the computation, even if the constraint is violated. Computing the above constraint stabilization equation yields

$$\dot{Q} + \beta Q = S_\theta \dot{v}_x - C_\theta \dot{v}_y + C(\theta, v_x, v_y, \omega) = 0, \tag{5.46}$$

where

$$C(\theta, v_x, v_y, \omega) = v_x C_\theta \omega + v_y S_\theta \omega + \beta(v_x S_\theta - v_y C_\theta).$$

Consequently, we have the following linear equation on $\dot{x}$, $\dot{y}$, $\dot{\theta}$, $\dot{v}_x$, $\dot{v}_y$, $\dot{\omega}$, and λ:

$$\begin{bmatrix} 1 & & & & & & \\ & 1 & & & & & \\ & & 1 & & & & \\ & & & m & & & -S_\theta \\ & & & & m & & C_\theta \\ & & & & & I & \\ & & & -S_\theta & C_\theta & & \end{bmatrix} \begin{bmatrix} \dot{x} \\ \dot{y} \\ \dot{\theta} \\ \dot{v}_x \\ \dot{v}_y \\ \dot{\omega} \\ \lambda \end{bmatrix} = \begin{bmatrix} v_x \\ v_y \\ \omega \\ -bv_x + f_x \\ -bv_y + f_y \\ -B\omega + \tau \\ C(\theta, v_x, v_y, \omega) \end{bmatrix}. \tag{5.47}$$

Note that the above linear equation is solvable because the matrix is regular, implying that we can compute $\dot{x}$, $\dot{y}$, $\dot{\theta}$, $\dot{v}_x$, $\dot{v}_y$, and $\dot{\omega}$. As a result, we can sketch x, y, θ, v_x, v_y, and ω using the Euler method or the Runge–Kutta method.

Assume that $m = 1000\,\text{kg}$, $I = 1500\,\text{kg m}^2$, $b = 200\,\text{N/(m/s)}$, and $B = 500\,\text{Nm/ (rad/s)}$. Let the initial values be $x(0) = 0\,\text{m}$, $y(0) = 0\,\text{m}$, $\theta(0) = \pi/6\,\text{rad}$, $v_x(0) = 0\,\text{m/s}$, $v_y(0) = 0\,\text{m/s}$, and $\omega(0) = 0\,\text{m/s}$. Assume that the direction of the driving force coincides with the orientation of the car, that is, $[\,f_x(t), f_y(t)\,] = f(t)\,[\cos\theta, \sin\theta\,]$, where $f(t)$ denotes the magnitude of the driving force. Then, $f(t)$ and $\tau(t)$ determine the motion of the car. Let us give $f(t)$ and $\tau(t)$ as listed in Table 5.1. Then, the motion of a car is given by Fig. 5.8. Figure 5.8a shows the position components $x(t)$ and $y(t)$. Figure

Table 5.1 Driving force and torque applied to a car

start time [s]	end time [s]	f [N]	τ [Nm]
0	20	3000	0
20	50	3000	50
50	60	0	0
60	90	−1200	0
90	100	0	0

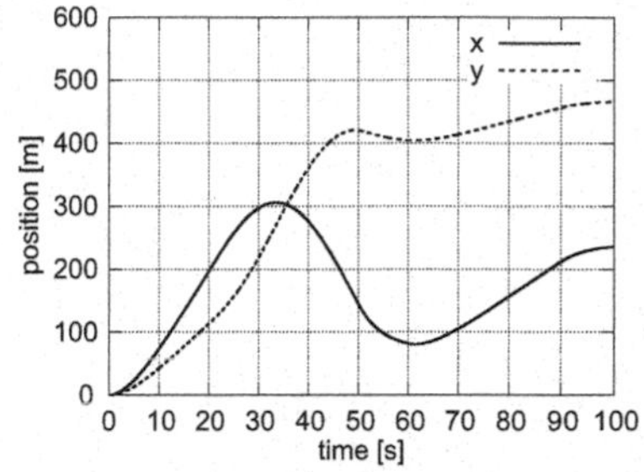

(**a**) Position (x, y)

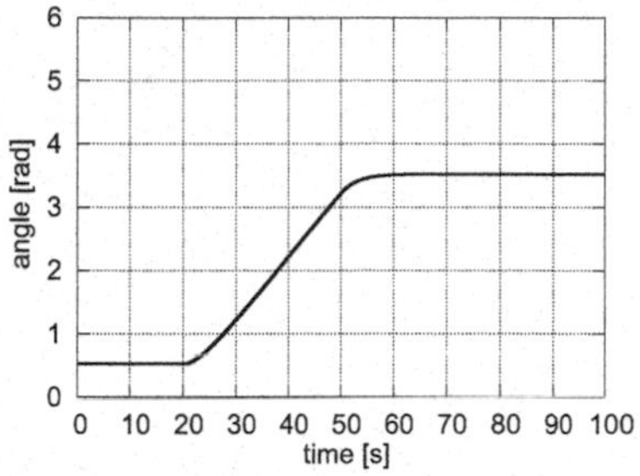

(**b**) Orientation θ

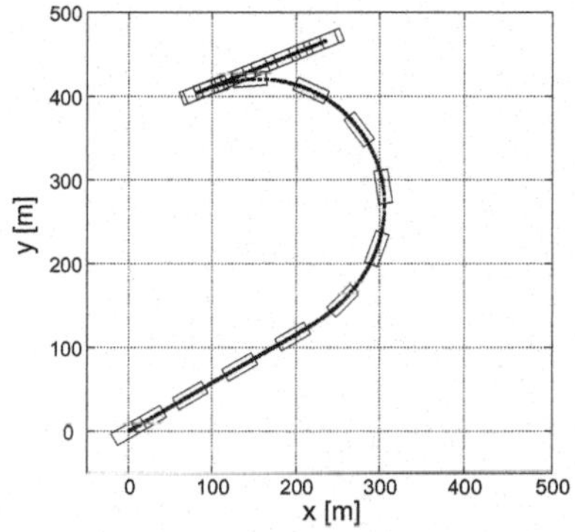

(**c**) Path (orientation every 5 s)

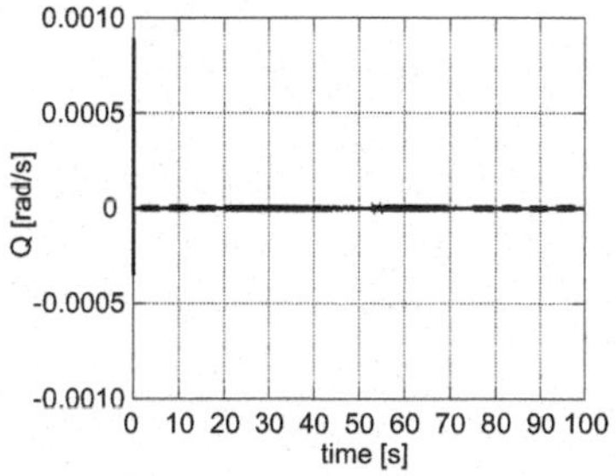

(**d**) Constraint Q

Fig. 5.8 Simulation result of the motion of a car

5.8b describes the orientation $\theta(t)$. The trajectory of a car in the horizontal plane is plotted in Fig. 5.8c. During the first 20 s, since the driving force is given but the torque is equal to zero, the car moves straight. When a positive torque is applied to the car during the following 30 s, the car turns to the left. Then, a negative driving force causes the car stop and move backward. Finally, the car stops again. Figure 5.8c proves that the direction of velocity of the car coincides with the direction of the car. The value of constraint Q is plotted in Fig. 5.8d, showing that constraint Q is almost satisfied during the computation. As shown above, we can compute the behavior of a system under nonholonomic Pfaffian constraints.

5.5 Concluding Remarks

As described in this chapter, the variational principles in mechanics provide an effective method to formulate mechanical systems under constraints. The following chapters employ these principles to formulate grasping and manipulation, which include normal and tangential constraints.

Chapter 6
Statics of Soft-fingered Grasping and Manipulation

6.1 Introduction

In Chap. 3, we formulated a simple 1D elastic model of hemispherical soft fingertips with a couple of variables: the contact orientation θ_{p} and the maximum finger displacement d, which was expressed as Eq. 3.15. In that chapter, the concept of the LMEE was introduced, and the elastic equilibrium of the fingers was revealed theoretically and experimentally. Unlike the analysis of the simple 1D model, the 2D model has six variables during a two-fingered manipulation process. In this chapter, we implement a static analysis of the overall elastic energy function including the gravitational force, which was improved as a 2D model described in Eq. 4.5. The existence of the LMEEwC (Fig. 4.9) will be ascertained through the numerical optimization method mentioned in Sect. 5.3.

6.2 Static Analysis Based on Force/Moment Equilibrium

6.2.1 Internal Energy Function

Let P_1 and P_2 be potential energies of two fingertips (Eq. 4.4), which are given by

$$P_1(d_{\mathrm{n1}}, d_{\mathrm{t1}}, \theta_{\mathrm{p1}}) = \pi E \left\{ \frac{d_{\mathrm{n1}}^3}{3\cos^2\theta_{\mathrm{p1}}} + d_{\mathrm{n1}}^2 d_{\mathrm{t1}} \tan\theta_{\mathrm{p1}} + d_{\mathrm{n1}} d_{\mathrm{t1}}^2 \right\}, \quad (6.1)$$

$$P_2(d_{\mathrm{n2}}, d_{\mathrm{t2}}, \theta_{\mathrm{p2}}) = \pi E \left\{ \frac{d_{\mathrm{n2}}^3}{3\cos^2\theta_{\mathrm{p2}}} + d_{\mathrm{n2}}^2 d_{\mathrm{t2}} \tan\theta_{\mathrm{p2}} + d_{\mathrm{n2}} d_{\mathrm{t2}}^2 \right\}. \quad (6.2)$$

In this section, we first define an internal energy function I using system variables $\boldsymbol{q}$, undetermined multipliers $\boldsymbol{\lambda}$, which physically correspond to constraint forces for each normal and tangential direction, the constraints vector $\boldsymbol{C}$, and Eq. 4.5 as follows:

$$I(\boldsymbol{q},\boldsymbol{\lambda}) \triangleq P_1 + P_2 - \boldsymbol{\lambda}^{\mathrm{T}}\boldsymbol{C}, \tag{6.3}$$

where

$$\boldsymbol{q} = [x_{\mathrm{obj}}, y_{\mathrm{obj}}, \theta_{\mathrm{obj}}, \boldsymbol{\theta}_{\mathrm{f}}^{\mathrm{T}}, \boldsymbol{d}_{\mathrm{n}}^{\mathrm{T}}, \boldsymbol{d}_{\mathrm{t}}^{\mathrm{T}}]^{\mathrm{T}} \in \mathcal{R}^{9\times 1}, \tag{6.4}$$

$$\boldsymbol{\lambda} = [\boldsymbol{\lambda}_n^{\mathrm{T}}, \boldsymbol{\lambda}_t^{\mathrm{T}}]^{\mathrm{T}} \in \mathcal{R}^{4\times 1}, \tag{6.5}$$

$$\boldsymbol{C} = [\boldsymbol{C}_n^{\mathrm{T}}, \boldsymbol{C}_t^{\mathrm{T}}]^{\mathrm{T}} \in \mathcal{R}^{4\times 1}. \tag{6.6}$$

Since it is difficult to analytically solve Eq. 6.3 and to find an LMEEwC, we apply numerical methods to search for a force/moment equilibrium point.

6.2.2 Numerical Analysis

The present study focuses on the static analysis of an object grasped by a set of soft fingers. We therefore perform the searching process when both rotational angles of the finger are maintained at an arbitrary angle under the condition that the grasping is performed successfully, as shown in Fig. 4.2. By repeating the computation process of Eq. 6.3 during the change of the joint angle, a sequential converged solution for the object can be searched, which is treated as a quasistatic problem. Note that the sequential solution can be found if Eq. 6.3 numerically converges to a certain value, which corresponds to the LMEEwC.

The method of numerical analysis used herein is based on the *Nelder–Mead method* [TVF+93] and is improved to solve constrained mechanical systems as static problems.

6.3 Simulation

6.3.1 Analysis Without Gravity

We first treat a nongravity environment so that the essence of physical stability of soft-fingered handling can be verified. Let us consider a quasicontinuous motion of both rotational fingers such that each finger moves in keeping with an operation manner as follows (Fig. 6.1):

1. The minimal incremental angle of the finger is 0.4°.

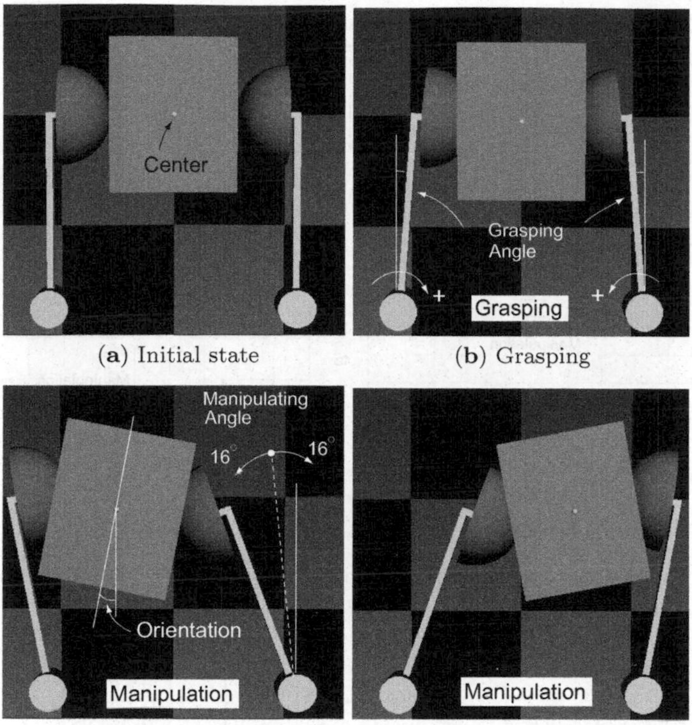

(**a**) Initial state (**b**) Grasping

(**c**) Toward left side (**d**) Toward right side

Fig. 6.1 Snapshots during soft-fingered manipulation

2. The motion starts from an initial state, in which both fingers and both sides of the object are parallel (Fig. 6.1a).
3. Grasp a planar object until the grasping angle reaches 3.6° (Fig. 6.1b).
4. Each finger rotates until $(\theta_{f1}, \theta_{f2}) = (19.6°, -12.4°)$ from operation 3 (Fig. 6.1c).
5. Each finger rotates until $(\theta_{f1}, \theta_{f2}) = (-12.4°, 19.6°)$ from operation 4 (Fig. 6.1d).
6. The search computation for an LMEEwC is implemented in every infinitesimal rotation of both fingers.

Here, we define the *grasping angle* as the angle where $\theta_{f1} = \theta_{f2}$, and each angle is positive inward. In addition, the *manipulating angle* is determined to be 16°, as illustrated in Fig. 6.1c. As shown in Table 6.1, let M_{obj} and M_{fi} be the masses of a grasped object and each finger, respectively, and let g be the acceleration of gravity. Note that the parameters M_{obj}, M_{fi}, and g will be used in the next section, in which we treat this manipulation problem under the gravitational effect.

Figure 6.2 shows the locus of object position and orientation when the above operation is performed on a grasped object. We know from these results

Table 6.1 Simulation parameters in static analysis

a	20 mm	L	76.2 mm
E	0.203 MPa	$2d_{\mathrm{f}}$	8 mm
W_{obj}	49 mm	$2W_{\mathrm{B}}$	97 mm
M_{obj}	0.2 kg	g	9.807 m/s^2

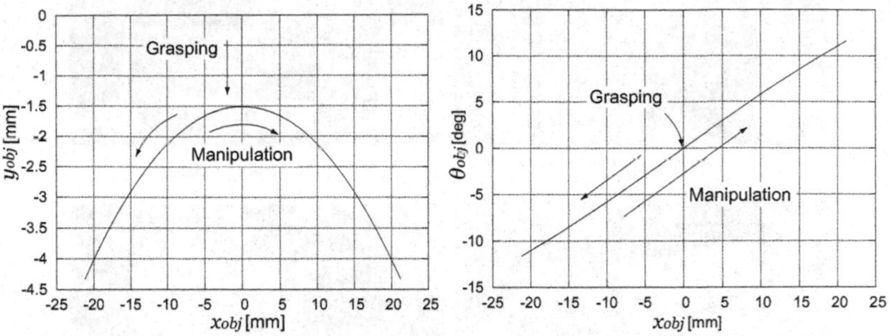

(**a**) Position (**b**) Orientation

Fig. 6.2 Locus of an object manipulated by two soft fingers when the grasping angle is 3.6°

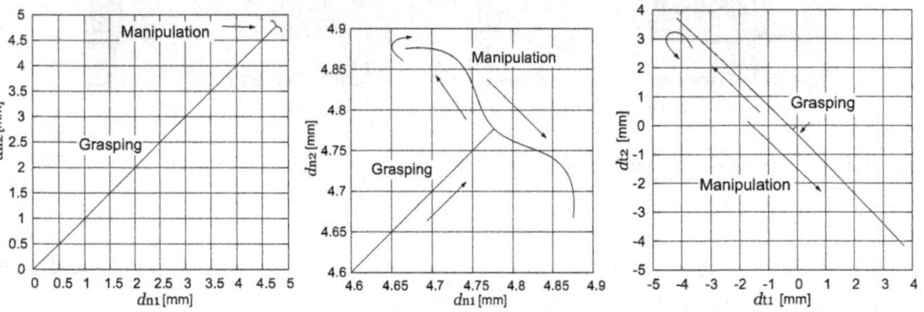

(**a**) d_{n1} vs. d_{n2} (**b**) Enlarged view of **a** (**c**) d_{t1} vs. d_{t2}

Fig. 6.3 Locus of $d_{\mathrm{n}i}$ and $d_{\mathrm{t}i}$ when the grasping angle is 3.6°

that the object moves straight downward in the grasping motion, and the locus of θ_{obj} exhibits a clear S-shaped curve in the manipulating motion. We infer from this result that the large restoring forces and moments from both fingertips, which are represented in Eq. 4.10, pull back the grasped object to the rotational direction as the rolling motion progresses.

Figure 6.3 shows the locus of $d_{\mathrm{n}i}$ and $d_{\mathrm{t}i}$ during manipulation when the grasping angle is 3.6°. We can see that both d_{n1} and d_{n2} first increase at the same rate during grasping, and then the locus of the object exhibits the characteristics of an axisymmetric arrangement with respect to $d_{\mathrm{n1}} = d_{\mathrm{n2}}$, as shown in Figs. 6.3a and 6.3b. In addition, Fig. 6.3c indicates that the tangential deflection of the soft fingertip never reaches the first quadrant. This means that each face of the soft fingers in contact with the object

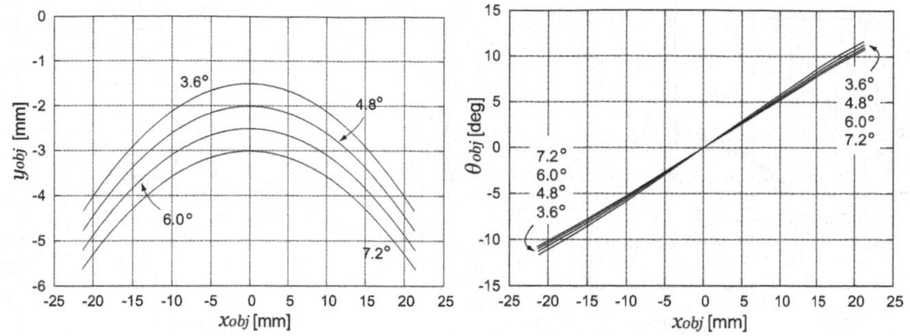

(a) Position (b) Orientation

Fig. 6.4 Each locus of the object as the grasping angle varies from 3.6° to 7.2°

deforms downward during the grasping motion, as illustrated in Fig. 6.1b, and that neither the right fingertip nor the left fingertip deforms upward.

Figure 6.4 shows a comparison of all the simulation results, in which all of the data of the object position and orientation are plotted for the cases in which the grasping angle is 3.6°, 4.8°, 6.0°, and 7.2°. As shown in Fig. 6.4a, the downward distance of the object in grasping becomes large. We find that both ends of the curves align vertically in the y-direction during the manipulating motion, where the x-value is approx. ± 21. This implies that x_{obj} can only be determined uniquely if the manipulating angle is decided independently of the grasping angle, that is, the x-position of the grasped object is independent of the grasping force that is equivalent to the elastic force due to the deformation of the soft fingertips. In addition, as shown in Fig. 6.4b, the characteristic of the S-shaped curve can be seen more clearly in the case of 3.6° than in the case of 7.2°.

6.3.2 Analysis Under Gravity

In order to clarify the stable grasping ability of the soft-fingered hand, we consider the gravitational effect on the system. We define another internal energy function I' in which the gravitational potential of a grasped object and both fingers are included as follows:

$$I'(\boldsymbol{q}, \boldsymbol{\lambda}) = I(\boldsymbol{q}, \boldsymbol{\lambda}) + M_{\mathrm{obj}}\, g\, y_{\mathrm{obj}} + \sum_{i=1}^{2} M_{\mathrm{f}i}\, g\, L \cos\theta_{\mathrm{f}i}, \tag{6.7}$$

where L denotes the distance between each finger joint and fingertip origin O_i, and $M_{\mathrm{f}i}$ denotes the mass of both fingers. As in the previous operation, we search an LMEEwC that is expected to be present in Eq. 6.7 and obtain

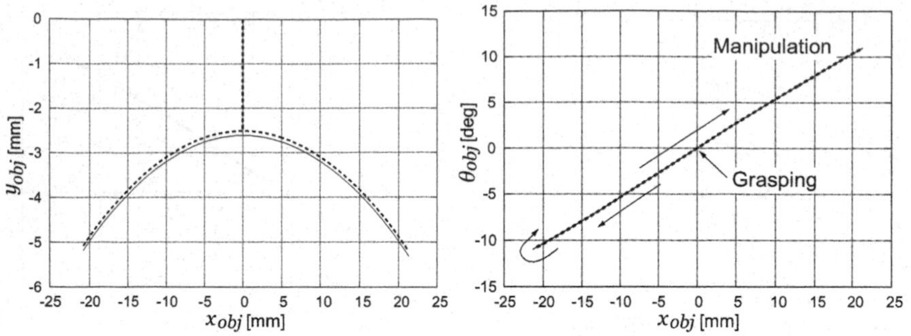

(**a**) Position (**b**) Orientation

Fig. 6.5 Locus comparison of $(x_{\mathrm{obj}}, y_{\mathrm{obj}}, \theta_{\mathrm{obj}})$ between the gravity-on condition (*continuous line*) and the gravity-off condition (*dotted line*) when each grasping angle is 6.0°

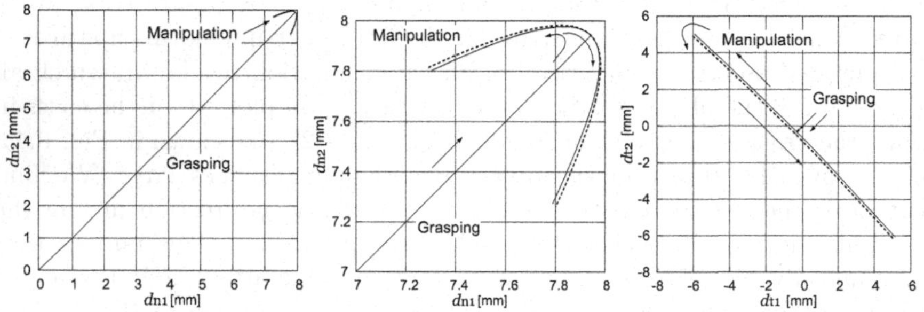

(**a**) $d_{\mathrm{n}1}$ vs. $d_{\mathrm{n}2}$ (**b**) Enlarged view of **a** (**c**) $d_{\mathrm{t}1}$ vs. $d_{\mathrm{t}2}$

Fig. 6.6 Locus comparison of $d_{\mathrm{n}i}$ and $d_{\mathrm{t}i}$ between the gravity-on condition (*continuous line*) and the gravity-off condition (*dotted line*) when each grasping angle is 6.0°

$(x_{\mathrm{obj}}, y_{\mathrm{obj}}, \theta_{\mathrm{obj}})$. Figure 6.5 shows that the object position is slightly lower for the gravitational case than the for the nongravitational case. The object orientation, however, does not change. Figure 6.6 compares the results for the locus of $d_{\mathrm{n}i}$ and $d_{\mathrm{t}i}$ in the cases with/without gravitational effect. From Fig. 6.6b we find that the paths of $d_{\mathrm{n}i}$ are consistent with each other in the grasping motion, but not in the manipulation operation. As in the case of the results of Fig. 6.3, axisymmetric properties with respect to $d_{\mathrm{n}1} = d_{\mathrm{n}2}$ and $d_{\mathrm{t}1} = d_{\mathrm{t}2}$ appear clearly throughout the entire process, as shown in Figs. 6.6b and 6.6c.

In addition, Fig. 6.7 shows the relationship between the constraint forces $\boldsymbol{\lambda}$ on each finger in the normal and tangential directions for the object. We can see that the tangential constraint force iteration increases significantly between the lower right and upper left directions (Figs. 6.7c and 6.7d), while in the manipulation process, the normal constraint force changes little (Figs. 6.7a and 6.7b). Likewise, there exists a difference in the locus between

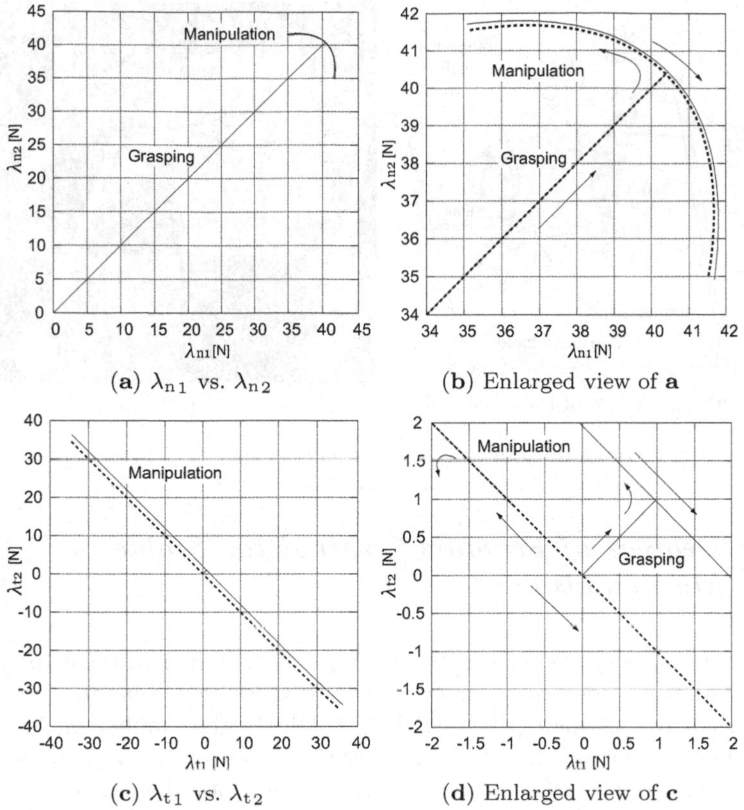

Fig. 6.7 Locus comparison of constraint forces, $\boldsymbol{\lambda}_n$ and $\boldsymbol{\lambda}_t$, between the gravity-on condition (*continuous line*) and the gravity-off condition (*dotted line*) when each grasping angle is 6.0°

the environments with/without gravitational force during the manipulation. These results imply that the tangential forces mainly affect the secure grasping under gravitational force (Fig. 6.8a). Thus, we can conclude that the concept of the LMEEwC can be easily applied to the 3D space. In summary, the object orientation angle in Fig. 6.5b remains consistently equivalent throughout the grasping and manipulation operations, despite the existence of the gravitational effect. This means that even a minimal-DOF robotic hand with soft fingertips may achieve posture control of a grasped object. The most important point is that the soft-fingered handling based on the physical phenomenon of the LMEEwC remains stable, even under gravitational force. That is, the soft-fingered robotic hand system does not require additional compensation inputs to compensate for gravity in actual manipulation tasks.

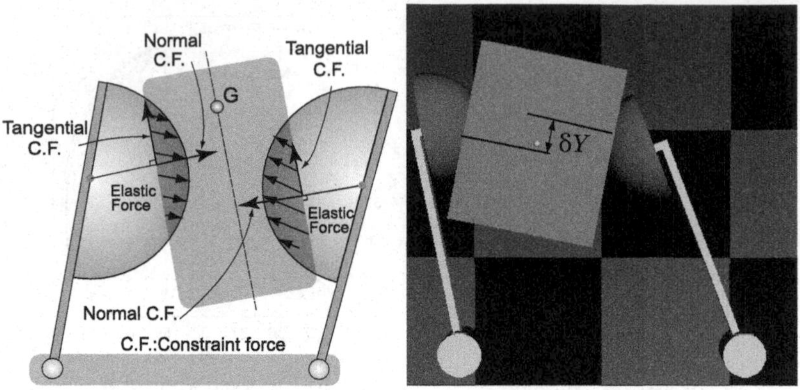

(**a**) The image of force balance (**b**) δY

Fig. 6.8 A conceptual diagram of force/moment balance on soft fingers

6.3.3 Degrees of Freedom Desired for Stable Manipulation

As an important characteristic of soft-fingered handling in terms of robust manipulation, the degrees of freedom required for stable manipulation can be reduced. In existing studies [ANH$^+$00, DF03], the number of degrees of freedom needed for soft-fingered manipulation has been reported to be three, so that the distance, δY, between the centers of contacting circles is zero, as shown in Fig. 6.8b. Those studies have proposed a control law to make δY zero so that unexpected moment of a couple does not occur between both fingers. However, the LMEEwC is achieved even though δY is not equal to zero, therefore, the control term for satisfying $\delta Y = 0$ need not be added to the control law. Attempting to input the control term for $\delta Y = 0$ to a soft-fingered robotic hand may result directly in a setback in desired tasks. Since the LMEEwC exists on soft-fingered manipulation due to the elastic deformation regardless of the number of links in the hand, stable manipulation does not depend on the number of links and can be sufficiently accomplished by means of a 2-DOF robotic hand.

6.4 Experiments

Figure 6.9 shows the experimental results of object position and orientation obtained using a CCD camera (Toshiba Teli, CS5111L) for tracking the locus of the object during manipulation (Fig. 6.10). In this experiment, the manipulating angle (Sect. 6.3.1) is set at 20°, and the grasping angle 2.4° is added. Figure 6.11 shows sequential binary images extracted from RGB signals by

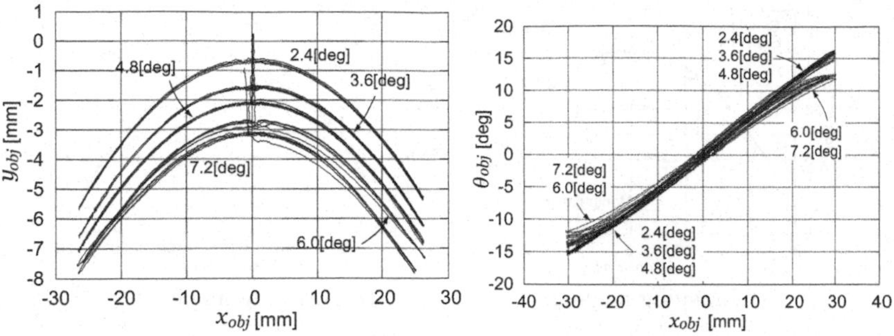

(**a**) Position (**b**) Orientation

Fig. 6.9 Experimental results when the manipulating angle is 20°

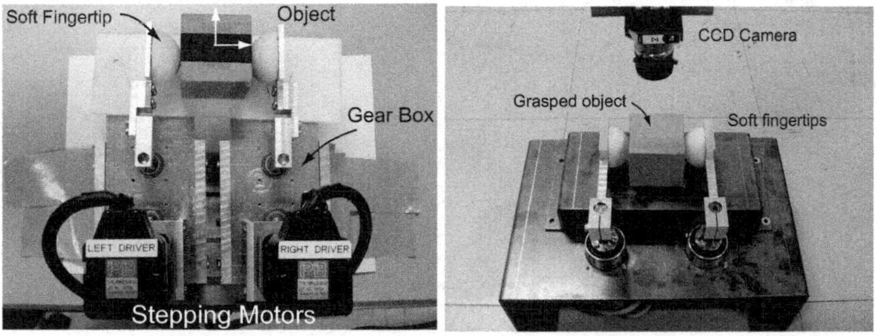

(**a**) Apparatus (**b**) Measurement with a CCD camera

Fig. 6.10 Experimental setup of grasping and manipulation processes

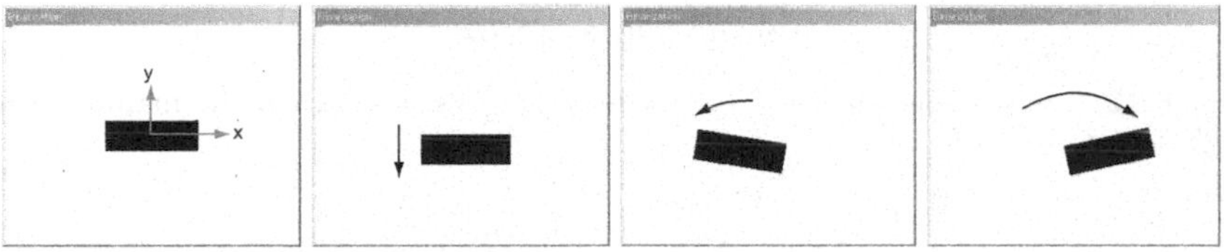

(**a**) Initial state (**b**) Operation 3 (**c**) Operation 4 (**d**) Operation 5

Fig. 6.11 Binary images obtained from a CCD camera

the CCD camera. A rectangular sheet of black paper was attached to the object. The object's center was taken to coincide with the luminance center of a binary image of the paper, which was obtained using a CCD camera above the robotic hand. The object orientation angle was measured by computing the second-order moment of the binary image.

As shown in Fig. 6.9a, the ends of all curves align vertically against the y-axis, as in the simulation results shown in Fig. 6.4a. In addition, an S-shaped curve can be seen when the grasping angle becomes larger, as shown in Fig. 6.9b. Figure 6.12 is a plot of the simulated and experimental results

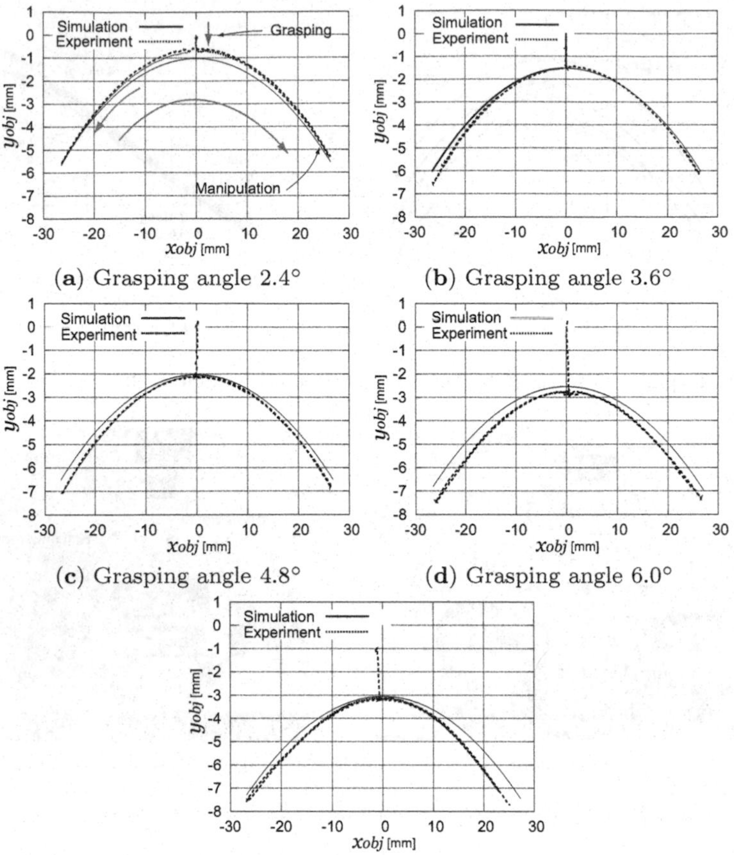

(a) Grasping angle 2.4°
(b) Grasping angle 3.6°
(c) Grasping angle 4.8°
(d) Grasping angle 6.0°
(e) Grasping angle 7.2°

Fig. 6.12 Comparison results of the locus of x_{obj} vs. y_{obj} when the manipulating angle is 20°

showing the path of the object's center $(x_{\mathrm{obj}}, y_{\mathrm{obj}})$ during grasping and manipulation processes. Note that we conducted this manipulation experiment on a 2D plane so that the gravitational force applied to the object can be ignored. The grasping angle is 2.4° at operation 3 in Fig. 6.12a and increases by 1.2 degrees per row, becoming 7.2° in the bottom graphs. The origin of each graph corresponds to the center of the black paper attached to the object, as shown in Fig. 6.10a and Fig. 6.11.

In the grasping motion, the position of the object shifts downward significantly. In the manipulating motion, we find that both resultant curves approach each other, as in the case of the grasping motion. Furthermore, the experimental curve deviates downward relative to the simulation curve as the grasping angle becomes large. This results from the large lateral deformation of the soft fingertip as the grasping angle increases. Thus, the grasped object moves downward in the experimental setup but not in the simulation.

6.5 Concluding Remarks

This chapter has provided conclusive evidence of the existence of a local minimum of elastic potential energy with constraints (LMEEwC), and the manipulation movement by a soft-fingered robotic hand was performed so as to satisfy the LMEEwC at all times. This fact is known from the consistency between the object loci obtained from numerical computation and experimental comparisons (Fig. 6.12). This also indicates that one hypothesis (*virtual trajectory hypothesis*) by Bizzi *et al.* and Hogan *et al.* [BAC+84, HH00] related to the virtual production of desired joint trajectories of the human arm, which was once completely denied by Kawato *et al.* [GK96], may resurge. This hypothesis is based on the idea that the transition between equilibrium states generated during arm movement should be solely a function of the contraction time of the motor units and the mechanical properties of the arm. The desired joint trajectory containing large allowable errors can be constructed as a continuously connected trajectory from the corresponding equilibrium states. Kawato insisted that the idea ignores the fact that human arm stiffness increases extremely during actual arm movements. However, in the case of soft-fingered manipulation, if the desired trajectory of finger joints in the present study can be produced without significant stiffness, the idea and findings about the LMEEwC may become a significant hypothesis that matches Bizzis' hypothesis [BAC+84].

Chapter 7
Dynamics of Soft-fingered Grasping and Manipulation

7.1 Introduction

This chapter provides formulations of the dynamics of soft-fingered manipulation on the basis of the parallel-distributed 2D fingertip model described in Chap. 4. Arimoto *et al.* have formulated the dynamics of soft-fingered manipulation based on the radially-distributed model [ANH+00, ADN+02]. Here, we apply the parallel-distributed model to the description of the dynamics and derive Lagrange's equations of motion of the soft-fingered manipulation using the constraint stabilization method (CSM) elaborated in Sect. 5.4.2. Furthermore, we verify the observations obtained in Chap. 2 by implementing dynamic simulations of constrained manipulation systems by a soft-fingered hand. We finally show the number of requisite DOFs for achieving robust manipulating motions without the use of object information. We refer to this ability of soft fingers as *conformable manipulation.*

7.2 Dynamics of Soft-fingered Grasping and Manipulation

Let us apply the extended parallel-distributed model (2D model) to a pair of 1-DOF fingers with soft hemispherical fingertips grasping and manipulating the rigid object in Fig. 4.2. Let θ_{f1} and θ_{f2} be the rotational angles of the right and left fingers, which have the same dimensions, and then the constraint equation facing normal to the grasped object is described using Eq. 4.17 as

$$C_{\mathrm{n}i} = (-1)^i(x_{\mathrm{obj}} - \mathrm{O}_{ix}) \cos\theta_{\mathrm{obj}} + (-1)^i(y_{\mathrm{obj}} - \mathrm{O}_{iy}) \sin\theta_{\mathrm{obj}} \\ -(a - d_{\mathrm{n}i}) - \frac{W_{\mathrm{obj}}}{2} - (-1)^i w = 0. \quad (7.1)$$

On the other hand, when we formulate tangential constraint equations for dynamic analysis, the previous equations derived as Eq. 4.23 should be transformed into a rigorous velocity-formed equation by differentiation with respect to time. This is because the velocity-formed equation can be easily incorporated into the computation procedure of dynamic systems. For instance, in this case, the entire rolling history of the object, $\mathrm{GQ}_i^{\mathrm{bef}}$, represented in Eq. 4.23 can be eliminated within the dynamic equations of motion. Thus, the tangential constraint equation is expressed by Eqs. 4.22 and 4.23 as

$$\dot{C}_{\mathrm{t}i} = \dot{\mathrm{GQ}}_i - a\dot{\theta_{\mathrm{p}i}} - \dot{d}_{\mathrm{t}i} = 0, \tag{7.2}$$

where $\dot{\theta_{\mathrm{p}i}} = \dot{\theta}_{\mathrm{f}i} + (-1)^i\dot{\theta}_{\mathrm{obj}}$. Hence we use Eq. 7.2 described as a velocity-formed rolling constraint, instead of Eq. 4.23, to maintain rigorous procedures mathematically. Note that Eq. 4.23 is not a nonholonomic constraint as it can be integrated and varies with a positional constraint equation.

The process of grasping and manipulation by a pair of 1-DOF rotational fingers with soft fingertips can be described by a set of nine generalized coordinates: object coordinates x_{obj} and y_{obj}, object orientation θ_{obj}, and, for the two fingers, respectively, rotational angles θ_{f1} and θ_{f2}, maximum normal deformations d_{n1} and d_{n2}, and tangential deformations d_{t1} and d_{t2}. Recall that two normal constraints (Eq. 7.1) and two tangential constraints (Eq. 7.2) are imposed during the process.

Let M_{obj} be the mass of the object, and let I_{obj} be the moment of inertia of the object around its center of gravity. Assume that the weights of both fingertips are negligible, and assume that the object and the two fingers move in a vertical plane, where gravitational forces act in the negative direction along the y-axis. The potential energy of the system is then given by the sum of the elastic potential energies of the two fingertips and the gravitational energy of the object using Eqs. 6.1 and 6.2 as follows:

$$P = P_1(d_{\mathrm{n1}}, d_{\mathrm{t1}}, \theta_{\mathrm{p1}}) + P_2(d_{\mathrm{n2}}, d_{\mathrm{t2}}, \theta_{\mathrm{p2}}) + M_{\mathrm{obj}}\, g\, y_{\mathrm{obj}} + \sum_{i=1}^{2} M_{\mathrm{f}i}\, g\, L\cos\theta_{\mathrm{f}i}. \tag{7.3}$$

Let $I_{\mathrm{f}i}$ be the moment of inertia of the ith finger around its rotational axis. Mass transfer due to the deformation of each fingertip varies during manipulation but is relatively small relative to the mass of the object. Thus, for the sake of simplicity, assume that mass transfers along normal and tangential directions of each fingertip are constant, denoted by m_{n} and m_{t}. Then, the kinetic energy of the system can be formulated as follows:

$$K = \frac{1}{2}M_{\mathrm{obj}}(\dot{x}_{\mathrm{obj}}^2 + \dot{y}_{\mathrm{obj}}^2) + \frac{1}{2}I_{\mathrm{obj}}\dot{\theta}_{\mathrm{obj}}^2 + \frac{1}{2}I_{\mathrm{f1}}\dot{\theta}_{\mathrm{f1}}^2 + \frac{1}{2}I_{\mathrm{f2}}\dot{\theta}_{\mathrm{f2}}^2 + \frac{1}{2}m_{\mathrm{n}}\dot{d}_{\mathrm{n1}}^2 + \frac{1}{2}m_{\mathrm{t}}\dot{d}_{\mathrm{t1}}^2 + \frac{1}{2}m_{\mathrm{n}}\dot{d}_{\mathrm{n2}}^2 + \frac{1}{2}m_{\mathrm{t}}\dot{d}_{\mathrm{t2}}^2. \tag{7.4}$$

From Eqs. 7.3 and 7.4 we can formulate the Lagrange equations of motion of a pair of fingers pinching a rigid object. The Lagrangian with holonomic constraints is described by

$$\mathcal{L} = K - P + \lambda_{\mathrm{n1}} C_{\mathrm{n1}} + \lambda_{\mathrm{n2}} C_{\mathrm{n2}}, \tag{7.5}$$

where λ_{n1} and λ_{n2} denote the Lagrange multipliers corresponding to the two holonomic constraints. Incorporating the rolling constraints expressed in Eq. 7.2, we have nine Lagrange equations of motion corresponding to the nine generalized coordinates:

$$\begin{aligned}
\frac{\mathrm{d}}{\mathrm{d}t}\frac{\partial \mathcal{L}}{\partial \dot{x}_{\mathrm{obj}}} - \frac{\partial \mathcal{L}}{\partial x_{\mathrm{obj}}} &= \frac{\partial}{\partial \dot{x}_{\mathrm{obj}}}\{\lambda_{\mathrm{t1}}\dot{C}_{\mathrm{t1}} + \lambda_{\mathrm{t2}}\dot{C}_{\mathrm{t2}}\},\\
\frac{\mathrm{d}}{\mathrm{d}t}\frac{\partial \mathcal{L}}{\partial \dot{y}_{\mathrm{obj}}} - \frac{\partial \mathcal{L}}{\partial y_{\mathrm{obj}}} &= \frac{\partial}{\partial \dot{y}_{\mathrm{obj}}}\{\lambda_{\mathrm{t1}}\dot{C}_{\mathrm{t1}} + \lambda_{\mathrm{t2}}\dot{C}_{\mathrm{t2}}\},\\
\frac{\mathrm{d}}{\mathrm{d}t}\frac{\partial \mathcal{L}}{\partial \dot{\theta}_{\mathrm{obj}}} - \frac{\partial \mathcal{L}}{\partial \theta_{\mathrm{obj}}} &= \frac{\partial}{\partial \dot{\theta}_{\mathrm{obj}}}\{\lambda_{\mathrm{t1}}\dot{C}_{\mathrm{t1}} + \lambda_{\mathrm{t2}}\dot{C}_{\mathrm{t2}}\},\\
\frac{\mathrm{d}}{\mathrm{d}t}\frac{\partial \mathcal{L}}{\partial \dot{\theta}_{\mathrm{f1}}} - \frac{\partial \mathcal{L}}{\partial \theta_{\mathrm{f1}}} &= \frac{\partial}{\partial \dot{\theta}_{\mathrm{f1}}}\{\lambda_{\mathrm{t1}}\dot{C}_{\mathrm{t1}} + \lambda_{\mathrm{t2}}\dot{C}_{\mathrm{t2}}\},\\
\frac{\mathrm{d}}{\mathrm{d}t}\frac{\partial \mathcal{L}}{\partial \dot{\theta}_{\mathrm{f2}}} - \frac{\partial \mathcal{L}}{\partial \theta_{\mathrm{f2}}} &= \frac{\partial}{\partial \dot{\theta}_{\mathrm{f2}}}\{\lambda_{\mathrm{t1}}\dot{C}_{\mathrm{t1}} + \lambda_{\mathrm{t2}}\dot{C}_{\mathrm{t2}}\},\\
\frac{\mathrm{d}}{\mathrm{d}t}\frac{\partial \mathcal{L}}{\partial \dot{d}_{\mathrm{n1}}} - \frac{\partial \mathcal{L}}{\partial d_{\mathrm{n1}}} &= \frac{\partial}{\partial \dot{d}_{\mathrm{n1}}}\{\lambda_{\mathrm{t1}}\dot{C}_{\mathrm{t1}} + \lambda_{\mathrm{t2}}\dot{C}_{\mathrm{t2}}\},\\
\frac{\mathrm{d}}{\mathrm{d}t}\frac{\partial \mathcal{L}}{\partial \dot{d}_{\mathrm{n2}}} - \frac{\partial \mathcal{L}}{\partial d_{\mathrm{n2}}} &= \frac{\partial}{\partial \dot{d}_{\mathrm{n2}}}\{\lambda_{\mathrm{t1}}\dot{C}_{\mathrm{t1}} + \lambda_{\mathrm{t2}}\dot{C}_{\mathrm{t2}}\},\\
\frac{\mathrm{d}}{\mathrm{d}t}\frac{\partial \mathcal{L}}{\partial \dot{d}_{\mathrm{t1}}} - \frac{\partial \mathcal{L}}{\partial d_{\mathrm{t1}}} &= \frac{\partial}{\partial \dot{d}_{\mathrm{t1}}}\{\lambda_{\mathrm{t1}}\dot{C}_{\mathrm{t1}} + \lambda_{\mathrm{t2}}\dot{C}_{\mathrm{t2}}\},\\
\frac{\mathrm{d}}{\mathrm{d}t}\frac{\partial \mathcal{L}}{\partial \dot{d}_{\mathrm{t2}}} - \frac{\partial \mathcal{L}}{\partial d_{\mathrm{t2}}} &= \frac{\partial}{\partial \dot{d}_{\mathrm{t2}}}\{\lambda_{\mathrm{t1}}\dot{C}_{\mathrm{t1}} + \lambda_{\mathrm{t2}}\dot{C}_{\mathrm{t2}}\},
\end{aligned} \tag{7.6}$$

where λ_{t1} and λ_{t2} denote Lagrange multipliers corresponding to the two velocity-formed constraints. Note that the treatment of the velocity-formed constraint must be done not in the Lagrangian (Eq. 7.5) but rather in the formulation process of the Lagrange equations (see Sect. 2.4 of [GPS02]). Let c_{n} and c_{t} be the viscous moduli of a fingertip along the normal and tangential directions. Then, we can formulate viscous terms, which are incorporated into the above Lagrange equations of motion detailed in Sect. 7.3.2.

Consequently, the dynamics of grasping and manipulation of an object by a pair of 1-DOF fingers with soft fingertips is formulated as a set of Lagrange equations of motion (Eq. 7.6) with four constraints, defined by Eqs. 7.1 and 7.2.

7.3 Simulation of Soft-fingered Grasping and Manipulation

7.3.1 Numerical Integration of Lagrange Equations of Motion Under Geometric Constraints

We can simulate the dynamic process of grasping and manipulation by a pair of fingers with soft fingertips by solving a set of Lagrange equations of motion (Eq. 7.6) under normal and tangential constraints (Eqs. 7.1 and 7.2). Let us convert these constraints into differential equations to incorporate them into the Lagrange equations of motion through the *constraint stabilization method* (CSM) [Bau72]. The CSM converts the two holonomic constraints (Eq. 7.1) into the following differential equations:

$$\begin{aligned} \ddot{C}_{\rm n1} + 2\alpha \dot{C}_{\rm n1} + \alpha^2 C_{\rm n1} &= 0, \\ \ddot{C}_{\rm n2} + 2\alpha \dot{C}_{\rm n2} + \alpha^2 C_{\rm n2} &= 0, \end{aligned} \tag{7.7}$$

where the parameter α is a sufficiently large constant. Note that the above equations describe critical damping, implying that the value of each holonomic constraint converges to zero, even if the constraint is violated due to numerical integration of the Lagrange equations of motion. The CSM converts the two velocity-formed constraints (Eq. 7.2) into the following differential equations:

$$\begin{aligned} \ddot{C}_{\rm t1} + \beta \dot{C}_{\rm t1} &= 0, \\ \ddot{C}_{\rm t2} + \beta \dot{C}_{\rm t2} &= 0, \end{aligned} \tag{7.8}$$

where the parameter β is a sufficiently large constant. Note that the above equations describe exponential damping, implying that the value of each constraint also converges to zero, even if the constraint is violated due to numerical integration of the Lagrange equations of motion.

A set of Lagrange equations of motion (Eq. 7.6) and differential equations (Eqs. 7.7 and 7.8) involves nine generalized coordinates and four multipliers, which are unknown variables. Thus, numerically solving 13 differential equations (Eqs. 7.6–7.8) yields 9 generalized coordinates and 4 multipliers. Let us introduce a collective vector consisting of generalized coordinates $\boldsymbol{q} = [x_{\rm obj}, y_{\rm obj}, \theta_{\rm obj}, \theta_{\rm f1}, \theta_{\rm f2}, d_{\rm n1}, d_{\rm n2}, d_{\rm t1}, d_{\rm t2}]^{\rm T}$ and a collective vector consisting of generalized velocities $\boldsymbol{p}$. Let $\boldsymbol{\lambda}_{\rm n} = [\lambda_{\rm n1}, \lambda_{\rm n2}]^{\rm T}$ be a vector consisting of Lagrange multipliers corresponding to holonomic constraints, and let $\boldsymbol{\lambda}_{\rm t} = [\lambda_{\rm t1}, \lambda_{\rm t2}]^{\rm T}$ be a vector consisting of Lagrange multipliers corresponding to velocity (rolling) constraints.

Let us substitute $\boldsymbol{p} = \dot{\boldsymbol{q}}$ into differential Eqs. 7.6–7.8 in order to move terms having a time derivative $\dot{\boldsymbol{p}}$ or Lagrange multiplier $\lambda_{\rm n}$ and $\lambda_{\rm t}$ to the left-hand side. The velocity vector $\boldsymbol{p}$, which is a set of system variables of Lagrange equations of motion (Eq. 7.6), and the corrected constraint Eqs. 7.7 and 7.8 to stabilize normal and tangential constraints (Eqs. 7.1 and 7.2) can be collectively described as follows [PH86, JE93]:

$$\begin{bmatrix} \boldsymbol{I} & \boldsymbol{O} & \boldsymbol{O} & \boldsymbol{O} \\ \boldsymbol{O} & \boldsymbol{M} & -\boldsymbol{\Phi}_{\mathrm{n}}^{\mathrm{T}} & -\boldsymbol{\Phi}_{\mathrm{t}}^{\mathrm{T}} \\ \boldsymbol{O} & -\boldsymbol{\Phi}_{\mathrm{n}} & \boldsymbol{O} & \boldsymbol{O} \\ \boldsymbol{O} & -\boldsymbol{\Phi}_{\mathrm{t}} & \boldsymbol{O} & \boldsymbol{O} \end{bmatrix} \begin{bmatrix} \dot{\boldsymbol{q}} \\ \dot{\boldsymbol{p}} \\ \boldsymbol{\lambda}_{\mathrm{n}} \\ \boldsymbol{\lambda}_{\mathrm{t}} \end{bmatrix} = \begin{bmatrix} \boldsymbol{p} \\ -\boldsymbol{f}_{\mathrm{p}} \\ \gamma_{\mathrm{n}} \\ \gamma_{\mathrm{t}} \end{bmatrix}, \tag{7.9}$$

where $\boldsymbol{I}$ denotes the 9×9 identity matrix and $\boldsymbol{O}$ denotes the 9×9 zero matrix. Matrix $\boldsymbol{M}$ is a 9×9 inertia matrix, which depends on a set of generalized coordinates $\boldsymbol{q}$. Matrices $\boldsymbol{\Phi}_{\mathrm{n}}$ and $\boldsymbol{\Phi}_{\mathrm{t}}$ are collectively referred to as the *constraint matrix* and are composed of 2×9 matrices, which depend on a set of generalized coordinates $\boldsymbol{q}$ and a set of generalized velocities $\boldsymbol{p}$. Nine-dimensional vectors $\boldsymbol{f}_{\mathrm{p}}$ associated with generalized potential forces produced by Eq. 7.6 and 2D vectors γ_{n} and γ_{t} all depend on a set of generalized coordinates $\boldsymbol{q}$ and a set of generalized velocities $\boldsymbol{p}$. See Sect. 7.3.2 for the computation of constraint stabilization. The derivation process of both γ_{n} and γ_{t} are shown in Sect. 7.3.2. Note that $\boldsymbol{f}_{\mathrm{p}}$ corresponds to the generalized force caused by the elastic energy of soft fingertips toward the individual generalized coordinate $\boldsymbol{q}$, which includes the moment around each joint of the fingers caused by the elastic force. In addition, let us add a viscous term $\boldsymbol{Dp}$, an external generalized force $\boldsymbol{f}_{\mathrm{ext}}$, and a control input $\boldsymbol{u}$ to this system. Equation 7.9 can then be given as a state-space description:

$$\begin{bmatrix} \boldsymbol{I} & \boldsymbol{O} & \boldsymbol{O} & \boldsymbol{O} \\ \boldsymbol{O} & \boldsymbol{M} & -\boldsymbol{\Phi}_{\mathrm{n}}^{\mathrm{T}} & -\boldsymbol{\Phi}_{\mathrm{t}}^{\mathrm{T}} \\ \boldsymbol{O} & -\boldsymbol{\Phi}_{\mathrm{n}} & \boldsymbol{O} & \boldsymbol{O} \\ \boldsymbol{O} & -\boldsymbol{\Phi}_{\mathrm{t}} & \boldsymbol{O} & \boldsymbol{O} \end{bmatrix} \begin{bmatrix} \dot{\boldsymbol{q}} \\ \dot{\boldsymbol{p}} \\ \boldsymbol{\lambda}_{\mathrm{n}} \\ \boldsymbol{\lambda}_{\mathrm{t}} \end{bmatrix} = \begin{bmatrix} \boldsymbol{p} \\ -\boldsymbol{Dp} - \boldsymbol{f}_{\mathrm{p}} + \boldsymbol{f}_{\mathrm{ext}} + \boldsymbol{u} \\ \gamma_{\mathrm{n}} \\ \gamma_{\mathrm{t}} \end{bmatrix}, \tag{7.10}$$

where Eq. 7.10$\in \mathcal{R}^{22\times 1}$ is satisfied. The coefficient matrix on the left-hand side of the above equation is regular, so $\dot{\boldsymbol{q}}$, $\dot{\boldsymbol{p}}$, $\boldsymbol{\lambda}_{\mathrm{n}}$, and $\boldsymbol{\lambda}_{\mathrm{t}}$ can be solved in the above linear equation. We can obtain $\boldsymbol{q}$ and $\boldsymbol{p}$ by applying a numerical integration of differential equations to this computation process.

7.3.2 Computation of Equations of Motion

In order to satisfy the condition that the constraints $C_{\mathrm{n}i}$ and $C_{\mathrm{t}i}$ are both equal to zero during numerical computation, we substitute each of Eqs. 7.1 and 7.2 into differential Eqs. 7.7 and 7.8, respectively. In Eqs. 7.7 and 7.8, α and β denote a pair of arbitrary constants associated with the speed of asymptotical stability. Equations 7.7 and 7.8 are then calculated as

$$\begin{aligned}\boldsymbol{\Phi}_{\mathrm{n}}\ddot{\boldsymbol{q}} = \boldsymbol{\Phi}_{\mathrm{n}}\dot{\boldsymbol{p}} &= -\sum_i\sum_j \frac{\partial^2 \boldsymbol{C}_{\mathrm{n}}}{\partial q_i \partial q_j}\dot{q}_i\dot{q}_j - 2\alpha\dot{\boldsymbol{C}}_{\mathrm{n}}(\boldsymbol{q},\dot{\boldsymbol{q}}) - \alpha^2\boldsymbol{C}_{\mathrm{n}}(\boldsymbol{q},\dot{\boldsymbol{q}}) \\ &= -\sum_i\sum_j \frac{\partial^2 \boldsymbol{C}_{\mathrm{n}}}{\partial q_i \partial q_j}p_i p_j - 2\alpha\dot{\boldsymbol{C}}_{\mathrm{n}}(\boldsymbol{q},\boldsymbol{p}) - \alpha^2\boldsymbol{C}_{\mathrm{n}}(\boldsymbol{q},\boldsymbol{p}) \\ &\triangleq -\gamma_{\mathrm{n}}(\boldsymbol{q},\boldsymbol{p}), \end{aligned} \tag{7.11}$$

$$\begin{aligned}\boldsymbol{\Phi}_{\mathrm{t}}\ddot{\boldsymbol{q}} = \boldsymbol{\Phi}_{\mathrm{t}}\dot{\boldsymbol{p}} &= -\sum_i\sum_j \frac{\partial^2 \boldsymbol{C}_{\mathrm{t}}}{\partial q_i \partial q_j}\dot{q}_i\dot{q}_j - \beta\dot{\boldsymbol{C}}_{\mathrm{t}}(\boldsymbol{q},\dot{\boldsymbol{q}}) \\ &= -\sum_i\sum_j \frac{\partial^2 \boldsymbol{C}_{\mathrm{t}}}{\partial q_i \partial q_j}p_i p_j - \beta\dot{\boldsymbol{C}}_{\mathrm{t}}(\boldsymbol{q},\boldsymbol{p}) \\ &\triangleq -\gamma_{\mathrm{t}}(\boldsymbol{q},\boldsymbol{p}). \end{aligned} \tag{7.12}$$

Note that Eq. 7.1 is substituted into Eq. 7.11, and Eqs. 4.23 and 7.2 are substituted into Eq. 7.12. By differentiating the constant term $\mathrm{GQ}_i^{\mathrm{bef}}$ that appeared in Eq. 4.23, we are able to prevent a formidable problem whereby the geometric constraint conditions may vary according to the parameter initialization. Thus, from Eqs. 7.11 and 7.12, the equations of motion of the total hand system (Eq. 7.10), which also contain external forces, viscous damping terms, and control inputs, can be completely described.

For fast numerical computation, we need only compute the following equation using the *Runge–Kutta method*:

$$\begin{bmatrix} \boldsymbol{M} & -\boldsymbol{\Phi}_{\mathrm{n}}^{\mathrm{T}} & -\boldsymbol{\Phi}_{\mathrm{t}}^{\mathrm{T}} \\ -\boldsymbol{\Phi}_{\mathrm{n}} & \boldsymbol{O} & \boldsymbol{O} \\ -\boldsymbol{\Phi}_{\mathrm{t}} & \boldsymbol{O} & \boldsymbol{O} \end{bmatrix} \begin{bmatrix} \dot{\boldsymbol{p}} \\ \boldsymbol{\lambda}_{\mathrm{n}} \\ \boldsymbol{\lambda}_{\mathrm{t}} \end{bmatrix} = \begin{bmatrix} -\boldsymbol{D}\boldsymbol{p} - \boldsymbol{f}_{\mathrm{p}} + \boldsymbol{f}_{\mathrm{ext}} + \boldsymbol{u} \\ \gamma_{\mathrm{n}} \\ \gamma_{\mathrm{t}} \end{bmatrix} \in \mathcal{R}^{13\times 1}. \tag{7.13}$$

The dynamic behavior of the total system, $(\boldsymbol{q},\boldsymbol{p})$, and the constraint forces, $(\boldsymbol{\lambda}_{\mathrm{n}},\boldsymbol{\lambda}_{\mathrm{t}})$, are finally obtained.

Next, let us calculate each element of the constraint matrix $(\boldsymbol{\Phi}_{\mathrm{n}},\boldsymbol{\Phi}_{\mathrm{t}})$. Since $\boldsymbol{\Phi}_{\mathrm{n}}$ corresponds to $\partial\boldsymbol{C}_{\mathrm{n}}/\partial\boldsymbol{q}$, the constraint matrix can be expressed as

$$\Phi_{\mathrm{n}ij} = \frac{\partial C_{\mathrm{n}i}}{\partial q_j}, \quad i = 1,2,\ j = 1,\cdots,9. \tag{7.14}$$

Recalling each element of the generalized coordinate q_j and the normal constraint $C_{\mathrm{n}i}$, we have

$$q_j = [x_{\mathrm{obj}}, y_{\mathrm{obj}}, \theta_{\mathrm{obj}}, \theta_{\mathrm{f1}}, \theta_{\mathrm{f2}}, d_{\mathrm{n1}}, d_{\mathrm{n2}}, d_{\mathrm{t1}}, d_{\mathrm{t2}}]^{\mathrm{T}}, \tag{7.15}$$

$$\begin{aligned} C_{\mathrm{n}i} &= (-1)^i(x_{\mathrm{obj}} - \mathrm{O}_{ix})\cos\theta_{\mathrm{obj}} + (-1)^i(y_{\mathrm{obj}} - \mathrm{O}_{iy})\sin\theta_{\mathrm{obj}} \\ &\quad - (a - d_{\mathrm{n}i}) - \frac{W_{\mathrm{obj}}}{2} - (-1)^i w = 0. \end{aligned} \tag{7.16}$$

Equation 7.14 can then be calculated as the following matrix:

$$\Phi_{nij} = \begin{bmatrix} -\cos\theta_{obj} & -\sin\theta_{obj} & A_{n1} & E_{n1} & 0 & 1\,0\,0\,0 \\ \cos\theta_{obj} & \sin\theta_{obj} & A_{n2} & 0 & E_{n2} & 0\,1\,0\,0 \end{bmatrix}, \tag{7.17}$$

where

$$A_{ni} = (-1)^{i+1}(x_{obj} - O_{ix})\sin\theta_{obj} + (-1)^i(y_{obj} - O_{iy})\cos\theta_{obj}, \tag{7.18}$$
$$E_{ni} = -L\cos\left\{\theta_{fi} + (-1)^i\theta_{obj}\right\} + d_f\sin\left\{\theta_{fi} + (-1)^i\theta_{obj}\right\}. \tag{7.19}$$

On the other hand, since Φ_t corresponds to $\partial \boldsymbol{C}_t/\partial \boldsymbol{q}$, the constraint matrix can be expressed as

$$\Phi_{tij} = \frac{\partial C_{ti}}{\partial q_j}, \quad i = 1, 2, \; j = 1, \cdots, 9. \tag{7.20}$$

Recalling the tangential constraint equation C_{ti}, we have

$$\begin{aligned} C_{ti} = &-(x_{obj} - O_{ix})\sin\theta_{obj} + (y_{obj} - O_{iy})\cos\theta_{obj} \\ &- a\left\{\theta_{fi} + (-1)^i\theta_{obj}\right\} - GQ_i^{bef} - d_{ti} = 0. \end{aligned} \tag{7.21}$$

Equation 7.20 can then be calculated as the following matrix:

$$\Phi_{tij} = \begin{bmatrix} -\sin\theta_{obj} & \cos\theta_{obj} & A_{t1} & F_{t1} & 0 & 0\,0\,1\,0 \\ -\sin\theta_{obj} & \cos\theta_{obj} & A_{t2} & 0 & F_{t2} & 0\,0\,0\,1 \end{bmatrix}, \tag{7.22}$$

where

$$A_{ti} = -(x_{obj} - O_{ix})\cos\theta_{obj} - (y_{obj} - O_{iy})\sin\theta_{obj} + (-1)^i a, \tag{7.23}$$
$$F_{ti} = L\sin\left\{\theta_{fi} + (-1)^i\theta_{obj}\right\} + d_f\cos\left\{\theta_{fi} + (-1)^i\theta_{obj}\right\} + a. \tag{7.24}$$

From Eq. 7.13, equations of motion with respect to $\boldsymbol{q}$ are represented in the following vector form:

$$\boldsymbol{M}\dot{\boldsymbol{p}} + \boldsymbol{D}\boldsymbol{p} + \boldsymbol{f}_p - \Phi_n^T\boldsymbol{\lambda}_n - \Phi_t^T\boldsymbol{\lambda}_t = \boldsymbol{f}_{ext} + \boldsymbol{u}. \tag{7.25}$$

In the present soft-finger handling system, we determine a viscous damping matrix as

$$\boldsymbol{D} = \mathrm{diag}(0, 0, 0, 0, 0, c_n, c_n, c_t, c_t), \tag{7.26}$$

which means that the corresponding viscous damping forces generated by elastic deformation are taken into account in the present system. Furthermore, the external force vector $\boldsymbol{f}_{ext}$ is assumed to be zero, and the control input vector $\boldsymbol{u}$ is equal to an input torque set that acts on each revolute joint of the two-fingered robotic hand, which is constructed using DC motors.

Using Eqs. 7.17, 7.22, and 7.25, the complete form of the equation of motion with respect to x_{obj} is finally expressed as

$$M_{\text{obj}}\ddot{x}_{\text{obj}} - \lambda_{\text{n}1}\Phi_{\text{n}11} - \lambda_{\text{n}2}\Phi_{\text{n}21} - \lambda_{\text{t}1}\Phi_{\text{t}11} - \lambda_{\text{t}2}\Phi_{\text{t}21} = 0, \tag{7.27}$$

and, therefore, we have

$$M_{\text{obj}}\ddot{x}_{\text{obj}} + \lambda_{\text{n}1}\cos\theta_{\text{obj}} - \lambda_{\text{n}2}\cos\theta_{\text{obj}} + \lambda_{\text{t}1}\sin\theta_{\text{obj}} + \lambda_{\text{t}2}\sin\theta_{\text{obj}} = 0. \tag{7.28}$$

Likewise, using Eqs. 7.17, 7.22, and 7.25, the complete form of the equation of motion with respect to y_{obj} is finally expressed as

$$M_{\text{obj}}\ddot{y}_{\text{obj}} - \lambda_{\text{n}1}\Phi_{\text{n}12} - \lambda_{\text{n}2}\Phi_{\text{n}22} - \lambda_{\text{t}1}\Phi_{\text{t}12} - \lambda_{\text{t}2}\Phi_{\text{t}22} + M_{\text{obj}}g = 0, \tag{7.29}$$

and, therefore, we have

$$M_{\text{obj}}\ddot{y}_{\text{obj}} + \lambda_{\text{n}1}\sin\theta_{\text{obj}} - \lambda_{\text{n}2}\sin\theta_{\text{obj}} - \lambda_{\text{t}1}\cos\theta_{\text{obj}} - \lambda_{\text{t}2}\cos\theta_{\text{obj}} + M_{\text{obj}}g = 0. \tag{7.30}$$

In addition to the above computations, the equation of motion with respect to θ_{obj} is obtained as

$$I_{\text{obj}}\ddot{\theta}_{\text{obj}} + \frac{2\pi E}{3}\left\{\frac{d_{\text{n}1}^3\sin\theta_{\text{p}1}}{\cos^3\theta_{\text{p}1}} - \frac{d_{\text{n}2}^3\sin\theta_{\text{p}2}}{\cos^3\theta_{\text{p}2}}\right\} + \pi E\left\{\frac{-d_{\text{n}1}^2 d_{\text{t}1}}{\cos^2\theta_{\text{p}1}} + \frac{d_{\text{n}2}^2 d_{\text{t}2}}{\cos^2\theta_{\text{p}2}}\right\}$$
$$-\lambda_{\text{n}1}A_{\text{n}1} - \lambda_{\text{n}2}A_{\text{n}2} - \lambda_{\text{t}1}A_{\text{t}1} - \lambda_{\text{t}2}A_{\text{t}2} = 0, \tag{7.31}$$

where I_{obj} denotes the moment of inertia of the object and $\theta_{\text{p}i}$ denotes $\theta_{\text{f}i} + (-1)^i\theta_{\text{obj}}$. Note that the second and third terms of Eq. 7.31 correspond to *restoring moments* that occur around the rotational axis of the grasped object due to the elastic energy P of both fingertips. That is, both of the terms corresponding to $\boldsymbol{f}_{\text{p}}$ can be derived by differentiating P with respect to q_j. They play an important role in making the grasped object come to rest in a stable posture in the manipulation processes. In addition, the equations of motion other than those for the grasped object in terms of $\theta_{\text{f}i}$, $d_{\text{n}i}$, and $d_{\text{t}i}$ are represented by differentiating P with respect to these variables as

$$I_{\text{f}i}\ddot{\theta}_{\text{f}i} - \frac{2\pi E d_{\text{n}i}^3\sin\theta_{\text{p}i}}{3\cos^3\theta_{\text{p}i}} + \frac{\pi E d_{\text{n}i}^2 d_{\text{t}i}}{\cos^2\theta_{\text{p}i}} - \lambda_{\text{n}i}E_{\text{n}i} - \lambda_{\text{t}i}F_{\text{t}i}$$
$$-M_{\text{f}i}gL\sin\theta_{\text{f}i} = u_i, \tag{7.32}$$

$$M_{\text{obj}}\ddot{d}_{\text{n}i} + \frac{\pi E d_{\text{n}i}^2}{\cos^2\theta_{\text{p}i}} + 2\pi E d_{\text{n}i}d_{\text{t}i}\tan\theta_{\text{p}i} + \pi E d_{\text{t}i}^2 - \lambda_{\text{n}i} + c_{\text{n}}\dot{d}_{\text{n}i} = 0, \tag{7.33}$$

$$M_{\text{obj}}\ddot{d}_{\text{t}i} + \pi E d_{\text{n}i}^2\tan\theta_{\text{p}i} + 2\pi E d_{\text{n}i}d_{\text{t}i} - \lambda_{\text{t}i} + c_{\text{t}}\dot{d}_{\text{t}i} = 0. \tag{7.34}$$

In Eq. 7.32, the first and second terms are equivalent to reaction moments against the moment exerted on the grasped object. $E_{\text{n}i}$ and $F_{\text{t}i}$ correspond to each moment arm related to the constraint forces $\lambda_{\text{n}i}$ and $\lambda_{\text{t}i}$. In Eq. 7.33,

Table 7.1 Simulation parameters in dynamic analysis

a	20 mm	L	76.2 mm
E	0.232 MPa	$2d_{\mathrm{f}}$	8 mm
$c_{\mathrm{n}}, c_{\mathrm{t}}$	300 Ns/m	$2W_{\mathrm{B}}$	98 mm
W_{obj}	50 mm	$I_{\mathrm{f1}}, I_{\mathrm{f2}}$	582 kg·mm^2
M_{obj}	0.3 kg	g	9.807 m/s^2
$M_{\mathrm{f1}}, M_{\mathrm{f2}}$	0.3 kg	α	10000
I_{obj}	41.7 kg·mm^2	β	10000

the second, third, and fourth terms are equal to the potential force on the deformable fingertips, which is directed normal to the grasped object. $c_{\mathrm{n}}\dot{d}_{\mathrm{n}i}$ is a viscous damping force along the normal direction of the object. In Eq. 7.34, the second and third terms correspond to the potential force to the tangent of the object surface. In addition to Eq. 7.33, $c_{\mathrm{t}}\dot{d}_{\mathrm{t}i}$ denotes a viscous damping force along the tangential direction of the object.

7.4 Simulation Results

Figure 7.1 shows a simulation, based on the parallel-distributed model (2D model), of a pair of 1-DOF fingers controlling the orientation of an object. We used the parameters listed in Table 7.1. Note that the Young modulus of Table 7.1 does not coincide with that of Table 6.1. This reason is that we remade a set of soft fingertips for dynamic experiments of robotic handling motion, which are presented in Sect. 7.5. This is why $E = 0.232$ MPa is applied and listed in Table 7.1. Figure 7.1a shows the initial contact position between the fingers and the object, in which the fingertip was not deformed, and Fig. 7.1b shows the initial grasping position, in which both fingertips are deformed in the same manner. In Fig. 7.1c, the fingers have both rotated counterclockwise and the object has rotated clockwise, while in Fig. 7.1d, the fingers have rotated clockwise and the object has rotated counterclockwise. The results of the simulation agree with the observations shown in Fig. 2.3. The simulation suggests that a pair of 1-DOF rotational fingers with soft fingertips can control the orientation of a pinched object.

Figure 7.2 compares the simulation and experimental results of the control of the orientation of a pinched object by the fingertips. In Fig. 7.2a, the manner in which the object's orientation angle θ_{obj} changes with its coordinate x_{obj} in the simulation was almost the same as that in the experiment. In addition, in Fig. 7.2b, the relative changes in the object's coordinates x_{obj} and y_{obj} in the simulation were almost the same as those in the experiment.

Figure 7.3 compares simulations based on the parallel model with and without tangential deformation of the fingertips. In Fig. 7.3a, the slope of the plot of the object's orientation angle θ_{obj} with respect to its coordinate

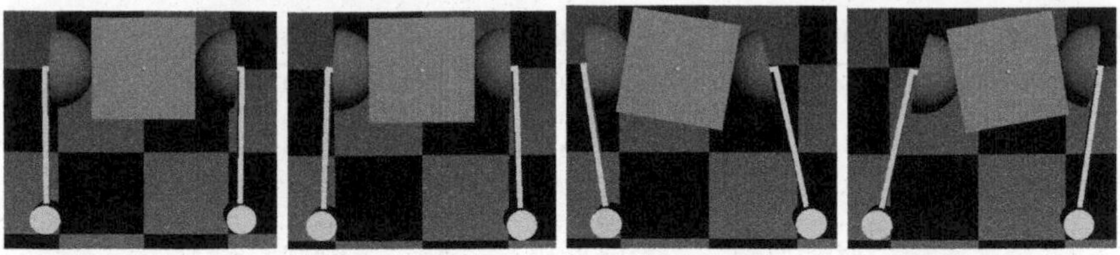

(a) Initial contact (b) Grasping (c) Left move (d) Right move

Fig. 7.1 Simulation of fingertips controlling the orientation of an object

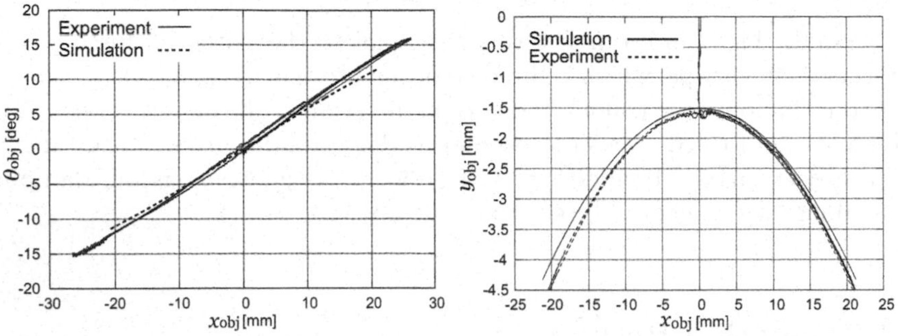

(a) Orientation (b) Position

Fig. 7.2 Comparison between simulation and experimental results

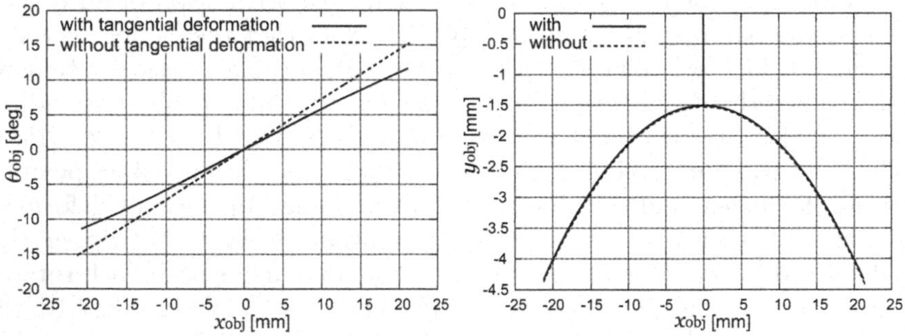

(a) Orientation (b) Position

Fig. 7.3 Comparison of simulation results with/without tangential deformation

x_{obj} is steeper without tangential deformation than with tangential deformation. Since the simulation results based on the parallel model with tangential deformation agree well with the experimental results, tangential deformation probably occurs in actual grasping and manipulation. In Fig. 7.2b, the path of the pinched object is the same regardless of the tangential deformation. Note that the center of the pinched object lies on a line between the two points of contact in the initial grasping position, as shown in Fig. 7.1a, which may be why the paths were the same.

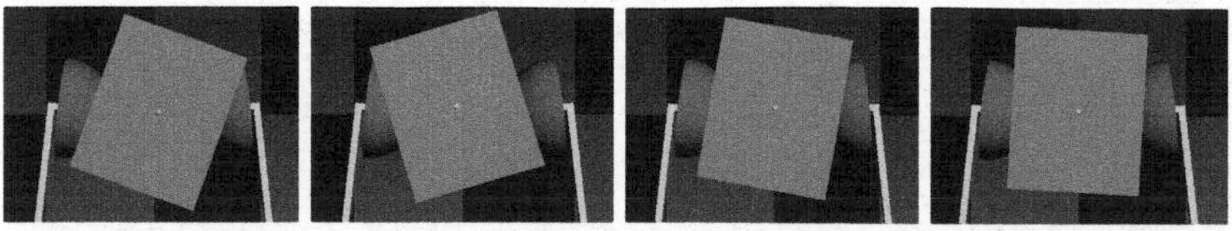

(a) Before release (b) oscillation 1 (c) oscillation 2 (d) oscillation 3

Fig. 7.4 Snapshots of the simulation of pinched object rotation by an external force

Table 7.2 Sequence of motions

Initial state	Both fingers grasp an object in parallel
Operation 1	$(\theta_{f1}^{d}, \theta_{f2}^{d}) = (6\,^\circ,\ 6\,^\circ)$
Operation 2	$(\theta_{f1}^{d}, \theta_{f2}^{d}) = (20\,^\circ,\ -10\,^\circ)$
Operation 3	$(\theta_{f1}^{d}, \theta_{f2}^{d}) = (-2\,^\circ,\ 13\,^\circ)$
Operation 4	$(\theta_{f1}^{d}, \theta_{f2}^{d}) = (-10\,^\circ,\ 20\,^\circ)$
Operation 5	$(\theta_{f1}^{d}, \theta_{f2}^{d}) = (-7\,^\circ,\ 17\,^\circ)$
Operation 6	$(\theta_{f1}^{d}, \theta_{f2}^{d}) = (17\,^\circ,\ -7\,^\circ)$
Operation 7	$(\theta_{f1}^{d}, \theta_{f2}^{d}) = (-15\,^\circ,\ 25\,^\circ)$
Operation 8	$(\theta_{f1}^{d}, \theta_{f2}^{d}) = (5\,^\circ,\ 5\,^\circ)$

Fig. 7.4 shows a simulation of the rotation of a pinched object by an external force. As shown in Fig. 7.4a, the pinched object rotates counterclockwise when the force is applied, though each angle of two joints is fixed at a certain angle. As shown in Fig. 7.4b-d, when the force is relaxed, the object rotates back to its initial orientation. The simulation results agreed well with the observations shown in Fig. 2.5.

Let us guide joint angles θ_{f1} and θ_{f2} to their desired values θ_{f1}^{d} and θ_{f2}^{d}. The input torques on the joints of the right and left fingers are u_1 and u_2, respectively. Now let us apply the following simple PID control laws to guide the joint angles to their desired values:

$$u_1 = -K_P(\theta_{f1} - \theta_{f1}^{d}) - K_D\dot{\theta}_{f1} - K_I \int_0^t \{\theta_{f1}(\tau) - \theta_{f1}^{d}\}\, d\tau, \tag{7.35}$$

$$u_2 = -K_P(\theta_{f2} - \theta_{f2}^{d}) - K_D\dot{\theta}_{f2} - K_I \int_0^t \{\theta_{f2}(\tau) - \theta_{f2}^{d}\}\, d\tau, \tag{7.36}$$

where K_P, K_D, and K_I denote proportional, differential, and integral gains, respectively. Let us apply the sequence of desired values in Table 7.2. Figure 7.5 shows the simulation results. We determined $K_P = 300\,\text{Nm/rad}$, $K_D = 14\,\text{Nm/(rad/s)}$, and $K_I = 0.1\,\text{Nm/(rad·s)}$, and the above control signals were made on the basis of the block diagram illustrated in Fig. 7.6. In Figs. 7.5a and b, respectively, the joint angles θ_{f1} and θ_{f2} of the right and left fingers converge to their desired values within one second. Figure 7.5c and d show the object's position, and Fig. 7.5e shows its orientation. These figures show

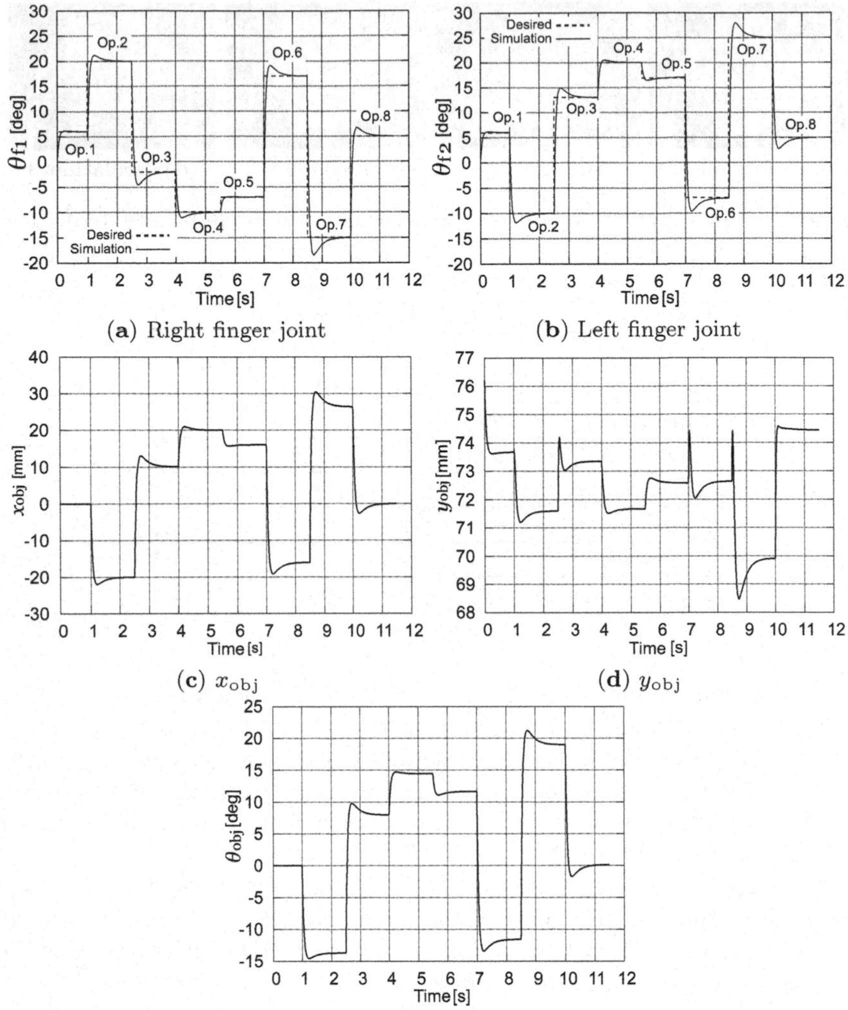

(a) Right finger joint

(b) Left finger joint

(c) x_{obj}

(d) y_{obj}

(e) θ_{obj}

Fig. 7.5 Simulation results of the joint angle control without using object information

that the object is stable once the joint angles have stabilized, which suggests that the motion of the object is inherently stable without any external sensor feedback associated with the object information. We refer to this intrinsic capability of soft-fingered grasps as *conformable manipulation.*

Based on the above simulations, we conclude that the proposed parallel-distributed model provides a good explanation of grasping and manipulation by soft hemispherical fingertips. Moreover, soft hemispherical fingertips enable the control law in grasping and manipulation to be simplified.

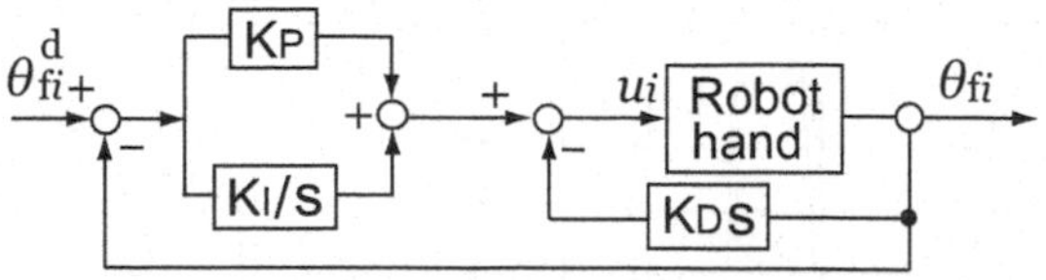

Fig. 7.6 Block diagram of the entire robotic hand system based on joint angle control

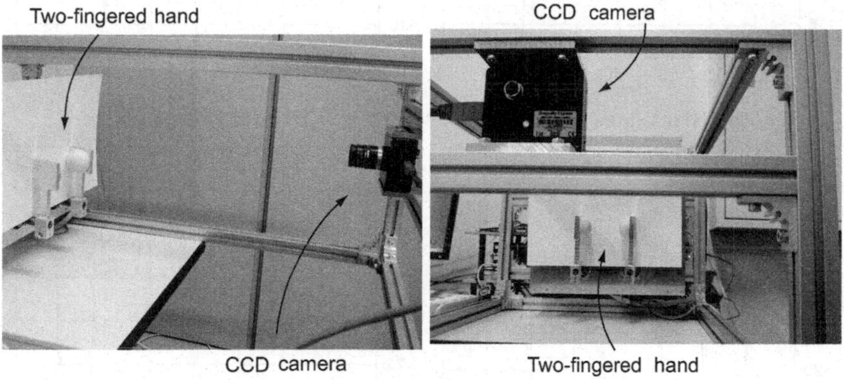

(**a**) Side view (**b**) Top view

Fig. 7.7 Experimental setup of a soft-fingered robotic hand

Table 7.3 Camera specification

Manufacturer and model	Point Gray Research Inc., Dragonfly Express
Interface	IEEE 1394b (FireWire)
Imaging sensor	Progressive scan CCD
Resolution	VGA Grayscale image
Frame Rates	200 (Format 7 mode), 120, 60, 30 fps

7.5 Experimental Results

To confirm the capability of the *conformable manipulation* in soft-fingered robotic hand, we implement an experimental task in which both joints of the robot are controlled by a PI controller that can be constructed by eliminating derivative terms from Eqs. 7.35 and 7.36. The experimental setup of a robotic hand is shown in Fig. 7.7, where a high-speed camera (200 Hz, Point Gray Research Inc.) is incorporated via a Linux computer that works as an AD/DA/encoder interface for DC motors of the robot and an image processing unit of captured object images. Specification of the camera is presented in Table 7.3. The sampling time of the vision system is approx. 25 ms that includes 5 ms-capturing duration of the camera and 20 ms-image processing process. The mass of a rectangular wood block (grasped object) and aluminum fingers is M_{obj}=90.0 g and $M_{\mathrm{f1}}, M_{\mathrm{f2}}$=88.2 g, respectively. Note that

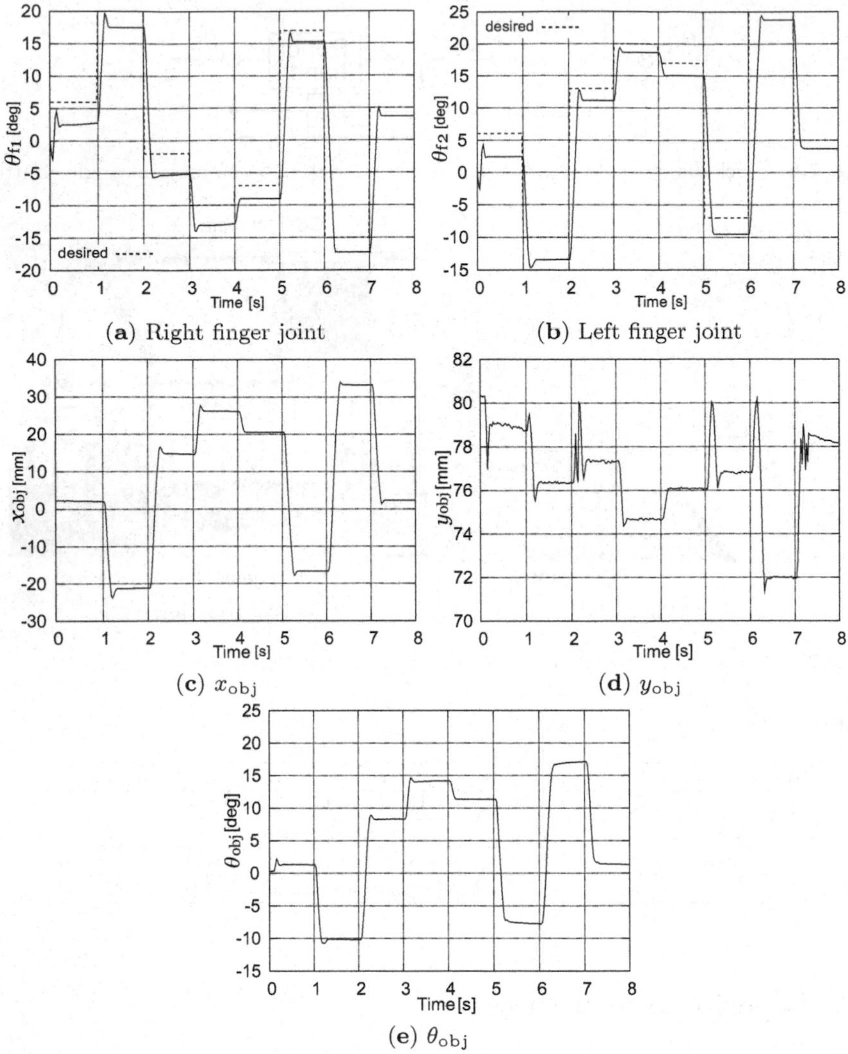

Fig. 7.8 Experimental results of joint angle control of a robotic hand. Feedback gains were given such that $K_P = 20$, $K_D = 0$, and $K_I = 0.1$

both parameters are not equal to those listed in Table 7.1, which are used in the simulation. Tasks given in this experiment are equal to the procedure described in Table 7.2.

Figure 7.8 shows experimental results where each trajectory of object variables, $(x_{obj}, y_{obj}, \theta_{obj})$, exhibits stable behavior despite any object information for visual feedback control is not applied to this system. On the other hand, the trajectories of both fingers did not fulfill each desired value in every step. In particular, both joint angles have large errors at the beginning of

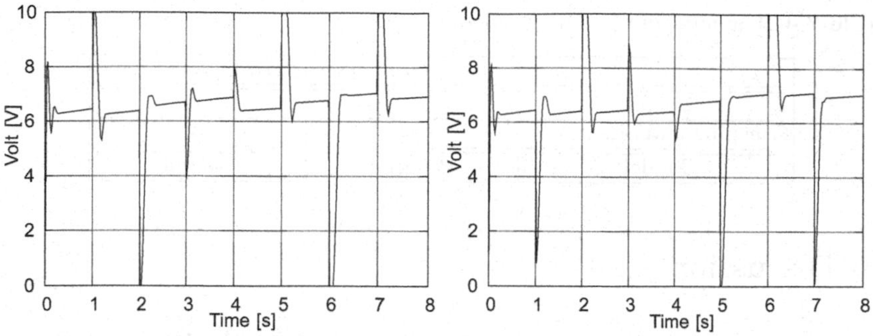

(**a**) Right finger voltage (**b**) Left finger voltage

Fig. 7.9 Change of voltage during the experiment. The neutral point of the motor drive is at 5 V, which means the motor does not move if no load is applied. 10 V voltage produces inward rotation of the finger, while 0 V produces outward rotation with no load, which directly results in a failure of grasp

operation 1, as shown in Fig. 7.8a and b. This large discrepancy stems from the fact that the joint torque of the individual finger, which can be induced by a DC motor, did not overcome the repulsion force due to the elastic deformation of a soft fingertip. This observation is reinforced with monitoring of voltage outputs of the DC motor, as shown in Fig. 7.9. Both joint angles have little change in the steady state (Figs. 7.8a and b) regardless of the increase in the voltage due to the integral controller (Eqs. 7.35 and 7.36). Even after operation 1, relatively-large errors remain in both joints, at the same time, the desired value of the joint is consistently above the actual joint angle in terms of the vertical axis, as shown in Figs. 7.8a and b. This reason is that the joint error arisen at the initial grasping (operation 1) remains throughout the task once the both fingers rotate to same direction after operation 2. As a result, stable grasping and manipulation by soft fingers can be achieved even if such a precipitous desired trajectory or the discontinuous step function of the joint is added to the system and if both joint angles remain large discrepancies. Comparison with the simulation result (Fig. 7.5) will allow us to know that both results show no oscillatory behavior except the y-trajectory. These findings are caused by the fact that the stiffness of the soft fingertip gradually increases as indicated in Fig. 3.3 as the grasping torque by both motors increase. Thus, we know that flexibility of the soft fingers contributes profoundly to dynamic pinching by robotic hands and compliant motion of a grasped object. This adaptable capability of a soft-fingered hand is named *conformable manipulation*.

Table 7.4 The number of DOFs needed.

	Soft-fingered hand	Hard-fingered hand
Grasping	1	2
Grasping and orientation	2	3
Model	Parallel	Radial

7.6 Discussion

Observations and simulations have shown that a pair of 1-DOF fingers with soft fingertips can regulate both grasping force and object orientation. The results agree with the observations described in Chap. 2. All of the dynamic behavior of the present system and for which control strategy (Eqs. 7.35 and 7.36) is used does not involve any object information, *i.e.*, $(x_{\rm obj}, y_{\rm obj}, \theta_{\rm obj})$, when manipulating the target object freely. Despite that, the manipulation movement is robustly and easily accomplished. Therefore, we define this type of manipulation performance as *conformable manipulation.* In addition, the number of DOFs needed for grasping is summarized in Table 7.4. It has been reported that, based on a radially-distributed model, for a pair of fingers with rigid fingertips to control the orientation of a rigid object, at least 3-DOFs are required [HAK+01]; otherwise, regardless of their relative orientation, the moments on the object would not be balanced and the orientation of the object must be given by sensor feedback. However, the present study shows that the moments are actually balanced inherently by a pair of soft fingertips with one less degree of freedom and that the orientation of the object can be controlled without sensor feedback. Consequently, the control law for the pair of 1-DOF soft fingertips is simpler.

Note that, in the parallel-distributed model, the elastic potential energy of a soft fingertip depends on its angle relative to the grasped object. This dependency is due to the fingertip being hemispherical and having a hard-back plate. This dependency does not exist for spherical fingertips because the shape does not vary when the spherical fingertips are rotated. In contrast, the hard plate behind the hemispherical soft fingertips imposes a boundary condition on their deformation and results in their elastic energy varying with the relative orientation. Consequently, the structure of a finger consisting of a soft fingertip and a hard fingernail enhances dexterity in grasping and manipulation.

7.7 Conclusion and Research Perspective

We modeled a pair of hemispherical soft fingertips with 1-DOF grasping and manipulation of a rigid object. We formulated the dynamics of grasping and manipulation performed by the pair of 1-DOF fingers with soft fingertips on the basis of a parallel-distributed model. We showed that the dynamics can be described by Lagrange equations of motion with holonomic and nonholonomic constraints due to the contact between each fingertip and the object. Finally, using our parallel-distributed model, we simulated the soft fingertips grasping and manipulating the object and showed that our model agrees well with the observations made in Chap. 2.

This chapter focused only on simple handling motions without using the object information through the two-fingered manipulation on the basis of the mechanical structure of a pair of 1-DOF opposed revolute joints. Chap. 8 therefore attempts to propose a sophisticated control method of object orientation from the viewpoint of sensory feedback strategies. In Chap. 10, we further regard not only the revolute joint but also the prismatic joint and translational movements of the finger joints of robotic hand, which can be seen in everyday movements of the distal phalanges of the human thumb and forefinger. Here we dealt with planar grasping and manipulation. Spatial grasping and manipulation will be formulated in Chap. 11.

Chapter 8
Control of Soft-fingered Grasping and Manipulation

8.1 Introduction

Soft-fingered grasping and manipulation problems have been discussed for some time. The first elastic-fingered robotic hand was proposed by Hanafusa and Asada [HA77a, HA77b], who derived an elastic potential function induced by the deviation of elastic fingers and revealed that the optimal prehension strategy could be constructed by computing multiple local minimum values of the elastic energy. Although their study mentioned stable grasping and the method of prehension, manipulation was not discussed. In order to treat soft-fingered manipulation along with hard-contact handling using a unified methodology, Salisbury introduced the *contact constraint matrix* $\boldsymbol{H}$, which is capable of evaluating contact types between fingers/manipulators and target objects [JJ82, MS85]. The $\boldsymbol{H}$ matrix works as an on-off indicator to transmit forces and moments applied by multiple fingers. The theory, however, is based only on the geometric configuration during contact. As such, the physical and mechanical characteristics of soft fingers cannot be treated [CK89, LHS89].

Recently, a mathematical model of the hemispherical soft fingertip was presented by Arimoto *et al.* [ANH$^+$00], satisfying the radial deployment of virtual springs within the fingertip. Based on their model, Doulgeri *et al.* [DFA02, DF03] proposed a linearly superimposed controller including a force feedforward term and a feedback term related to orientation control of a grasped object to meet the stable grasping and posture control simultaneously. Their approach, however, complies with traditional control methods based on hard-contact manipulation analyses in the sense that it includes a control term to cancel out an unexpected moment of a couple, which is acting on a grasped object. These studies could not incorporate the intrinsic property of soft fingers, which is intimately related to secure manipulation motions. The most important characteristic of soft-fingered manipulation is softness and flexibility during manipulation operations. By importing the physical advantage offered by flexibility into the elastic model of the soft fingertips, it is expected that a straightforward control scheme could be derived relative to conventional control strategies.

The objective is to present a new and straightforward control scheme in terms of soft-fingered manipulation that is capable of accomplishing posture control of an object grasped by a soft-fingered hand. Here we deal with a two-fingered robotic hand on which a set of hemispherical soft fingers is mounted, and a rigid rectangular object is grasped by the hand. Based on a parallel-distributed model, the Lagrangian of this system and its equations of motion are presented in a general and straightforward vector form, including nine generalized coordinates with respect to the reference coordinate system. In addition, we newly show a serially-coupled two-phased controller associated with the posture control of an object grasped by the robotic hand. Finally, we clarify that the posture control of the grasped object can be achieved by means of a minimal-degree-of-freedom (DOF) soft-fingered hand.

8.2 Equations of Motion of the Two-fingered Hand

In this section, we derive the equations of motion of a two-fingered hand system shown in Fig. 4.2 as a general vector form. This derivation is of an energy-based approach, that is, we describe the handling motions by making use of the elastic energy induced by the deformation of the soft fingers. Since the present soft-fingered manipulation system and a grasped rigid object comprise a closed-link mechanism maintaining a plane contact between the object and two fingers, four geometric constraints related to the normal and tangential directions with respect to the object surface are represented. Among the constraint equations, the tangential constraints should basically be a velocity form and a nonholonomic relationship because the effective rolling radius on the soft fingertip varies steadily according to the rolling motion of the grasped object, despite the formulation in the 2D plane. In the presnt study, we assume that the rolling radius is constant and equivalent to the original radius a of the hemispherical fingertip, based on Chang's observations [CC95] that the rolling radius varies during the rolling motion, but the change is very small. Therefore, we again use Eqs. 7.1 and 7.2 for both normal and tangential constraints.

A pair of 1-DOF robotic hands has nine system variables along with the displacements of the soft fingertip, $\boldsymbol{q} = [x_{\rm obj}, y_{\rm obj}, \theta_{\rm obj}, \boldsymbol{\theta}_{\rm f}^{\rm T}, \boldsymbol{d}_{\rm n}^{\rm T}, \boldsymbol{d}_{\rm t}^{\rm T}]^{\rm T}$, in which $\boldsymbol{d}_{\rm n}$ and $\boldsymbol{d}_{\rm t}$ are both dependent on the other variables. The Lagrangian of the system can be expressed using Eqs. 7.3 and 7.4 as

$$\mathcal{L} = K - P + \sum_{i=1}^{2} \lambda_{{\rm n}i} C_{{\rm n}i}. \tag{8.1}$$

As a result, the equations of motion of the system are finally described from Eq. 8.1 as

$$\frac{\mathrm{d}}{\mathrm{d}t}\frac{\partial \mathcal{L}}{\partial \dot{\boldsymbol{q}}} - \frac{\partial \mathcal{L}}{\partial \boldsymbol{q}} = \sum_{i=1}^{2} \lambda_{\mathrm{t}i} \frac{\partial \dot{C}_{\mathrm{t}i}}{\partial \dot{\boldsymbol{q}}} + \boldsymbol{f}_{\mathrm{ext}} + \boldsymbol{u}, \tag{8.2}$$

in which a generalized external force and a control input are added. Note that the first term on the right-hand side of the above equation denotes the constraint forces facing the grasped object tangentially. Thus, the equation of motion of the two-fingered robotic hand is finally expressed as the following equation:

$$\boldsymbol{M}\ddot{\boldsymbol{q}} + \boldsymbol{D}\dot{\boldsymbol{q}} + \boldsymbol{f}_{\mathrm{p}} - \boldsymbol{\Phi}^{\mathrm{T}}\boldsymbol{\lambda} = \boldsymbol{f}_{\mathrm{ext}} + \boldsymbol{u}, \tag{8.3}$$

where $\boldsymbol{M}$ denotes the inertia matrix of the system and $\boldsymbol{f}_{\mathrm{p}}$ denotes the generalized potential force caused by P. In addition, $\boldsymbol{\Phi}$ corresponds to the *constraint matrix* obtained by differentiating $C_{\mathrm{n}i}$ with respect to $\boldsymbol{q}$ and $\dot{C}_{\mathrm{t}i}$ with respect to $\dot{\boldsymbol{q}}$. In Eq. 8.3, we add the effect of viscous damping, $\boldsymbol{D}\dot{\boldsymbol{q}}$, of both fingertips. Since Eq. 8.3 corresponds to a nonlinear simultaneous equation with four constraints in terms of the nine variables $\boldsymbol{q}$, a numerical computation algorithm *constraint stabilization method* (CSM) [Bau72] is applied for the dynamic behavior of the constrained system by recalling Sect. 5.4.2. The useful state-space description of Eq. 8.3 for numerical computation to apply the CSM was presented in Eq. 7.10. A complete description of the system was also detailed in Sect. 7.3.2.

8.3 Simulations I: Posture Control of a Grasped Object

The posture control of a grasped object by means of two-fingered robotic hands has been reported to require a total of at least three joints in the hand system [HAK+01]. In this section, we show that the object posture control can be realized even in the minimal-DOF robotic hand with two rotational joints.

8.3.1 Serially-coupled Two-phased Object Orientation Controller

We first propose a two-phased controller capable of determining the orientation of a grasped object as follows:

Table 8.1 Simulation parameters

α	20000	W_{obj}	50 mm
K_{P}	300	$M_{\mathrm{obj}}, M_{\mathrm{f}i}$	0.3 kg
K_{D}	14	$m_{\mathrm{n}}, m_{\mathrm{t}}$	0.3 kg
K_{I}	0.1	I_{obj}	41.7 kg·mm^2
$c_{\mathrm{n}}, c_{\mathrm{t}}$ (Eq. 7.26)	300 Ns/m	$I_{\mathrm{f1}}, I_{\mathrm{f2}}$	582 kg·mm^2
L	76.2 mm	d_{f}	4 mm
$2W_{\mathrm{B}}$	98 mm	E	0.232 MPa
a	20 mm	τ_{b}	30 Nm

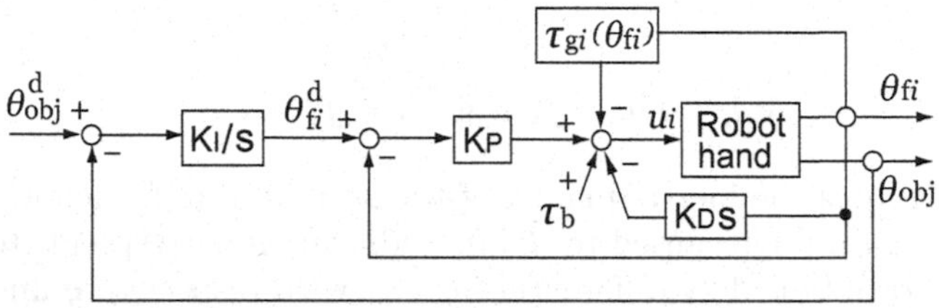

Fig. 8.1 Block diagram of the proposed control method. The characteristics of the control method are that the desired joint angle of the hand is produced dynamically by the integral controller with respect to the object orientation

$$\theta_{\mathrm{f}i}^{\mathrm{d}} = -(-1)^i K_{\mathrm{I}} \int_0^t (\theta_{\mathrm{obj}} - \theta_{\mathrm{obj}}^{\mathrm{d}})\,\mathrm{d}\tau, \tag{8.4}$$

$$u_i = -K_{\mathrm{P}}(\theta_{\mathrm{f}i} - \theta_{\mathrm{f}i}^{\mathrm{d}}) - K_{\mathrm{D}}\dot{\theta}_{\mathrm{f}i} + \tau_{\mathrm{b}} - \tau_{\mathrm{g}i}(\theta_{\mathrm{f}i}). \tag{8.5}$$

At the first stage of this controller, a virtual desired joint angle of the finger, $\theta_{\mathrm{f}i}^{\mathrm{d}}$, is produced by the integral controller of the object orientation, as described in Eq. 8.4. Furthermore, a PD controller using the previously mentioned desired joint angle and a constant torque, τ_{b}, are given at the second stage, as represented in Eq. 8.5. The constant torque works to generate the torque for sustaining the stable grasping even if both the proportional and derivative terms simultaneously become zero during the convergence of the system in the steady state. Figure 8.1 shows a block diagram of the system based on the proposed control method.

Figure 8.2 shows the simulation results when the desired angle of the object is set to be a step input such that

$$\theta_{\mathrm{obj}}^{\mathrm{d}} = \begin{cases} 3^\circ \ (0 \le t < 1), \\ 8^\circ \ (1 \le t < 2), \\ -5^\circ \ (2 \le t < 3), \end{cases} \tag{8.6}$$

where the mechanical parameters and other values used for the numerical simulation are listed in Table 8.1. In addition, we set the constant force at $\tau_{\mathrm{b}} = 30$ Nm and assume that this manipulating motion is implemented on a

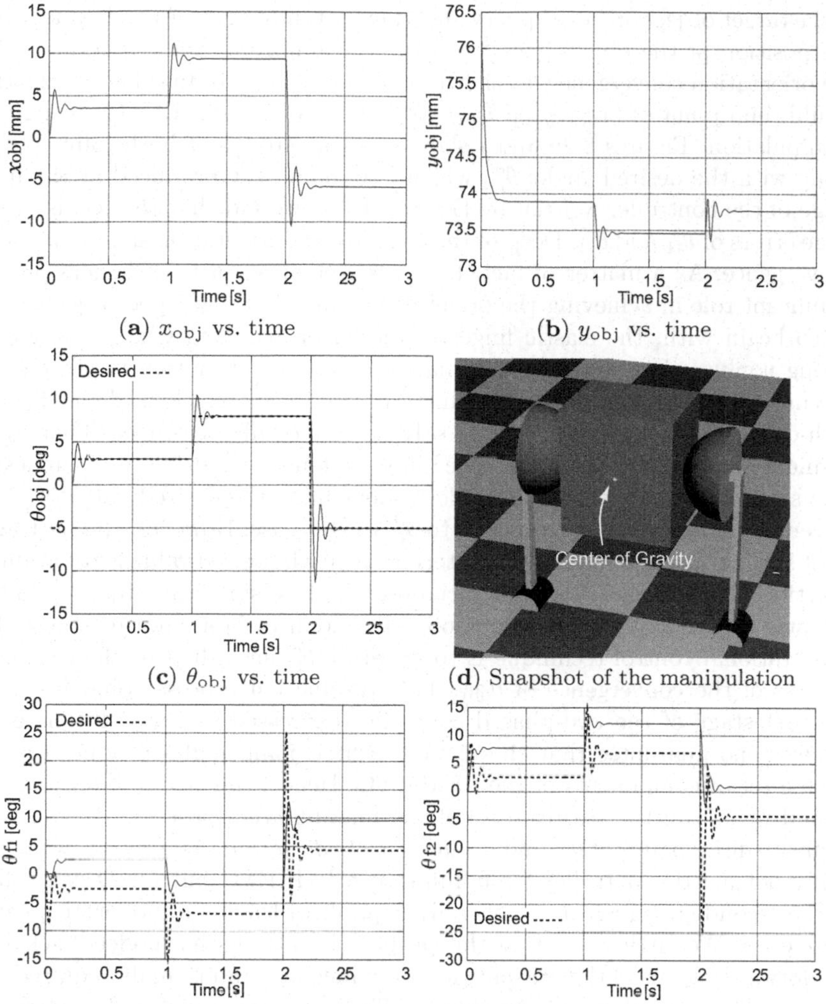

(a) x_{obj} vs. time (b) y_{obj} vs. time

(c) θ_{obj} vs. time (d) Snapshot of the manipulation

(e) θ_{f1} vs. time (f) θ_{f2} vs. time

Fig. 8.2 The simulation results show that the posture control of an object grasped by a minimal-DOF robotic hand can be achieved, although each of the joint angles contains the steady-state error

vertical plane with the gravitational force (Fig. 8.2d). Figures 8.2a-c depict the trajectory of the grasped object $(x_{\mathrm{obj}}, y_{\mathrm{obj}}, \theta_{\mathrm{obj}})$ with respect to time. We can see that the two-phased orientation controller works well so that θ_{obj} robustly converges to the desired step trajectory (Fig. 8.2c). In particular, it is important in Eq. 8.4 to produce the desired finger angles that satisfy equivalent positive and negative values of each finger. This simple control structure of the first stage takes advantage of the natural rolling movement of the target object on both spherical surfaces of the fingers. At the same time, the position of the object goes to a stable equilibrium point together with the orientation convergence, as shown in Figs. 8.2a and b. In other words, the equilibrium point of $(x_{\mathrm{obj}}, y_{\mathrm{obj}}, \theta_{\mathrm{obj}})$ corresponds to an LMEEwC during the manipulation. Figures 8.2e and f show the trajectories of both joint angles along with the desired angle, $\theta^{\mathrm{d}}_{\mathrm{f}i}$, which is dynamically produced at the first stage of the controller written by Eq. 8.4. Here we find that there exist very large errors of $\theta_{\mathrm{f}1}$ and $\theta_{\mathrm{f}2}$. Despite that, the object orientation shows an exact convergence. As a matter of fact, these discrepancies of both fingers play a significant role in achieving the orientation control of a grasped object.

To begin with, the elastic fingertip model proposed in Chap. 4 exhibits strong nonlinearity, and yet the dynamic equations of motion implicitly involving a complicated potential function (Eq. 7.3), which is defined as a high-order nonlinear model, can be solved analytically no longer. Therefore, a unique solution set of LMEEwC cannot be found, resulting in an undesirable situation where we are not able to determine the desired value of each system variable, which corresponds to $\theta^{\mathrm{d}}_{\mathrm{f}i}$ in this case. However, the best and most straightforward method for determining each desired value is to produce it artificially from other system variables. That is, since the object orientation was chosen as a target task among all system variables in this study, the most rational control technique is to generate $\theta^{\mathrm{d}}_{\mathrm{f}i}$ according to the dynamic process of the convergence of θ_{obj}. This production process coincides with the first stage of the two-phased controller expressed as Eq. 8.4. There is, however, no guarantee that the virtual desired joint angles produced comprise a set of components of the LEMMwC. Hence, the errors of $\theta_{\mathrm{f}1}$ and $\theta_{\mathrm{f}2}$ should remain in the steady-state of the manipulation process. We call this requisite discrepancy of $\theta_{\mathrm{f}1}$ and $\theta_{\mathrm{f}2}$ *admissible deviation*. As a result, we must apply not a PID controller for fulfilling the LMEEwC but a PD controller at the second stage, which is generally capable of leaving a moderate steady state error. We further find that this proposed controller has no feedback and feedforward signals of the grasping forces, which are traditionally required for the control of articulated robotic hands. The latter generally needs a rigorous mathematical model of a dynamic system with soft fingertips, and we must solve the inverse dynamics of a closed-loop system. That is, the successful convergence of an object under these conditions indicates that the proposed controller design requires neither the fingertip models nor the real-time computing of the feedforward torque and is based on a complete sensory-based control.

8.3.2 Examples of Failure

Here, we show an example of failure when a PID controller relating to the joint angles is applied to a system. Usually, the PID controller contributes

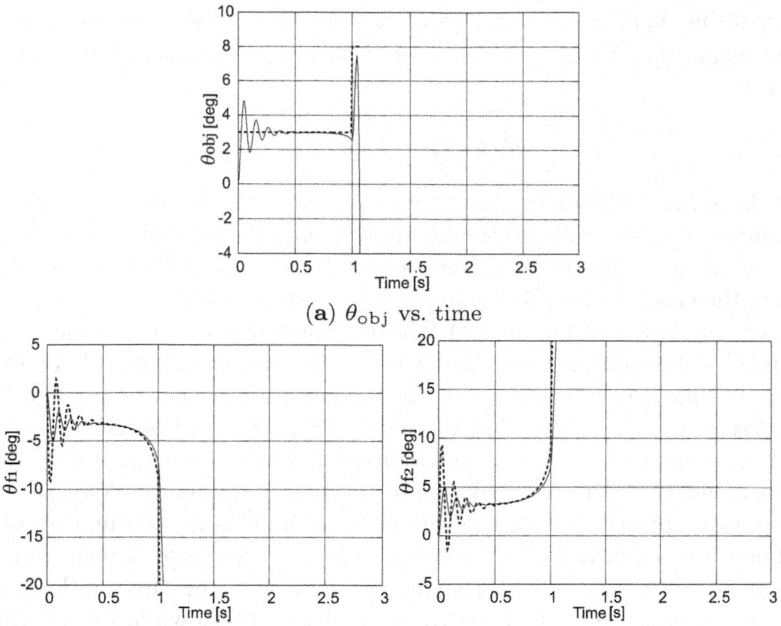

(**a**) θ_{obj} vs. time

(**b**) θ_{f1} vs. time

(**c**) θ_{f2} vs. time

Fig. 8.3 Simulation results in failure when a PID controller related to $\theta_{\mathrm{f}i}$ (Eq. 8.7) is applied to a system

to a robust and secure robotic system if allowing some delays in response, whereas it may invoke unexpected errors to the system in the use of our *serially-coupled two-phased controller*. In this case, the PID controller is given from Eq. 8.5 by

$$u_i = -K_{\mathrm{P}}(\theta_{\mathrm{f}i} - \theta_{\mathrm{f}i}^{\mathrm{d}}) - K_{\mathrm{D}}\dot{\theta}_{\mathrm{f}i} - K_{\mathrm{I}} \int_0^t (\theta_{\mathrm{f}i} - \theta_{\mathrm{f}i}^{\mathrm{d}})\mathrm{d}\tau + \tau_{\mathrm{b}} - \tau_{\mathrm{g}i}(\theta_{\mathrm{f}i}). \quad (8.7)$$

Also in this case, $\theta_{\mathrm{f}i}^{\mathrm{d}}$ is dynamically produced at the first stage of the two-phased controller. Generally, the PID controller enables one to eliminate steady-state errors of a target variable, but in this case, it becomes a crucial cause of the failure as shown in Fig. 8.3. In this control procedure, Eq. 8.7 as a matter of course tries to reduce individual error of both joint angles over time. θ_{f1} and θ_{f2} converge at a certain value around 0.7 s. However, the error increases exponentially, and, as a result, the trajectories of the joint angles diverge rapidly. These results indicate that a control law to eliminate the steady-state error of the joint angle should not be applied to soft-fingered manipulation. The reason for this stems from the fact that the LMEEwC determines a unique equilibrium point during the manipulation. If $\theta_{\mathrm{obj}}^{\mathrm{d}}$, predetermined in this process, is $\theta_{\mathrm{obj}}^{\star}$, that is, a component of the LMEEwC, then a stable equilibrium point $(\theta^{\star}_{\mathrm{obj}}, \theta^{\star}_{\mathrm{f1}}, \theta^{\star}_{\mathrm{f2}})$ that exists uniquely does not always equal $(\theta^{\star}_{\mathrm{obj}}, \theta^{\mathrm{d}}_{\mathrm{f1}}, \theta^{\mathrm{d}}_{\mathrm{f2}})$. That is, the following relationships are generally satisfied:

$$\theta^{\mathrm{d}}_{\mathrm{f1}} \neq \theta^{\star}_{\mathrm{f1}}, \quad \theta^{\mathrm{d}}_{\mathrm{f2}} \neq \theta^{\star}_{\mathrm{f2}}. \tag{8.8}$$

Since the actual joint angles must not converge to the desired value, the PID controller of the joint induces the failure of manipulation tasks. Note that the actual joint angles of the fingers correspond to the LMEEwC in the steady state in the case of successful manipulation shown in Fig. 8.2.

Next, we show another case of failure obtained when zero biased torque is applied to the control law in order to clarify the significance of the biased torque. In this case, we again apply the proposed control law based on the PD control scheme expressed as Eqs. 8.4 and 8.5. Figure 8.4 shows that the object orientation once appears to converge to a desired value for 0.4 to 0.7 s. The PD control alone, however, is not enough to produce a minimally required torque to overcome the strong moment induced by the elastic deformation of the soft fingertips. Eventually, the deficiency of the torque provokes the outward rotation of both fingers, resulting in the fact that the tips of the fingers disengage from the surface of the grasping object, as shown in Figs. 8.4d and e, in which the fingertip deformation of the right finger becomes a negative value after approx. 0.43 s. Thus, the biased torque inputs correspond to an *open-loop control* on joint torque space and are able to determine the prehension strength of each finger without any use of grasping force sensings, excluding environments in contact with the robot or the target object.

8.3.3 Available Range of the Biased Torque

Here we have a query containing the upper and lower bounds of the biased torque for achieving stable grasping and manipulation. To make clear the available range of the torque, we should solve analytically the closed-loop dynamics of a soft-fingered robotic hand. However, because of the complexity of a nonlinear dynamic system, the analytical approach cannot reveal torque limitations. We therefore present a static analysis that can be described as a steady state condition of a closed-loop system; in particular, the static relationship of the right finger is written using Eqs. 7.3, 7.19, 7.24, and 8.5 as

$$E_{\mathrm{n1}}\lambda_{\mathrm{n1}} + F_{\mathrm{t1}}\lambda_{\mathrm{t1}} - \frac{\partial P}{\partial \theta_{\mathrm{f1}}} - K_{\mathrm{P}}(\theta_{\mathrm{f1}} - \theta^{\mathrm{d}}_{\mathrm{f1}}) + \tau_{\mathrm{b}} - \tau_{\mathrm{g1}} = 0. \tag{8.9}$$

When the dynamic behavior of the system remains stable during the manipulation, Eq. 8.9 must be satisfied in the steady state.

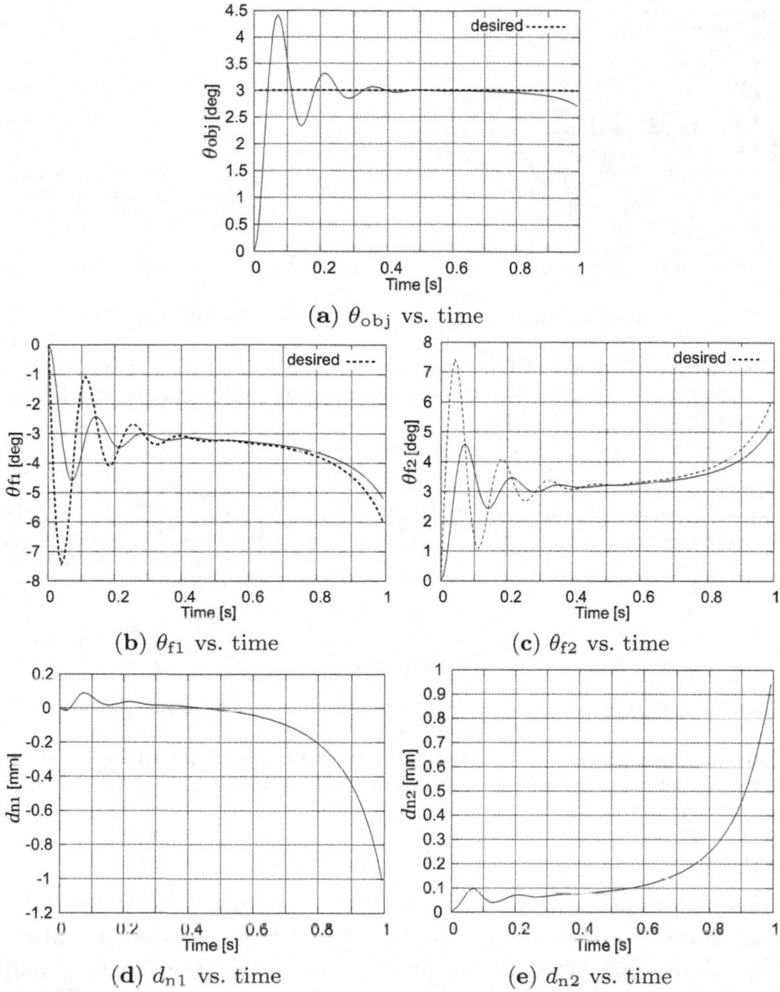

(**a**) θ_{obj} vs. time

(**b**) θ_{f1} vs. time

(**c**) θ_{f2} vs. time

(**d**) d_{n1} vs. time

(**e**) d_{n2} vs. time

Fig. 8.4 Simulation result when $\tau_{\mathrm{b}} = 0$ is added to the system. Once the object orientation appears to converge to a desired value, but the orientation deviates from the value and immediately moves away due to insufficient torque

Figure 8.5 shows the case of failure where a large biased torque $\tau_{\mathrm{b}} = 51$ Nm is applied. This result depends on the mechanical and numerical parameters of this simulation, and in this case we configured them according to an actual experimental setup that will be presented in Sect. 8.5. Consequently, the mechanical parameters in this result obey the dimension of the actual robot hand mechanism, and they are modified such that

$$M_{\mathrm{obj}} = 86\,\mathrm{g},\ M_{\mathrm{f}i} = 88\,\mathrm{g},\ I_{\mathrm{obj}} = 12\,\mathrm{kg\,mm}^2,\ I_{\mathrm{f}i} = 171\,\mathrm{kg\,mm}^2. \quad (8.10)$$

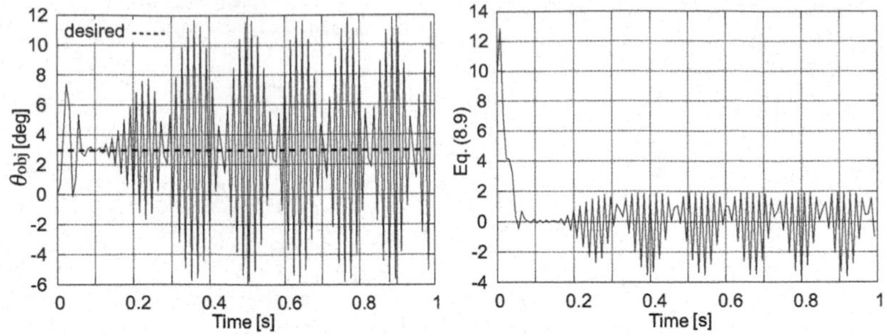

(**a**) θ_{obj} vs. time (**b**) Eq. 8.9

Fig. 8.5 Simulation result of failure obtained when a large biased torque $\tau_{\mathrm{b}} = 51$ Nm is applied to the system, where the gains are assumed to be $K_{\mathrm{P}} = 50$, $K_{\mathrm{D}} = 1$, and $K_{\mathrm{I}} = 0.01$

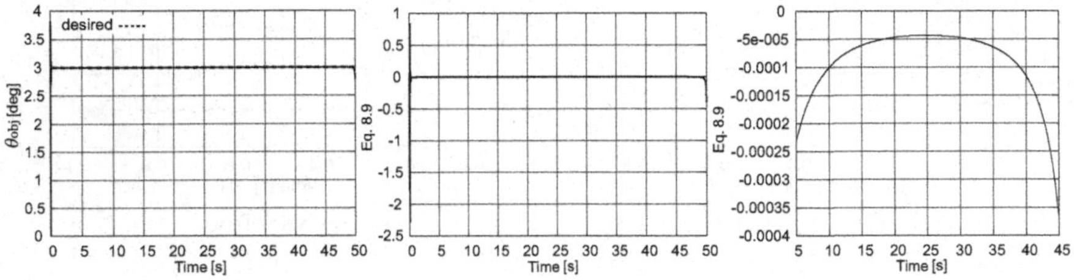

(**a**) θ_{obj} vs. time (**b**) Eq. 8.9 (**c**) Enlarged view of (b)

Fig. 8.6 Simulation result of failure obtained when a small biased torque $\tau_{\mathrm{b}} = 0.37$ Nm is applied to the system, where the gains are assumed to have the same values and parameters as Fig. 8.5

Note that the upper and lower limits of the biased torque specified in the following analysis are therefore confined only to this section. Returning to the Fig. 8.5, it shows that the trajectory of θ_{obj} oscillates acutely and the steady-state relation of Eq. 8.9 does not fulfill the zero condition. This result stems from the fact that the fine adjustment of the remaining torque produced at the proportional controller as an *admissible deviation* can be inhibited by the relatively large torque of τ_{b}. On the other hand, when the biased torque is $\tau_{\mathrm{b}} = 50$ Nm, the simulation of the object orientation is successfully performed without any oscillation of the trajectory. Thus, the upper limit of the torque can be determined such that $\tau_{\mathrm{b}} \approx 50$ Nm.

In addition, Fig. 8.6 shows an opposite case of failure obtained when the biased torque is set at $\tau_{\mathrm{b}} = 0.37$ Nm. The object orientation θ_{obj} appears to converge immediately to 3° and its stable condition is maintained for a long time. However, just before 50 s, the steady-state relationship of Eq. 8.9 is abruptly disturbed by some unexpected factor together with the deviation of the trajectory of θ_{obj}, as shown in Figs. 8.6a and b. In fact, the apparent convergences of θ_{obj} and Eq. 8.9 were not truly achieved permanently, as is clearly seen in an enlarged view shown in Fig. 8.6c. This result stems from an antithetical reason, unlike the case of large torque input, where the torque regulation by the PD controller of Eq. 8.5 largely affects the production of u_i in comparison with the magnitude of τ_b. When τ_b takes a large value, its torque and the restoring moment of fingertip deformation counterbalance each other in a high-energy state (high torque). In that case, the manipulating motion results in a large oscillatory behavior, as shown in Fig. 8.5, because of the simple open-loop control structure of τ_b. On the other hand, when τ_b takes a small value, the PD controller acts effectively as the torque producer within u_i. Hence, because of owing to the feedback as opposed to an open-loop control structure of the system, the trajectories of θ_{obj} and Eq. 8.9 deviate gradually as shown in Fig. 8.6, unlike the results of Fig. 8.5. Note that the simulation result of $\tau_b = 0.38$ Nm was successfully completed. Therefore, a presumable condition of the upper and lower limits of the biased torque based on a parameter example of Eq. 8.10 can be derived as $0.38\,\mathrm{Nm} \leq \tau_b \leq 50\,\mathrm{Nm}$.

8.3.4 Passivity Analysis

Recalling the Lagrange equations of motion (Eq. 8.2) of the soft-fingered hand system including the dynamic constraints, the equations of motion are rewritten using a viscous damping matrix of the soft fingertip and zero external forces as

$$\frac{\mathrm{d}}{\mathrm{dt}}\left(\frac{\partial \mathcal{L}}{\partial \dot{\boldsymbol{q}}}\right) - \frac{\partial \mathcal{L}}{\partial \boldsymbol{q}} - \left[\frac{\partial \dot{\boldsymbol{C}}_{\mathrm{t}}}{\partial \dot{\boldsymbol{q}}}\right]^{\mathrm{T}} \boldsymbol{\lambda}_{\mathrm{t}} + \boldsymbol{D}\dot{\boldsymbol{q}} = \begin{bmatrix} \mathbf{0}_3 \\ \boldsymbol{u} \\ \mathbf{0}_4 \end{bmatrix} \in \mathcal{R}^{9\times 1}, \tag{8.11}$$

where $\mathbf{0}_i$ denotes the i-dimensional zero vector, and $\boldsymbol{u}$ means a 2D vector that is redescribed from Eq. 8.5 as

$$\boldsymbol{u} = -K_{\mathrm{P}}(\boldsymbol{\theta}_{\mathrm{f}} - \boldsymbol{\theta}_{\mathrm{f}}^{\mathrm{d}}) - K_{\mathrm{D}}\dot{\boldsymbol{\theta}}_{\mathrm{f}} + \boldsymbol{\tau}_{\mathrm{b}} + \boldsymbol{\tau}, \tag{8.12}$$

where $\boldsymbol{\tau}$ is assumed to be a virtual zero-torque signal that is needed for the following analyzing procedure for descriptive purposes, and $\boldsymbol{\theta}_{\mathrm{f}}^{\mathrm{d}}$ denotes the vector form of Eq. 8.4. In this analysis, the gravity compensator is assumed to be eliminated in Eq. 8.12 because it is not indispensable for controlling the target variable, *i.e.*, object orientation in the proposed two-phased controller. This fact will be verified in the subsequent simulations of Sect. 8.4 and Chap. 10. To investigate the passivity of the system [Ari96], computing the inner product of $\dot{\boldsymbol{q}}$ on both sides of Eq. 8.11 and summing up both sides, we obtain $\dot{\boldsymbol{\theta}}_{\mathrm{f}}^{\mathrm{T}}\boldsymbol{\tau}$ as follows:

$$\dot{\boldsymbol{\theta}}_{\mathrm{f}}^{\mathrm{T}}\boldsymbol{\tau} = \dot{\boldsymbol{q}}^{\mathrm{T}}\left\{\frac{\mathrm{d}}{\mathrm{dt}}\left(\frac{\partial \mathcal{L}}{\partial \dot{\boldsymbol{q}}}\right)\right\} - \dot{\boldsymbol{q}}^{\mathrm{T}}\frac{\partial \mathcal{L}}{\partial \boldsymbol{q}} + \dot{\boldsymbol{q}}^{\mathrm{T}}\left[\frac{\partial \dot{\boldsymbol{C}}_{\mathrm{t}}}{\partial \dot{\boldsymbol{q}}}\right]^{\mathrm{T}}\boldsymbol{\lambda}_{\mathrm{t}}$$
$$+ \dot{\boldsymbol{q}}^{\mathrm{T}}\boldsymbol{D}\dot{\boldsymbol{q}} + \dot{\boldsymbol{q}}^{\mathrm{T}}\begin{bmatrix}\boldsymbol{0}_3 \\ K_{\mathrm{P}}(\boldsymbol{\theta}_{\mathrm{f}} - \boldsymbol{\theta}_{\mathrm{f}}^{\mathrm{d}}) + K_{\mathrm{D}}\dot{\boldsymbol{\theta}}_{\mathrm{f}} - \boldsymbol{\tau}_{\mathrm{b}} \\ \boldsymbol{0}_4\end{bmatrix}. \quad (8.13)$$

Here, considering that $\boldsymbol{D}$ relates only to the fingertip variable and $\boldsymbol{C}_{\mathrm{n}}$ and $\dot{\boldsymbol{C}}_{\mathrm{t}}$ can be represented respectively as a linear combination of each of $\boldsymbol{q}$ and $\dot{\boldsymbol{q}}$, Eq. 8.13 is then expressed as

$$\dot{\boldsymbol{\theta}}_{\mathrm{f}}^{\mathrm{T}}\boldsymbol{\tau} = \frac{\mathrm{d}K}{\mathrm{dt}} + \frac{\mathrm{d}P}{\mathrm{dt}} - \dot{\boldsymbol{C}}_{\mathrm{n}}^{\mathrm{T}}\boldsymbol{\lambda}_{\mathrm{n}} - \dot{\boldsymbol{C}}_{\mathrm{t}}^{\mathrm{T}}\boldsymbol{\lambda}_{\mathrm{t}} + c_{\mathrm{n}}\dot{\boldsymbol{d}}_{\mathrm{n}}^{\mathrm{T}}\dot{\boldsymbol{d}}_{\mathrm{n}} + c_{\mathrm{t}}\dot{\boldsymbol{d}}_{\mathrm{t}}^{\mathrm{T}}\dot{\boldsymbol{d}}_{\mathrm{t}}$$
$$+ K_{\mathrm{P}}\dot{\boldsymbol{\theta}}_{\mathrm{f}}^{\mathrm{T}}(\boldsymbol{\theta}_{\mathrm{f}} - \boldsymbol{\theta}_{\mathrm{f}}^{\mathrm{d}}) + K_{\mathrm{D}}\dot{\boldsymbol{\theta}}_{\mathrm{f}}^{\mathrm{T}}\dot{\boldsymbol{\theta}}_{\mathrm{f}} - \dot{\boldsymbol{\theta}}_{\mathrm{f}}^{\mathrm{T}}\boldsymbol{\tau}_{\mathrm{b}}. \quad (8.14)$$

Furthermore, from the relationships of $\dot{\boldsymbol{C}}_{\mathrm{n}} = \boldsymbol{0}$ and $\dot{\boldsymbol{C}}_{\mathrm{t}} = \boldsymbol{0}$, Eq. 8.14 can finally be represented as

$$\dot{\boldsymbol{\theta}}_{\mathrm{f}}^{\mathrm{T}}\boldsymbol{\tau} = \frac{\mathrm{d}E}{\mathrm{dt}} + c_{\mathrm{n}}\dot{\boldsymbol{d}}_{\mathrm{n}}^{\mathrm{T}}\dot{\boldsymbol{d}}_{\mathrm{n}} + c_{\mathrm{t}}\dot{\boldsymbol{d}}_{\mathrm{t}}^{\mathrm{T}}\dot{\boldsymbol{d}}_{\mathrm{t}}$$
$$+ K_{\mathrm{P}}\dot{\boldsymbol{\theta}}_{\mathrm{f}}^{\mathrm{T}}(\boldsymbol{\theta}_{\mathrm{f}} - \boldsymbol{\theta}_{\mathrm{f}}^{\mathrm{d}}) + K_{\mathrm{D}}\,\dot{\boldsymbol{\theta}}_{\mathrm{f}}^{\mathrm{T}}\dot{\boldsymbol{\theta}}_{\mathrm{f}} - \dot{\boldsymbol{\theta}}_{\mathrm{f}}^{\mathrm{T}}\boldsymbol{\tau}_{\mathrm{b}}, \quad (8.15)$$

where $E = K + P$ denotes an entire energy function of the total system. In addition, integrating Eq. 8.15 with respect to time, we can obtain the following integral equation:

$$\int_0^T \dot{\boldsymbol{\theta}}_{\mathrm{f}}^{\mathrm{T}}\boldsymbol{\tau}\,\mathrm{dt} = E(T) - E(0) + c_{\mathrm{n}}\int_0^T \dot{\boldsymbol{d}}_{\mathrm{n}}^{\mathrm{T}}\dot{\boldsymbol{d}}_{\mathrm{n}}\,\mathrm{dt} + c_{\mathrm{t}}\int_0^T \dot{\boldsymbol{d}}_{\mathrm{t}}^{\mathrm{T}}\dot{\boldsymbol{d}}_{\mathrm{t}}\,\mathrm{dt}$$
$$+ K_{\mathrm{P}}\int_0^T \dot{\boldsymbol{\theta}}_{\mathrm{f}}^{\mathrm{T}}(\boldsymbol{\theta}_{\mathrm{f}} - \boldsymbol{\theta}_{\mathrm{f}}^{\mathrm{d}})\,\mathrm{dt} + K_{\mathrm{D}}\int_0^T \dot{\boldsymbol{\theta}}_{\mathrm{f}}^{\mathrm{T}}\dot{\boldsymbol{\theta}}_{\mathrm{f}}\,\mathrm{dt} - \int_0^T \dot{\boldsymbol{\theta}}_{\mathrm{f}}^{\mathrm{T}}\boldsymbol{\tau}_{\mathrm{b}}\,\mathrm{dt}. \quad (8.16)$$

Generally, if the above equation, which refers to an energy balance of a closed-loop system, satisfies the subsequent relationship, it is said that the corresponding dynamic system fulfills the passivity condition:

$$\int_0^T \dot{\boldsymbol{\theta}}_{\mathrm{f}}^{\mathrm{T}}\boldsymbol{\tau}\,\mathrm{dt} \geq c_{\mathrm{n}}\int_0^T \dot{\boldsymbol{d}}_{\mathrm{n}}^{\mathrm{T}}\dot{\boldsymbol{d}}_{\mathrm{n}}\,\mathrm{dt} + c_{\mathrm{t}}\int_0^T \dot{\boldsymbol{d}}_{\mathrm{t}}^{\mathrm{T}}\dot{\boldsymbol{d}}_{\mathrm{t}}\,\mathrm{dt} + K_{\mathrm{D}}\int_0^T \dot{\boldsymbol{\theta}}_{\mathrm{f}}^{\mathrm{T}}\dot{\boldsymbol{\theta}}_{\mathrm{f}}\,\mathrm{dt}. \quad (8.17)$$

However, we notice that the success and failure of Eq. 8.17 depend on the joint angles and their initial conditions, which can be seen in Eq. 8.16, and that the change of $\boldsymbol{\tau}_{\mathrm{b}}$ plays an important role in achieving the passivity condition. That is, even if we start manipulation movements after determining the viscous damping, control gains, and initial conditions preliminarily, we can adjust the system stability/instability by regulating the biased torque $\boldsymbol{\tau}_{\mathrm{b}}$.

As a result, stability can be maintained by implementing the real-time computation of $\boldsymbol{\tau}_{\text{b}}$ during the robot motions and applying it to a corresponding control law. This study, however, employs a preliminary offline computation of $\boldsymbol{\tau}_{\text{b}}$ to avoid the burden on the CPU performance, whose upper and lower limits were already presented in Sect. 8.3.3.

On the other hand, we are generally not able to compensate the convergence performance of the target variable even though we prove the internal stability of the closed-loop system using passivity analysis. That is, since the second stage of the two-phased controller is added to the robot as a PD controller of joint angles, the angles do not converge to the desired value as long as a gravity compensator in not introduced to the joints. However, if the joint angles do not need to be converged during the manipulation, then the gravity compensator is not indispensable for the control system. In this case, the idea is based on a peculiar control technology that can permit an unexpected steady-state discrepancy of the joint angles if only the target variable (object orientation) successfully converges to the desired trajectory by interlocking the PD controller directly with the target variable to be precisely controlled. The part that works as the controller of the target variable corresponds to the first-stage controller that can be serially connected to the second-stage controller for the finger joints on the basis of the inventive concept.

8.4 Simulations II: Responses for Time Delay

It is said that a biological motor control system involves a large time delay along with neural information transfer processes [Hol90, Dev00]. That delay of the spinal reflex system is much larger than the sampling time for control of traditional robot systems, resulting in an unstable state of the robot. Furthermore, it is reported that the delay of human visual system that contains image processing at retina and transmission on visual pathways to visual cortex where high-level perception takes place increases largely than the neuromuscular transmission on the afferent nerve pathway [ATÖ+06].

In this section, we first present several simulation results obtained when the delay of sampling time of the visual feedback loop that corresponds to the first-stage controller occurs during the orientation control task. In particular, we indicate that the two-phased object orientation controller is capable of improving the degradation of the control performance, which is caused by the delay of sampling time of the visual feedback. Note that, in what follows of this chapter, the gravity compensator expressed in Eq. 8.5 is eliminated because it never affect the convergence performance of a target object, but we consider 2D vertical-planar operations to which the gravitational force applies. The second stage controller can, therefore, be rewritten by

$$u_i = -K_{\text{P}}(\theta_{\text{f}i} - \theta_{\text{f}i}^{\text{d}}) - K_{\text{D}}\dot{\theta}_{\text{f}i} + \tau_{\text{b}}. \tag{8.18}$$

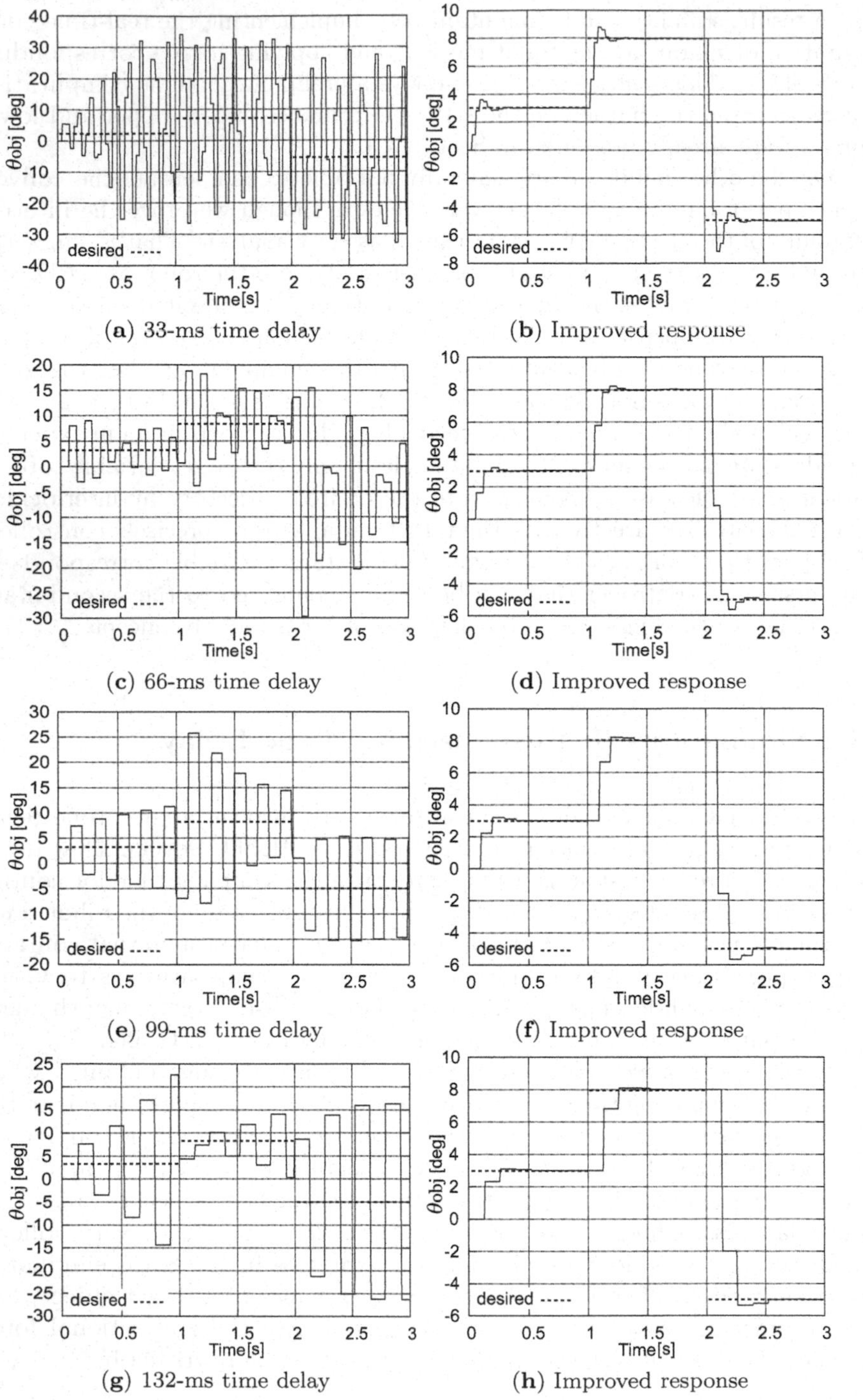

(**a**) 33-ms time delay (**b**) Improved response

(**c**) 66-ms time delay (**d**) Improved response

(**e**) 99-ms time delay (**f**) Improved response

(**g**) 132-ms time delay (**h**) Improved response

Fig. 8.7 Improvement effects of oscillatory phenomena caused by the *visual-updating delay* (Fig. 8.10) of the orientation controller (first stage) can be seen when the integral gain decreases. **a-b** K_{I} decreases from 0.01 to 0.002. **c-d** K_{I} decreases from 0.002 to 0.001. **e-f** K_{I} decreases from 0.0027 to 0.0008. **g-h** K_{I} decreases from 0.002 to 0.0006

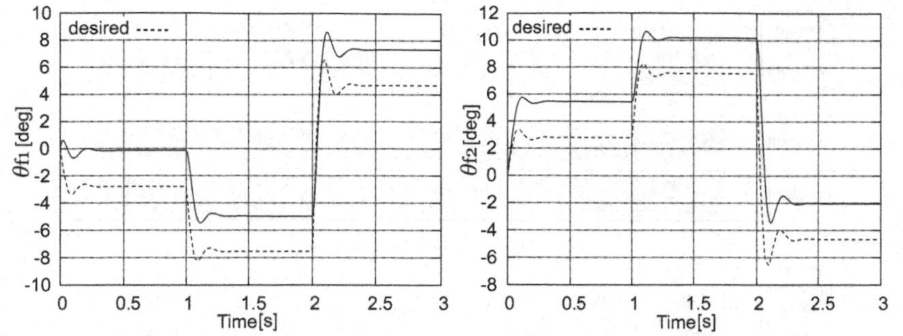

(**a**) θ_{f1} and $\theta_{\mathrm{f1}}^{\mathrm{d}}$ (**b**) θ_{f2} and $\theta_{\mathrm{f2}}^{\mathrm{d}}$

Fig. 8.8 Admissible discrepancies of the joint angles can be seen also in this case where the delay of the sampling time is 33 ms and $K_{\mathrm{I}} = 0.002$

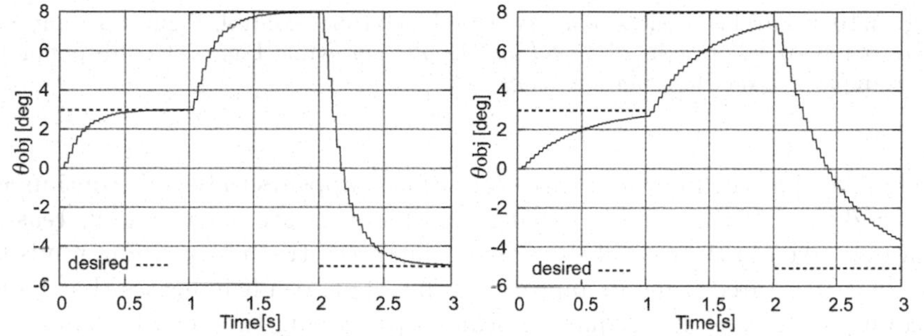

(**a**) $K_{\mathrm{I}} = 0.0005$ (**b**) $K_{\mathrm{I}} = 0.0002$

Fig. 8.9 Further degradation of response can be seen when the integral gain K_{I} decreases more and more. In this case, the delay of the sampling time is 33 ms

In general, visual sensing methods using a commercially available CCD camera have a specific architectural limitation of the image processing speed, *i.e.*, periodic 33-ms processing time is inevitable for the practical use of an industrial camera. Therefore, we intentionally alter the sampling time of the visual sensing in the simulation such that the time delays become 33 ms (30 Hz), 66 ms (15 Hz), 99 ms (10 Hz), and 132 ms (7.5 Hz). That is, the timing of acquisition of camera images is assumed to be the above four patterns. Hence, the time delay of the image information can be ignored for this example. These four patterns were chosen because in our future works we plan to consider a high-resolution camera (*e.g.*, 2 megapixels or more) for capturing handling motions, whose maximal frame rate is approx. 15 Hz or less.

Figure 8.7a shows an oscillatory movement of a manipulated object in simulation, which is obtained when the time delay is 33 ms and $K_{\mathrm{I}} = 0.01$. This oscillatory phenomenon can easily be improved dramatically by decreasing K_{I} slightly, as shown in Fig. 8.7b. Likewise, the results of 66 ms, 99 ms, and 132 ms are all improved clearly by the slight change in K_{I}. Also, we can

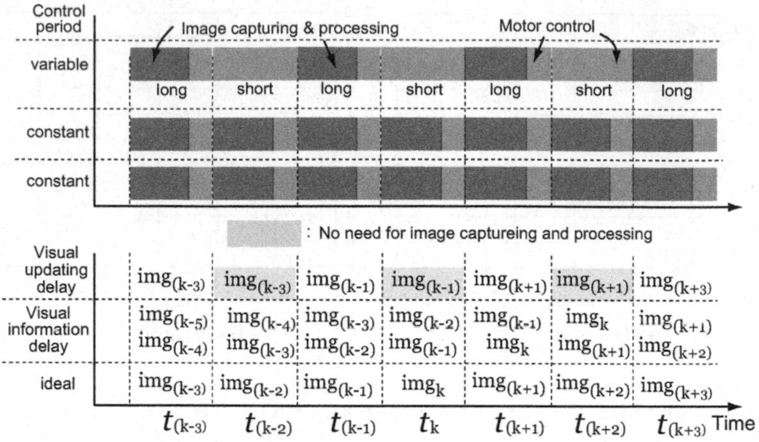

Fig. 8.10 Control sequences of a visual feedback robot control. Figures 8.7 and 8.9 corresponds to an example of *visual-updating delay*, while Fig. 8.11 corresponds to an example of *visual-information delay*

easily find that the improved trajectory of θ_{obj} appears to be a discontinuous line and the rise time of θ_{obj} expands gradually as the delay time increases. However, even in the case of the 132-ms delay, the trajectory of θ_{obj} tends to follow each desired value in approx. 250 ms. This result indicates that high-resolution cameras that require enormous processing time can be employed for robotic manipulation tasks by a soft-fingered hand. In addition, all the improved results become a steplike trajectory according to the time delay, and then the width of the step increases steadily together with the change in the delay. Furthermore, the onset of the trajectory movement on every step has a specific time delay. For example, we find that a certain invariant state of the trajectory at 1 s exists in every improved response more than expected. These results stem from the fact that the integral multiple of each sampling time (*i.e.*, 33, 66, 99, and 132 ms) do not coincide with that of 1000 ms, that is, the beginning for control in each step no longer occurs at 1000 ms or 2000 ms. Despite that, the trajectory of the object robustly converges to a desired value. In addition, Fig. 8.8 shows admissible discrepancies of the joint angles as well as the simulation results with no delay, which were presented in Fig. 8.2. Both of these joint trajectories are obvious continuous lines, unlike the object orientation shown in Fig. 8.7. This is because the PD controller for the joints of the second stage works within continuous real-time processing with no delay. On the other hand, the excessive decrease in K_{I} may provoke an undesirable further degradation of dynamic response, as shown in Fig. 8.9. Experimental results associated with the lower limit of the integral gain will be presented at Sect. 8.6. Thus, we notice the existence of an optimal value of K_{I} that is capable of enhancing the control performance significantly even though the sampling time of the visual feedback is delayed significantly.

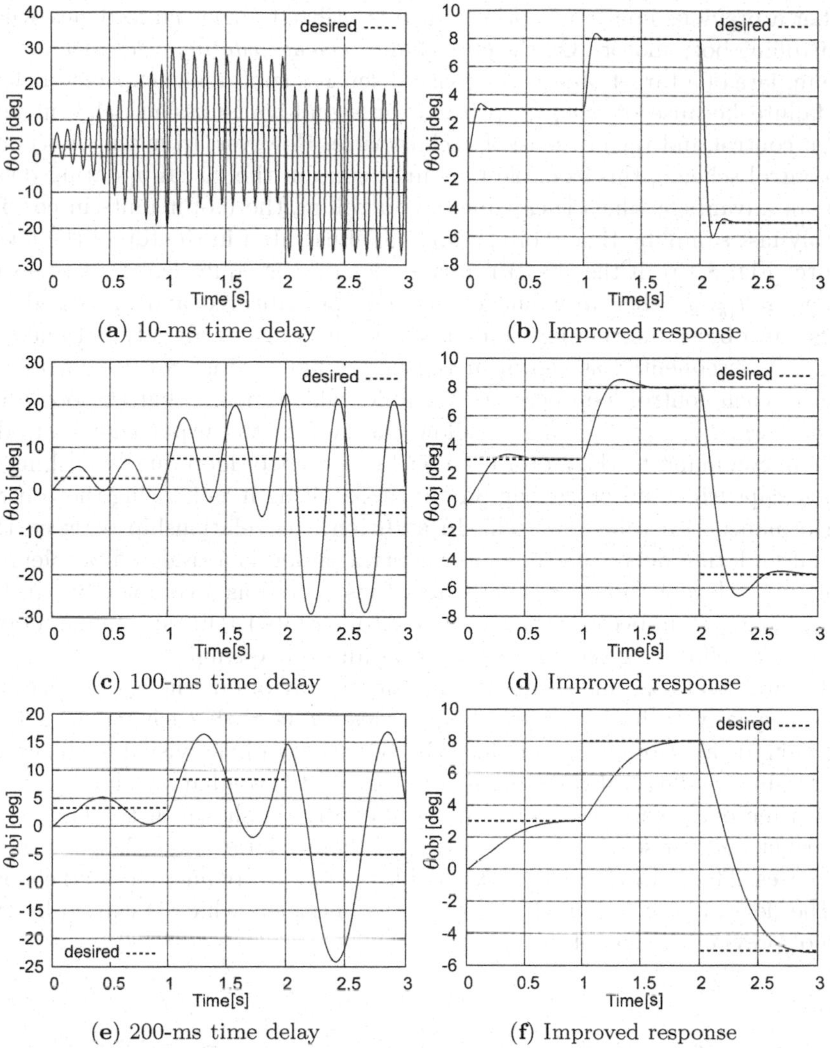

(**a**) 10-ms time delay (**b**) Improved response

(**c**) 100-ms time delay (**d**) Improved response

(**e**) 200-ms time delay (**f**) Improved response

Fig. 8.11 Improvement effects of oscillatory phenomena caused by the *visual-information delay* (Fig. 8.10) of the orientation controller (first stage) can be seen when the integral gain decreases. **a-b** K_{I} decreases from 0.01 to 0.002. **c-d** K_{I} decreases from 0.0015 to 0.0005. **e-f** K_{I} decreases from 0.0008 to 0.0002

Next, we show another simulation result in which the time delay defined as the transmission latency of human optical nerve pathways is considered [HK03, Bur81]. From the viewpoint of control strategy, the time delay means that the previous (old) visual information in a time series is used for the orientation controller of the first stage, which corresponds to *visual-information delay* shown in Fig. 8.10. That is, the middle of the lower figure indicates that either a previous image or a prior image is utilized for visual feedback when controlling both motors. On the other hand, *visual-updating delay* means that the update of a target object using visual sensing contains unexpected delays or failure because of, *e.g.*, drop frame or signal-transmission delay in network control and remote sensing architecture. However, from the viewpoint of control scheme, this example that may contain the visual-updating delay has an advantage which is capable of controlling the motor units in considerably fast sampling time. For example, as illustrated in the top of the lower figure (Fig. 8.10), if the $(k-3)$th image at $t_{(k-3)}$ is equivalent to a $(k-3)$th image at $t_{(k-2)}$, the latter image need not be captured and processed because an object information computed by an image is already obtained at $t_{(k-3)}$. Consequently, as shown in the upper figure (Fig. 8.10), control period in each control loop becomes variable, which means that the posterior loops $(e.g., t_{(k-2)}, t_k, t_{(k+2)}, \cdots)$ allow the system to output short-periodic control command to the motor even within a control loop due to nothing of image capturing and processing processes, resulting in enhancing the control performance of a robotic system. In addition, the relationship between the visual-updating delay and the motor control period is a trade-off problem as long as a relevant robotic system is not constructed as a completely-parallel processing system because the motor control and its feedback command does not work well during the image capturing and processing.

Figures 8.11a, c, and e show oscillatory behaviors of the $\theta_{\rm obj}$ trajectory that respectively have different frequencies along with each magnitude of the time delay. However, we reduce the sensitivity of the visual feedback by decreasing the integral gain, and as a result, all the oscillatory movements at each time delay can be suppressed significantly, as shown in the figures on the right. Of course, the rise time of $\theta_{\rm obj}$ becomes large when the time delay increases. Thus, the decrease in $K_{\rm I}$ enables one to overcome the rapid change in the deviation of $\theta_{\rm obj}$ in the orientation controller, which is caused by the use of previous image information.

8.5 Experiments I: Posture Control of a Grasped Object

To consolidate our two-phased object orientation control scheme, we implement several experiments by means of a soft-fingered robotic hand that consists of a pair of rotational fingers on which are mounted rubberlike soft fingertips made of polyurethane gel. Figure 8.12 shows an apparatus of the robotic hand system that has a CCD camera for visual feedback control, whose technical specification is listed in Table 8.2. As shown in Fig. 8.12, this experiment was conducted in a vertical 2D plane including the gravitational force, in which no compensator for the gravity was introduced in the controller design. An interface board of AD/DA conversions to directly produce a DC motor current is able to output the voltage from 0 to 10 V,

Table 8.2 Camera specification

Manufacturer and model number	Toshiba Teli CS5111L
TV system	NTSC (30 Hz)
Image sensor	2:1 Interline Color CCD
Total pixels	768(H)×494(V); 0.38 megapixel
Scanning lines	525 lines
Aspect ratio	4:3

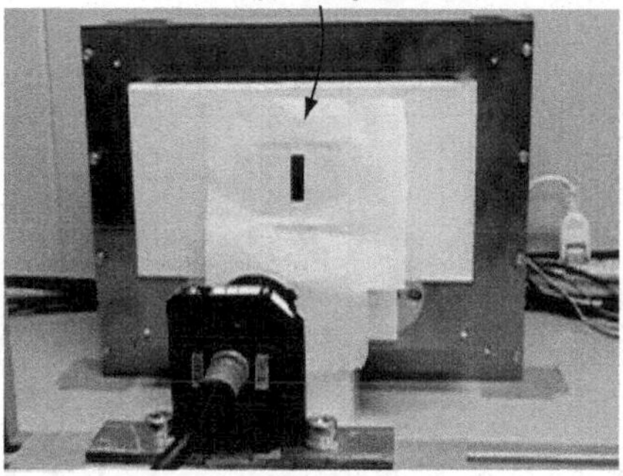

Fig. 8.12 Experimental setup

which means that 0 V and 10 V correspond to the maximal torque for inward/outward rotation and 5 V corresponds to the neutral state, meaning zero torque output.

8.5.1 Object Orientation Control Under Constant Biased Torque (Exp. 1)

First, we implement an experiment of object orientation control where the biased torque remains constant. The designated trajectory of θ_{obj} is identical to that of the simulation. Unlike the configuration of the simulation, each step of the manipulation task is set at 2 s in this experiments.

Figure 8.13a shows an exact convergence of the object orientation while keeping the rise time to approx. 500 ms in every step. This result indicates that our proposed two-phased controller works extremely well regardless of the very simple structure based on the traditional PD controller design. Figure 8.13a also exhibits influences of the delay of the video frame rate (33 ms). That is, the trajectory of the orientation becomes a discontinuous step-like function according to the time delay of the video frame rate. In addition, Fig. 8.13b exhibits an important function of the biased torque capable of pre-

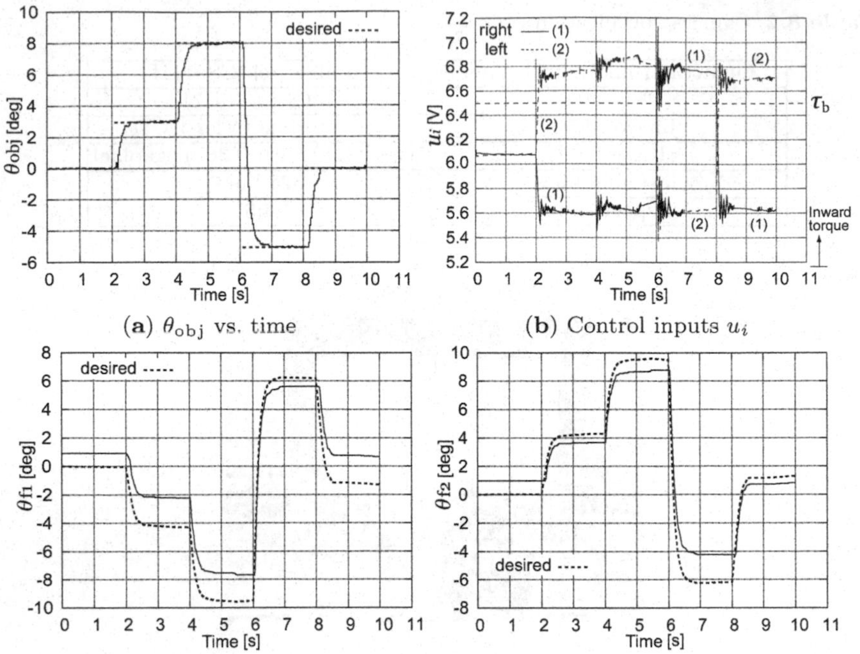

(**a**) θ_{obj} vs. time (**b**) Control inputs u_i

(**c**) θ_{f1} vs. time (**d**) θ_{f2} vs. time

Fig. 8.13 Experimental results of the orientation control of a grasped object when the biased torque is constant such that $\tau_{\mathrm{b}} = 0.41$ Nm (1.5 V), where $K_{\mathrm{P}} = 162$, $K_{\mathrm{D}} = 15$, and $K_{\mathrm{I}} = 0.028$

venting outward rotation of the fingers, which can be understood from the fact that even the lowest point in the voltage of left finger at 6 s lies above the neutral voltage (5 V), which cannot produce any motor torque for the left finger. We further find that the trajectory of the control input u_i is depicted in an axisymmetric manner because the biased torque remains constant during the manipulation process. This axisymmetric behavior of the torque implies the important fact that either the right or the left finger rotates in the direction of increasing grasping forces and the remaining opposite finger rotates in the direction of decreasing grasping forces. This phenomenon stems from a characteristic of the underlying control structure in the first-stage of the two-phased controller. That is, each integral controller for the object orientation produces contrary plus and minus values of the virtual desired joint angle such that $\theta_{\mathrm{f1}}^{\mathrm{d}} + \theta_{\mathrm{f2}}^{\mathrm{d}} = 0$. Note that τ_{b} is plotted at 6.5 V as a dotted line because its initial value (1.5 V) is predetermined so that the fingers can rotate inward. Figures 8.13c and d show the trajectory of both finger angles, which contains an admissible discrepancy that never influences the manipulation stability. As mentioned in Sect. 8.3.1, we find that the biased torque τ_{b} is not the one obtained from the real-time computation of the inverse dynamics of the total system and is rather a constant torque that is merely predetermined in the experiment. That is, the feedback and feedforward signals of the grasping forces measured by a force sensor unit are not indispensable for the posture control of the object.

8.5.2 Open-loop Control of Biased Torque (Exp. 2)

This section verifies that the biased torque can be altered arbitrarily while keeping the object orientation at a certain position, that is, we demonstrate that the open-loop control of the biased torque can be achieved during soft fingered manipulation based on the proposed control method. Here, the change of the biased torque is equivalent to that of the strength of grasping, and eventually we show that the two-phased orientation controller is capable of dealing the object orientation and the biased torque independently.

In this experiment, first we set the object orientation at 5°. After that we augment the biased torque continuously so that the increasing ratio becomes linear, as shown in Fig. 8.14a. We can find that the trajectory of the orientation spectacularly converges to the desired angle regardless of the rapid increase of the torque, as shown in Fig. 8.14c. Interestingly, as shown in Fig. 8.14b, both final torques u_i are subequal to each other starting at approx. 4.5 s, though these values are different largely before the increase in the biased torque. In other words, u_1 and u_2 are equal even though neither the object position nor orientation is not situated in a neutral state ($x_{\mathrm{obj}} = y_{\mathrm{obj}} = \theta_{\mathrm{obj}} = 0$). This result may suggest the possibility that equivalent strength of the grasping torque on both fingers is enough to manipulate the target object as long as the object orientation does not change. Figures 8.14d and e show the trajectories of both joint angles when the biased torque increases continuously. In both figures, each error of the joint angles augments gradually; in particular, the angle of the left finger rises though the artifical joint angle that is virtually produced as a desired trajectory deviates largely. This indicates that the discrepancies of the joints never affect the stability and performance of the manipulation, and the object orientation is only necessary to converge to a commanded orientation.

8.5.3 Object Orientation Control Under Variable Biased Torque (Exp. 3)

In this section, we implement a combined manipulation task performed by incorporating the object orientation control under constant torque and the open-loop control of the biased torque, which are both presented in the preceding sections. The main task is the object orientation control as well

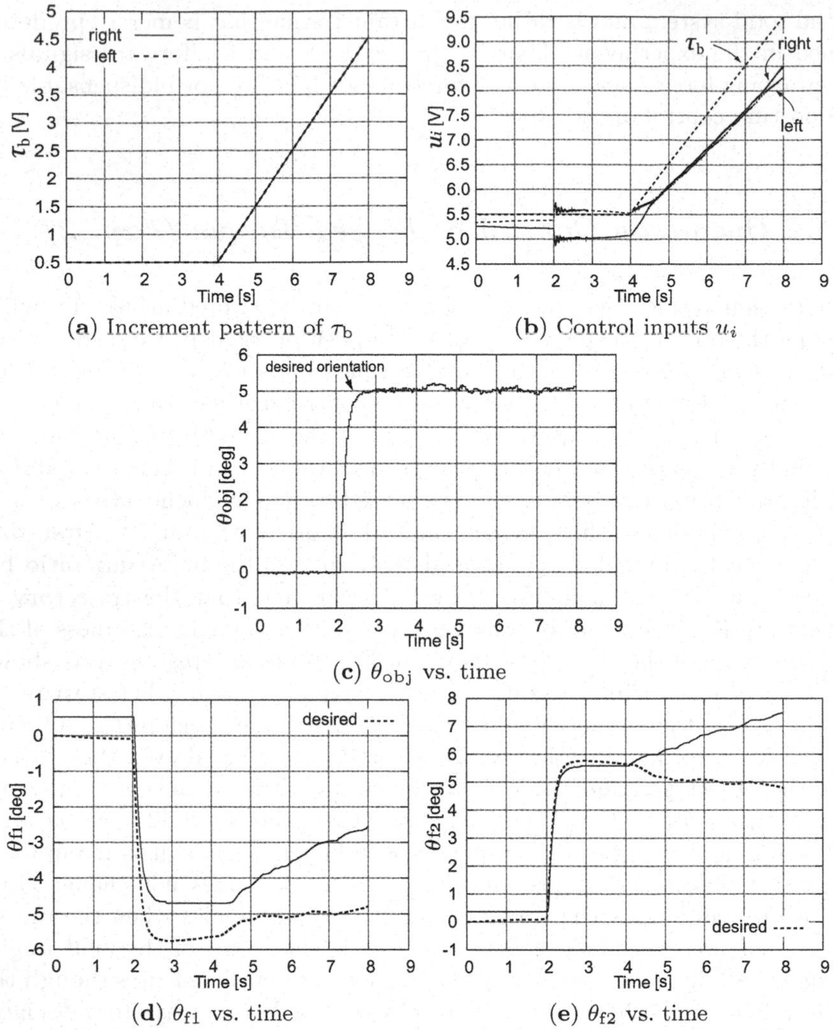

(a) Increment pattern of τ_b (b) Control inputs u_i

(c) θ_{obj} vs. time

(d) θ_{f1} vs. time (e) θ_{f2} vs. time

Fig. 8.14 Experimental results of the posture control of a grasped object during the increase in biased torque τ_b

as Fig. 8.13, while we try to increase the biased torque linearly as well as Fig. 8.14.

Figure 8.15a shows an increasing pattern of the torque, and Fig. 8.15c presents a successful convergence of the object orientation, indicating that the simultaneous control of both the object information and the biased torque that contributes to the secure grasping can be achieved because the torque relates directly to the strong and light grasping forces. Also in this case, the time delay due to the structural sampling cycle of the CCD camera

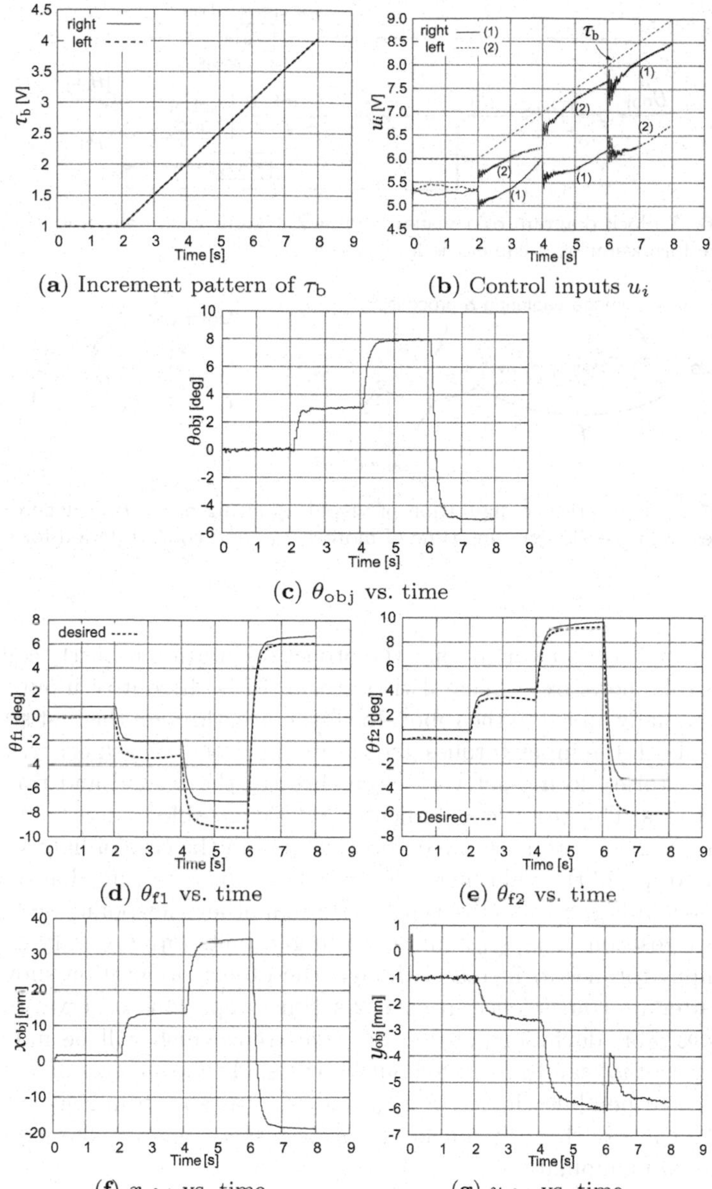

Fig. 8.15 Experimental results of the posture control of a grasped object when the biased torque τ_b varies (biased-torque control)

appears. Note that, as a matter of fact, the result of the figure contains both the time delay and another one due to the discrepancy between the multiple number of 33 ms and each instant of time (*i.e.*, 2000, 4000, $\cdots$ ms).

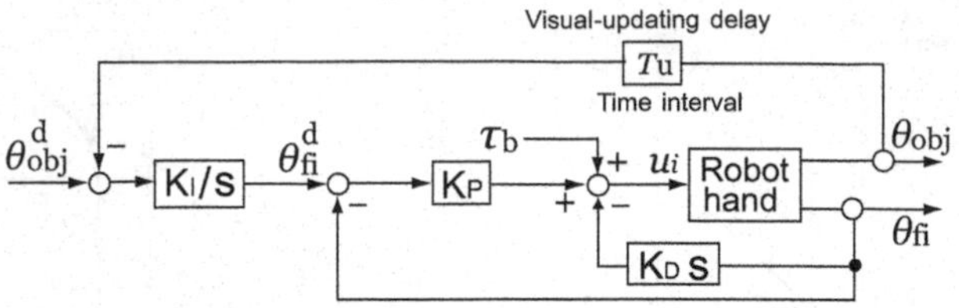

Fig. 8.16 A block diagram of a visual feedback system where the updating timing of visual information by a camera is largely delayed

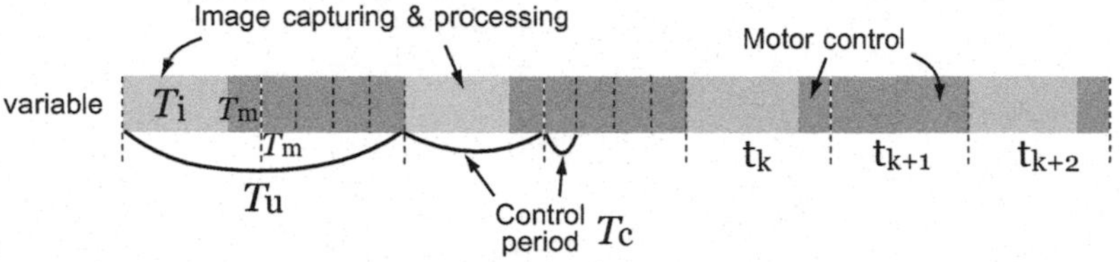

Fig. 8.17 A time series configuration of visual-updating delay T_u, image capturing and processing time T_i, sampling time of motors T_m, and control period for a system T_c

Nevertheless, it is noteworthy that the object orientation robustly converges. Figure 8.15b shows experimental trajectories of both control input torques that are actually added to the robot. In this figure, through the first step from 2 s to 4 s, both the input torques get gradually closer to each other, whereas these trajectories do not get very close through the second and third steps from 4 s to 8 s. This result may indicate that we are able to determine almost the same grasping torques between both fingers in the case of light grasping, while the torque of the right finger doubles in comparison with that of the left finger at 8 s in Fig. 8.15b. Observing it from another viewpoint, we find that the object position (x_{obj}, y_{obj}) and both finger joints $(\theta_{f1}, \theta_{f2})$ vary explicitly as the input torques u_i increase, though the object orientation shows little change after its convergence in each 2-s time step. This observation infers that undesirable deviations of θ_{obj} on each step, which will be induced by the increase of u_i, can be canceled out by the slight movements of θ_{f1} and θ_{f2} and the corresponding change of x_{obj} and y_{obj}, as shown in Figs. 8.15f and 8.15g. We can say that this phenomenon is just a superior characteristic of soft fingered manipulation.

8.6 Experiments II: Responses for Time Delay

In this section, we show experimental results of object orientation control using the *serially-coupled two-phased controller* (Eqs. 8.4 and 8.18), where the visual-updating delay is considered unlike the previous section. Therefore,

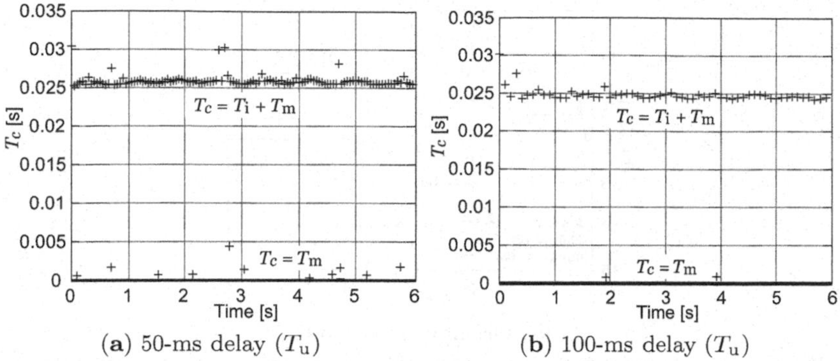

Fig. 8.18 A wide gap in the control period, T_c, of a loop between the conditions of $T_c = T_i + T_m$ (*image-capturing on*) and $T_c = T_m$ (*image-capturing off*). T_i is approx. 25 ms for which the test bed shown in Fig. 7.7 is utilized, while T_m becomes extremely short sampling time such that it ranges from 20 μs up to 30 μs dynamically

the concept of all experiments presented in this section obeys the simulation studies of Figs. 8.7, 8.8, and 8.9. Configurations of a robot and a camera device is equal to Table 7.3 and Fig. 7.7. Figure 8.16 shows a block diagram of a system including the visual-updating delay. In this control loop, as shown in Fig. 8.17, T_u is defined as the time interval of visual-updating delay, and let T_i be the time spent in capturing and processing a camera image, T_m be the sampling time for motor commands to both fingers, and T_c be the control period required for this control loop. Therefore, the control period can be expressed as $T_c = T_i + T_m$ only if the image capturing and processing process is performed within the control loop, otherwise it is represented as $T_c = T_m$. That is, T_c is a variable parameter in a situation where the visual-updating delay exists. In other words, fast-sampling intervals for the motor control, which are determined according to the condition of $T_c = T_m$, successfully conduces to the fine movement of a robotic hand. Figure 8.18, in fact, shows a wide gap between the conditions of $T_c = T_i + T_m$ and $T_c = T_m$, which is obtained by a fundamental identification test (Fig. 7.7 and Table 7.3) where the visual-updating delay is set at $T_u = 50$ ms and 100 ms. Figure 8.18b, in particular, indicates clear 100-ms updating intervals in every time step. Note that the sampling time for motor control, T_m, is measured in the approx. 20 μs to 30 μs range. In this experiment, T_u is set at each of 25 ms, 50 ms, 75ms, and 100 ms at most.

Figure 8.19 shows an experimental result that includes the desired trajectory of a grasped object, a delayed response caused by low integral gain $K_I = 1.2$ of the first stage (Eq. 8.4), and an improved response obtained by increasing the integral gain up to $K_I = 12.0$. This tenfold-increase operation in visual feedback gain dramatically improved the languid trajectory of delayed response and produced fine convergence, and resulting in a fast rising time less than approx. 200 ms. In addition, Figs. 8.19b and c clearly indicate

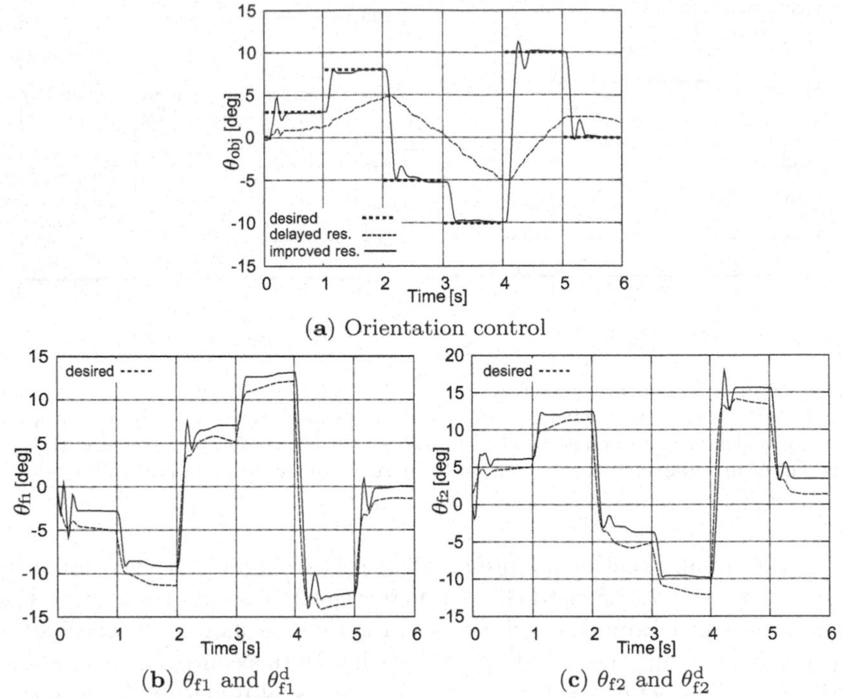

Fig. 8.19 Experimental results of object orientation control where $T_u = 25$ ms, $K_P = 40$, $K_D = 0.001$, $K_I = 12.0$, and $\tau_b = 2$ V (0.55 Nm) are given at the improved response, while $K_I = 1.2$ is given at the delayed response. **a** It contains a desired step trajectory, a delayed response, and an improved response

admissible deviations between the dynamically-produced desired joint angles and actual joint trajectories, which never affect the robust convergence of a target object. This fact may confirm the "virtual trajectory hypothesis" proposed by Hogan *et al.* [BAC+84, HH00]. According to the hypothesis, *the CNS controls simple large movements by specifying only a final equilibrium point, and the details of movement trajectory are determined by the inherent and viscoelastic properties of the limb and the muscles* [BAC+84]. In addition, that hypothesis proposed that *the CNS could define a final limb position by setting the spring constants of agonist and antagonist muscles even in the absence of peripheral feedback* [BAC+84]. In comparison with this hypothesis, as stated in Sect. 6.5, the convergence point of θ_{obj}, θ_{f1}, and θ_{f2} (Fig. 8.19), which is a set of LMEEwC even in the dynamic manipulation, corresponds to a *final equilibrium point.* The details of the trajectory of θ_{obj}, θ_{f1}, and θ_{f2} other than the final equilibrium point (LMEEwC) are determined by the intrinsic physical natures of soft fingertips. The hypothesis can furthermore be interpreted such that transient trajectories in limb movement are given by, in other words, unknown internal models within the human

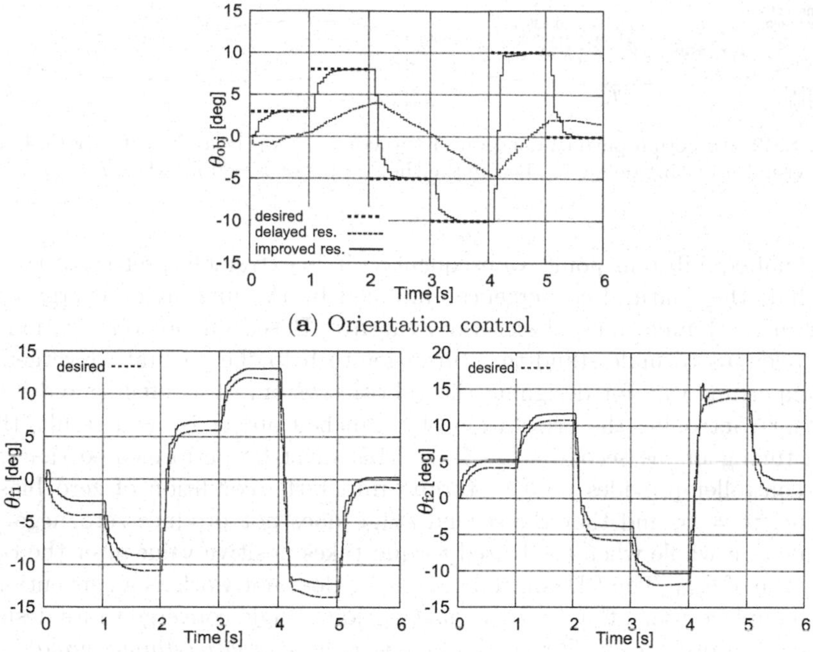

(**a**) Orientation control

(**b**) θ_{f1} and θ_{f1}^{d}

(**c**) θ_{f2} and θ_{f2}^{d}

Fig. 8.20 Experimental results of object orientation control where $T_u = 50$ ms, $K_P = 60$, $K_D = 0.001$, $K_I = 0.02$, and $\tau_b = 2$ V (0.55 Nm) are given at an improved response, while $K_I = 0.002$ is given at a delayed responce. **a** It contains a desired step trajectory, a delayed response, and an improved response

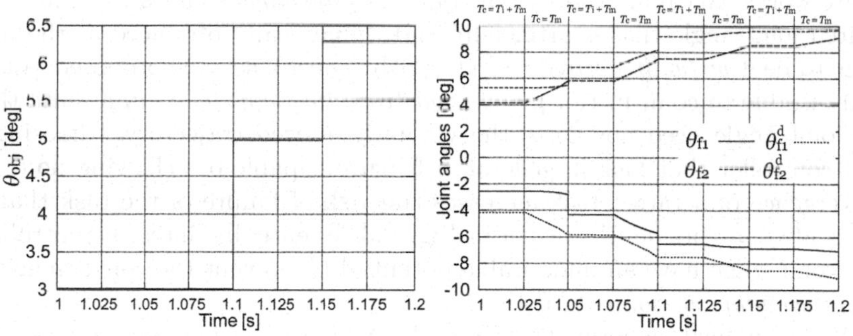

(**a**) θ_{obj}

(**b**) θ_{f1}, θ_{f1}^{d}, θ_{f2}, and θ_{f2}^{d}

Fig. 8.21 Enlarged view of Fig. 8.20. **b** It includes all trajectories of θ_{f1}, θ_{f1}^{d}, θ_{f2}, and θ_{f2}^{d}

body (*i.e.*, inherent and viscoelastic properties). The trajectories, in addition, converge with moving proximally along a contour or curvilinear surface of potential energy, as long as the movement of a target comes to rest at

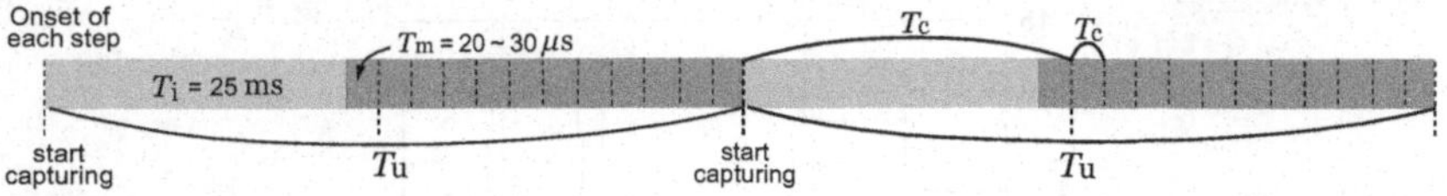

Fig. 8.22 An actual pattern diagram of the control loop in a present robotic hand. The visual-updating delay, T_u, becomes either 25 ms or 20–30 μs when $T_\mathrm{u} > T_\mathrm{i} + T_\mathrm{m}$

the final equilibrium point. Consequently, it is extremely preferable not to preclude the "natural convergence" induced by the underlying energy contour in the human musculoskeletal structure. Based on this observation, it will allow us to understand that, in order to help the natural convergence, an important way for designing the robotic system of a soft-fingered hand and for controlling the orientation of a pinched object is, as a result, that gain tuning in the second stage (Eq. 8.18) must be performed so that the PD controller provides positive torque under the condition of zero biased torque, $\tau_\mathrm{b} = 0$, and that the second stage does not produce zero/negative torque as a whole when the biased torque takes positive value after the convergence of θ_obj. The PD controller, that is, does not work as a conventional commonplace usage that usually makes a joint angle converge to its desired value given previously. This methodology of *zero/negative-torque prohibition* is very simple and control designers for the robot need not to regard whether the no-error convergence of the joint was successfully conducted in second stage controller because the joint error does not influence the goal task given as an object-orientation control in this case. In that sense, more simple and straightforward control laws might be introduced in the second stage in our future works. Recalling the *virtual trajectory hypothesis* and if assuming the desired joint angle that is artificially and dynamically produced at the first stage to be a *virtually-desired trajectory*, the control scheme proposed in this study is able to confirm Hogans' *virtual trajectory hypothesis* in a sense that the joint angle need not go to the virtually-desired trajectory. Choosing a PID controller that has, in general, a feature capable of achieving no-error convergence of a target task increases the risk of failure of the task than a PD control, as previously shown in Fig. 8.3. Eventually, integral controllers in second stage have an undesirable potential to prevent the convergence to a *final equilibrium point* (LMEEwC).

We show next experiment where the visual-updating period is set at $T_\mathrm{u} = 50$ ms. Figure 8.20 indicates a further delayed response of the object orientation, where the visual feedback gain is given as $K_\mathrm{I} = 0.002$. Due to the large delay time of updating, the trajectory of the delayed response appears to become a step-like function slightly and never reach half of the desired orientation in part. However, as well as the result of Fig. 8.19, the tenfold-increase operation in the visual feedback gain from $K_\mathrm{I} = 0.002$ to $K_\mathrm{I} = 0.02$ dramatically improved the response of object orientation. In addition, it is clearly indicated that the object trajectory does not change for

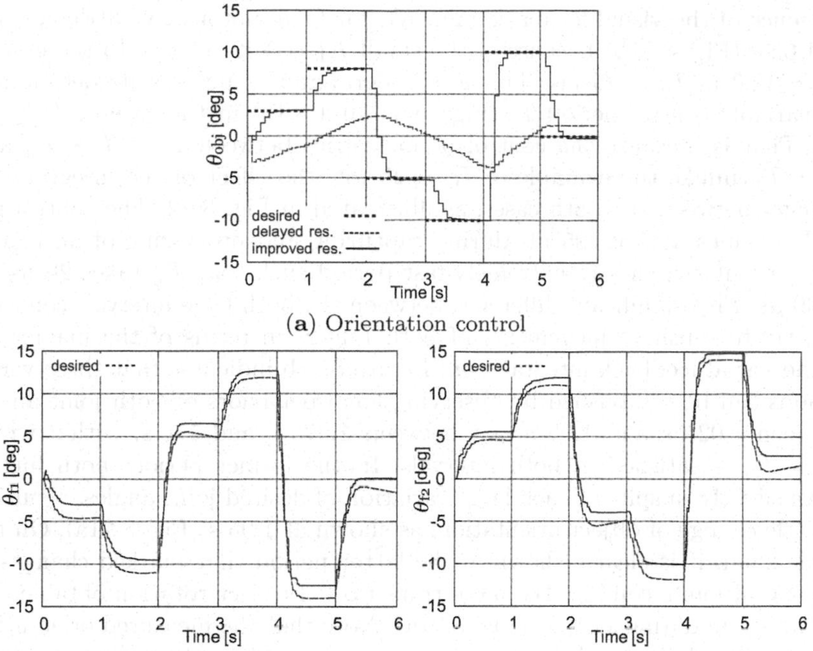

(a) Orientation control

(b) $\theta_{\rm f1}$ and $\theta_{\rm f1}^{\rm d}$

(c) $\theta_{\rm f2}$ and $\theta_{\rm f2}^{\rm d}$

Fig. 8.23 Experimental results of object orientation control where $T_{\rm u} = 75$ ms, $K_{\rm P} = 60$, $K_{\rm D} = 0.001$, $K_{\rm I} = 0.01$, and $\tau_{\rm b} = 2$ V (0.55 Nm) are given at an improved response, while $K_{\rm I} = 0.001$ is given at a delayed responce. **a** It contains a desired step trajectory, a delayed response, and an improved response

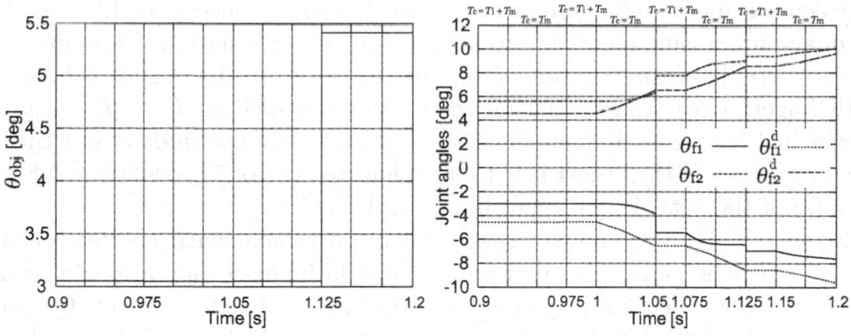

(a) $\theta_{\rm obj}$

(b) $\theta_{\rm f1}$, $\theta_{\rm f1}^{\rm d}$, $\theta_{\rm f2}$, and $\theta_{\rm f2}^{\rm d}$

Fig. 8.24 Enlarged view of Fig. 8.23. **b** It includes all trajectories of $\theta_{\rm f1}$, $\theta_{\rm f1}^{\rm d}$, $\theta_{\rm f2}$, and $\theta_{\rm f2}^{\rm d}$

approx. 100 ms at the beginning of each time step, *i.e.*, 1 s, 2 s, $\cdots$, 5 s despite the visual-updating delay is 50 ms. We refer to this as *double-delayed phenomenon.* This reason can be found from Figs. 8.21 and 8.22 that the influence of the visual feedback gain, K_I, for final motor drive little appears at 1.05 s (Fig. 8.21b) because of the small $K_\mathrm{I} = 0.02$ relative to the case of $K_\mathrm{I} = 12.0$ in $T_\mathrm{u} = 25$ ms. This low integral gain is necessarily determined by variable control period T_c (Figs. 8.17 and 8.18) in the case of $T_\mathrm{u} = 50$ ms. That is, because the control period varies between $T_\mathrm{c} = T_\mathrm{i} + T_\mathrm{m}$ and $T_\mathrm{c} = T_\mathrm{m}$ unlike the example of $T_\mathrm{u} = 25$ ms, the effect of the integral gain changes between the both cases, as illustrated in Fig. 8.22. The control period becomes approx. 25 ms during capturing and processing of an image, otherwise it switches to extremely-fast period such that T_m takes 20 μs up to 30 μs. This significant difference between the both time intervals connects directly to sensitive movements of both fingers in terms of the magnitude of the visual-feedback integral gain. Figure 8.21b indicates that these verifications can be established by observing local behaviors of both joint angles between 1.025 s and 1.05 s and between 1.075 s and 1.1 s, within which $T_\mathrm{c} = T_\mathrm{m}$ is satisfied in both intervals. In the former period, both fingers move slightly in spite of each large variation of desired joint angles, resulting in little change of object orientation, as shown at 1.05 s (Fig. 8.21a). On the other hand, it is clearly shown in the latter period that marked changes of the actual joints conduce the large transition of object rotation obtained by the image-updating at 1.1 s (Fig. 8.21a). Note that the measured orientation θ_obj and its desired value $\theta^\mathrm{d}_\mathsf{obj}$ do not change consistently for 100 ms from 1 s, therefore, the virtually-desired trajectories of both fingers exhibit linear increase and same angle of gradient in the both time intervals on the basis of the structure of visual-feedback controller expressed as Eq. 8.4. In addition, every time interval containing the image capturing and processing process, $T_\mathrm{c} = T_\mathrm{i} + T_\mathrm{m}$ ($T_\mathrm{i} \gg T_\mathrm{m}$), is totally occupied with the time for communication between a camera and a computer for direct memory access (DMA), and for an image processing algorithm to acquire object information. Consequently, all the virtually-desired joints, actual joints, and even the output voltage on both fingers never change during those intervals such as $T_\mathrm{c} = T_\mathrm{i} + T_\mathrm{m}$, as shown in Fig. 8.21b. Furthermore, note that we should understand that a true trajectory of the object is not equivalent to that of Figs. 8.20a and 8.21a because of the actual time spent for T_u and T_i.

Likewise, Fig. 8.23 shows a successful improvement with no steady-state errors, which is caused by performing tenfold-increase operation in visual feedback gain. The final trajectory of the grasped object, however, becomes clearly visible step-like behavior due to significantly large updating delay. Also in this case, we find that the admissible deviation associated with the virtually-desired joint angle does not affect the successful convergence, as shown in Figs. 8.23b and c. Figure 8.24a shows an enlarged view of Fig. 8.23a, in which the trajectory of the object orientation does not change for 225 ms at most despite the updating delay is set at $T_\mathrm{u} = 75$ ms. In this case, since the visual-updating delay is assumed to be $T_\mathrm{u} = 75$ ms, each updating time at every 75-ms intervals no longer coincides with the switching time of the desired orientation, $\theta^\mathrm{d}_\mathsf{obj}$, at *e.g.,* 1 s, 2 s, 4 s, and 5 s. For example, the last

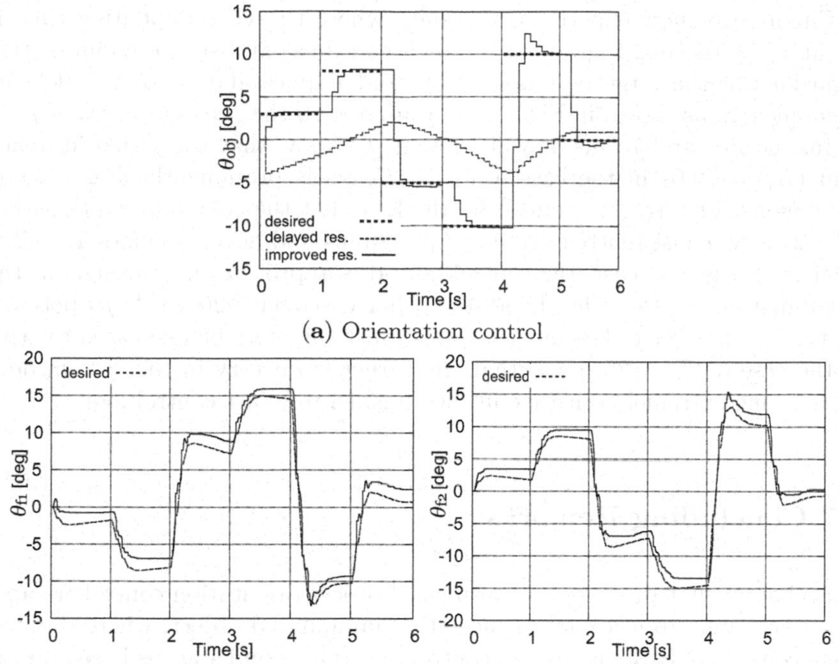

(a) Orientation control

(b) θ_{f1} and θ_{f1}^{d}

(c) θ_{f2} and θ_{f2}^{d}

Fig. 8.25 Experimental results of object orientation control where $T_u = 100$ ms. **a** It contains a desired step trajectory, a delayed response, and an improved response

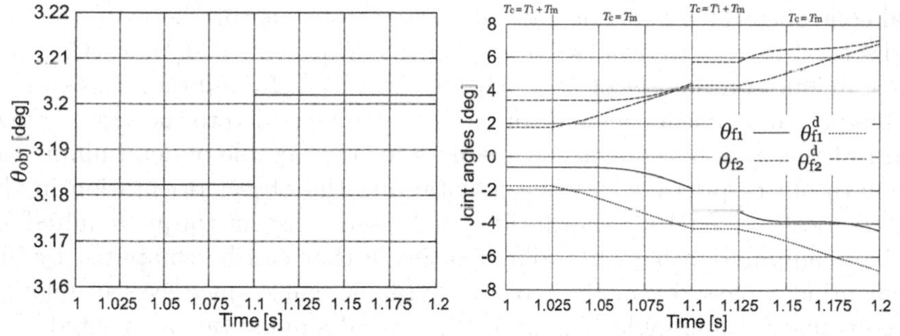

(a) θ_{obj}

(b) θ_{f1}, θ_{f1}^{d}, θ_{f2}, and θ_{f2}^{d}

Fig. 8.26 Enlarged view of Fig. 8.25. **b** It includes all trajectories of θ_{f1}, θ_{f1}^{d}, θ_{f2}, and θ_{f2}^{d}

update before 1 s occurs at 0.975 s, and continuously next update appears at 1.05 s and 1.125 s, as shown in Fig. 8.24b. Due to the simultaneous generation of this update discrepancy and the *double-delayed phenomenon* explained in Fig. 8.21, the trajectory of object orientation appears to become constant for 225 ms at most.

Finally, we show experimental results where the visual-updating time is set at $T_u = 100$ ms. Figure 8.25a indicates robust and secure manipulation regardless of the large updating delay, and implies 200-ms double-delayed phenomenon, as shown in Fig. 8.26. Compared to the previous cases of $T_u =$ 25 ms, 50 ms, and 75 ms, it will allow us to know that the visual feedback gain ($K_I = 12.0$) in the case of $T_u = 25$ ms is prominently large among these gains. This reason stems from the fact that the control period satisfies $T_c = T_i + T_m$ consistently in case of $T_u = 25$ ms and never becomes $T_c = T_m$. That is, in the first case, the control period is approx. 25 ms throughout the manipulation, on the other hand, the other cases have 20- or 30-μs periodic control loops in part. The integral gains must therefore be decreased relative to the case of $T_u = 25$ ms so that the target trajectory in the system does not become extremely sensitive in the considerably fast control loop.

8.7 Concluding Remarks

This chapter first presented a two-phased object orientation controller capable of achieving robust convergence of a manipulated object, where the first stage consists only of an integral control and the second stage is based on the traditional PD control scheme associated with the finger joints of a robot. The most outstanding characteristic of the controller is that desired joint angles used in the proportional controller of the second stage are directly connected to the first-stage controller and are virtually produced by an integral controller. This dynamically coupled control structure between the first and second stages can avoid solving specific solutions of the LMEEwC, which has a unique solution set of $(x_{obj}, y_{obj}, \theta_{obj}, \theta_{f1}, \theta_{f2})$. In addition, this control technique can define a constant torque that can be added to the second-stage controller as an independent linear sum, which is capable of determining the strength of grasping forces while maintaining the object orientation in the desired posture. We have shown that the biased constant torque required for stable manipulation has an available latitude that can be computed by implementing a passivity analysis of the dynamic constrained system relating to soft-fingered manipulation. Through several simulations associated with object orientation control, we have validated the availability and significant performances of the two-phased controller and revealed that discrepancies of joint angles are totally acceptable during dynamic manipulation. We have indicated that the soft-fingered manipulation is robust and secure even in an environment where there exist the structural delay of sampling time of a visual system and the information delay of visual feedback that human optical nerve pathways have. Secure manipulation under large time delays can be achieved by regulating the integral gain of the first-stage controller. Finally, we have demonstrated by implementing several robotic experiments that the proposed two-phased controller can independently accomplish both the object orientation control and the open-loop control of the constant biased torque.

Observing our proposed control scheme from the viewpoint of *virtual trajectory hypothesis*, it is natural to consider that attempting a tracking *virtually-desired trajectory* of both fingers is no longer impossible or rather not necessary for control designers if even the desired trajectory, as well as actual trajectory, is produced from ambiguous body model involving musculoskeletal structure and neurophysiological features. It is further obvious that tracking the virtually-desired trajectory of the joint with no steady-state error should be prevented because the desired trajectory does not coincide with a final equilibrium point anymore. We can conclude that above observations are equivalent to the *virtual trajectory hypothesis* that cannot be explained in terms of traditional control mechanics.

In addition, we attempted to consider a large time-delay system using a soft-fingered robotic hand, which is introduced in terms of large transmission duration due to optic nerve pathways of the human. A lot of simulations and experiments indicated that stable pinching and dexterous manipulation can be accomplished even if large time delays more than 100 ms exist in a robotic system. These new indications are all based on our serially-coupled two-phased controller for object manipulating operations.

Chapter 9
Geometric and Material Nonlinear Elastic Model

9.1 Introduction

The elastic force and elastic potential energy equations derived in Chap. 3 enable us to perform an analytical observation of the control design and the stability problem in soft-fingered handling. However, application fields exist that require a numerical-analysis approach, rather than an analytical approach. For example, numerical analysis uses a contact deformation model between multiple soft objects and its elastic force formulation in virtual space [MT96, CB04]. If numerical analysis is applied, it is not necessary to analytically solve the elliptical area integration, which is expressed as Eqs. 3.4, 3.9, and 3.13. That is, we are able to deal with the integrand of these equations in more complicated forms. Numerical analysis allows us to incorporate the material properties of soft objects such as rubber materials. By applying the material nonlinearity that is obvious in elastomer materials, we extend the elastic force formula, Eq. 3.9, previously discussed in Chap. 3, to a more adequate model that more closely models real fingertip deformation.

In this chapter, we clarify the stress-strain relationship of polyurethane gel used as a hemispherical soft fingertip and then obtain an approximated nonlinear Young modulus. By substituting the nonlinear Young's modulus into Eq. 3.9, we formulate a more accurate force model by means of numerical analysis.

9.2 Hertzian Contact and Kao's Elastic Model

In 1881, Hertz proposed a contact theory for two elastic objects having arbitrary curved surfaces [Joh85]. He showed that a normal contact force generated between an elastic sphere and a plane having a Young modulus of infinity can be expressed as

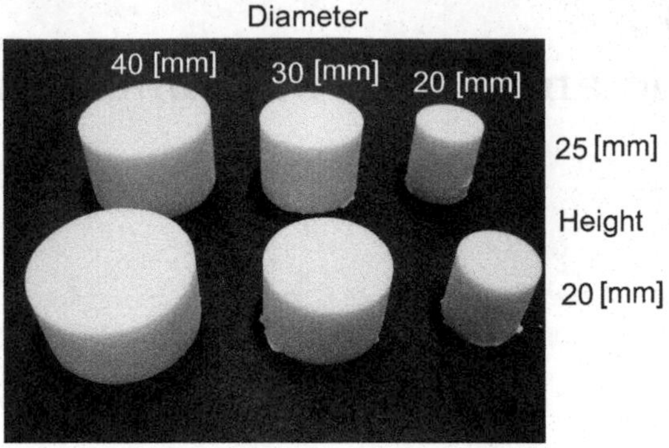

Fig. 9.1 Cylindrical specimens

$$F = \frac{4\sqrt{R}}{3}\left(\frac{E}{1-\nu^2}\right)d^{\frac{3}{2}}, \tag{9.1}$$

where R is the radius, E the Young modulus of the object, ν the Poisson ratio, and d the maximum displacement of the sphere. Since the above equation is useful from a practical viewpoint, it has been widely used for computing the contact stress between, for example, a wheel and a rail, a roll and material, and a retainer and a ball in a bearing. However, in Hertzian contact, it is assumed that both elastic objects are open elliptic paraboloids with an arbitrary radius of curvature. Consequently, no boundary conditions are used in the Hertzian contact model.

Kao *et al.* defined the parameter c_d corresponding to a material and geometric nonlinearity [KY04] and transformed Eq. 9.1 into

$$F = c_d d^{\zeta}. \tag{9.2}$$

They conducted a vertical compression test using a hemispherical soft fingertip and estimated the parameter c_d empirically using a weighted least-squares method. It has been shown that ζ is approx. 2.3 or 1.75 when the rate of deformation of the finger is above or below 20%, respectively. In other words, the parameter ζ is not identical to 3/2 in the contact model of soft fingertips. Therefore, the Hertzian contact theory cannot be adopted for deriving an elastic model of the hemispherical soft fingertip.

9.3 Identification of Nonlinear Young's Modulus

We obtain a nonlinear Young's modulus by compressing several cylindrical soft specimens, which are made of the same material as the soft fingertip. Specifically, three cylinders were 20 mm in height and 20, 30, and 40 mm

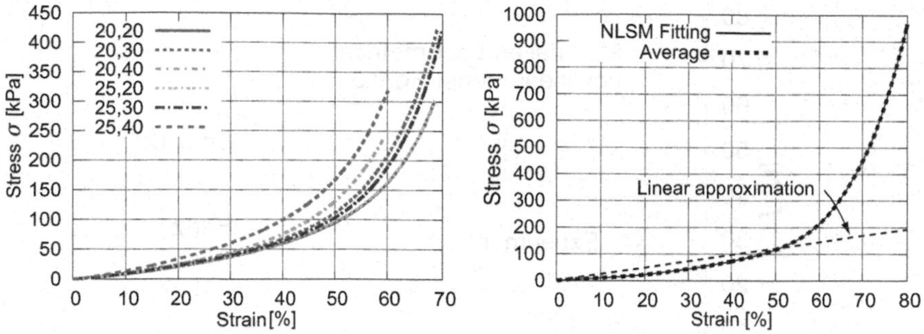

(a) Experiments (b) Average and fitting

Fig. 9.2 Stress-strain diagram

in diameter, and three were 25 mm in height and 20, 30, and 40 mm in diameter, as shown in Fig. 9.1. Note that we constructed several specimens for measuring the nonlinear Young's modulus, and these specimens are not identical to those shown in Fig. 3.6a.

Figure 9.2a shows the stress-strain diagram of an individual specimen, and Fig. 9.2b shows a result obtained by applying the nonlinear least-squares method (NLSM) to an average value of the diagram, on which a linearly approximated curve for 50% compression strain is plotted together with the nonlinearly fitted curve. We apply the linear Young's modulus, $E = 0.232$ MPa, to Eqs. 3.12 and 9.1, in which the modulus corresponds to the slope of the linearly approximated curve depicted in Fig. 9.2b.

On the other hand, performing a quintic nonlinear approximation (see Appendix A.4) of the stress with respect to the strain as shown in Fig. 9.2b, we obtain the following result:

$$\begin{aligned}\sigma(\epsilon) \cong\ & 1.829\epsilon - 1.455 \times 10^{-1}\epsilon^2 + 8.778 \times 10^{-3}\epsilon^3 \\ & - 1.908 \times 10^{-4}\epsilon^4 + 1.548 \times 10^{-6}\epsilon^5. \end{aligned} \tag{9.3}$$

Defining the nonlinear Young modulus as the slope of the nonlinear stress-strain diagram, the Young modulus can then be calculated by differentiating Eq. 9.3 with respect to the strain ϵ as

$$\begin{aligned}E(\epsilon) \cong\ & 1.829 - 2.910 \times 10^{-1}\epsilon + 2.633 \times 10^{-2}\epsilon^2 \\ & - 7.632 \times 10^{-4}\epsilon^3 + 7.740 \times 10^{-6}\epsilon^4. \end{aligned} \tag{9.4}$$

We then obtain an elastic force formula including the material nonlinearity together with the geometric nonlinearity of the hemispherical soft fingertip:

$$F = \int_{ell} \left\{ \int E(\epsilon)\, \mathrm{d}\epsilon \right\} \mathrm{d}S = \int_{ell} \sigma(\epsilon)\, \mathrm{d}S, \tag{9.5}$$

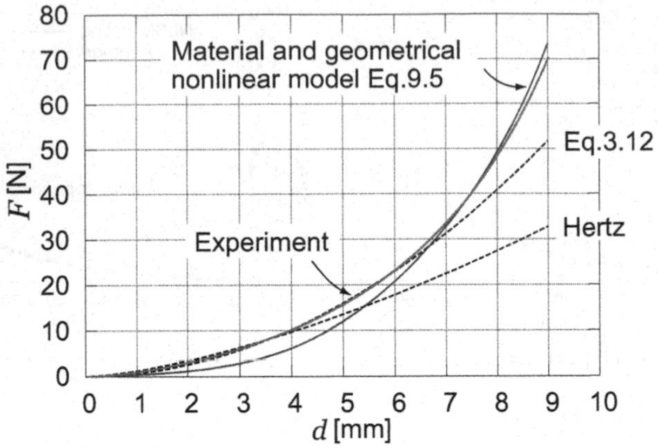

Fig. 9.3 Comparison between the material nonlinear elastic force model and conventional force models (Eq. 3.12 and Hertzian model) when $\theta_p = 0$ and $E = 0.232$ MPa

where ϵ is hidden in Eqs. 3.2 and 3.3 as

$$\epsilon = 1 - \frac{a - d - x \sin\theta_p}{\cos\theta_p \sqrt{a^2 - (x^2 + y^2)}}. \tag{9.6}$$

As mentioned above, it is sufficient to use Eq. 9.5 instead of Eq. 3.12 as the elastic force formula when we do not need to analytically solve the equations of motion of a grasped object in soft-fingered manipulation.

9.4 Comparison with Hertzian Contact

Figure 9.3 shows a comparison result in which the elastic force value with respect to the displacement d is plotted when a hemispherical soft fingertip having a radius of 20 mm is compressed vertically, as shown in Fig. 3.6b. Here, the reason for choosing a vertical contact between the object and the fingertip is that the Hertzian contact does not have the contact orientation, θ_p. Since it is impossible to compare Eqs. 3.12 and 9.5 with the Hertzian contact theory expressed as Eq. 9.1 when θ_p has a certain value, we substitute $\theta_p = 0$ into Eqs. 3.12 and 9.5. Finally, we obtain a good adjusted result relative to the material linear elastic model expressed as Eq. 3.12. The vertically oriented spring model is more suitable for deriving an elastic force over the entire deformation range because the proposed model contains a geometric nonlinearity due to the hemispherical shape and the material nonlinearity of the fingertip. In addition, the Hertzian contact theory cannot be applied to formulate the elastic forces on the soft contact material.

Soft materials exhibit nonlinear characteristics, even for infinitesimal deformations. In fact, Tatara derived a nonlinear Young modulus with respect to compressive strain [Tat91]. Furthermore, the concept of the contact angle of the object is not incorporated into the Hertzian contact theory. Although the Hertzian contact theory can be used for a simple contact pattern corresponding to normal contact, no contact at any other arbitrary angle or rolling contact can be defined. On the other hand, the elastic models proposed herein cover any contact angle of the object, and these models can therefore be used to analyze grasping and manipulating motions of various contact forms by a soft-fingered robotic hand.

Note that we experimentally measured the elastic forces on a soft fingertip by means of a load cell with a compression machine made by INSTRON. In addition, we used $E = 0.232$ MPa to measure the elastic forces.

9.5 Force Comparison

In this section, we compare the geometric and material nonlinear elastic force expressed as Eq. 9.5 with Eq. 3.12 when the orientation angle θ_{p} becomes 2.5° to 30.0° at the interval of 2.5°.

As shown in Fig. 9.4, we obtained preferable results even in the large deformation range, despite the change in the object orientation. However, while the experimental data become larger than the theoretical data in the small deformation range, the theoretical data become larger than the experimental data in the large deformation range. This phenomenon is caused by the intrinsic elastomer characteristics corresponding to a physical property whereby the rate of increase in the fingertip stiffness is different between the large deformation and the small deformation [KY04]. In addition, the theoretical data shift toward the left side in the figures according to the increase in the object orientation angle θ_{p}. In order to clarify the phenomenon, we show the comparison results of the experiment and simulation in Fig. 9.5. In the figure, the results of each elastic force curve are shown when θ_{p} changes from 0° to 30° at an interval of 5°.

In the simulation results shown in Fig. 9.5a, the rate of change in the force through the increase of the contacting angle is relatively large, and such a trend can be seen even in the small deformation area. On the other hand, in the case of the experimental results shown in Fig. 9.5b, the slope of the force changes little up to approx. d=5 mm.

In addition, as shown in Fig. 9.6, a local minimum point of the elastic force appears when the normal contact force with several orientation angles is applied to the soft fingertip. For the large orientation angle, we can find a large discrepancy, especially at ± 30°. This discrepancy is caused by the fact that while the fingertip easily deforms along the lateral direction in actual compression, such a complicated motion of the fingertip cannot occur in a

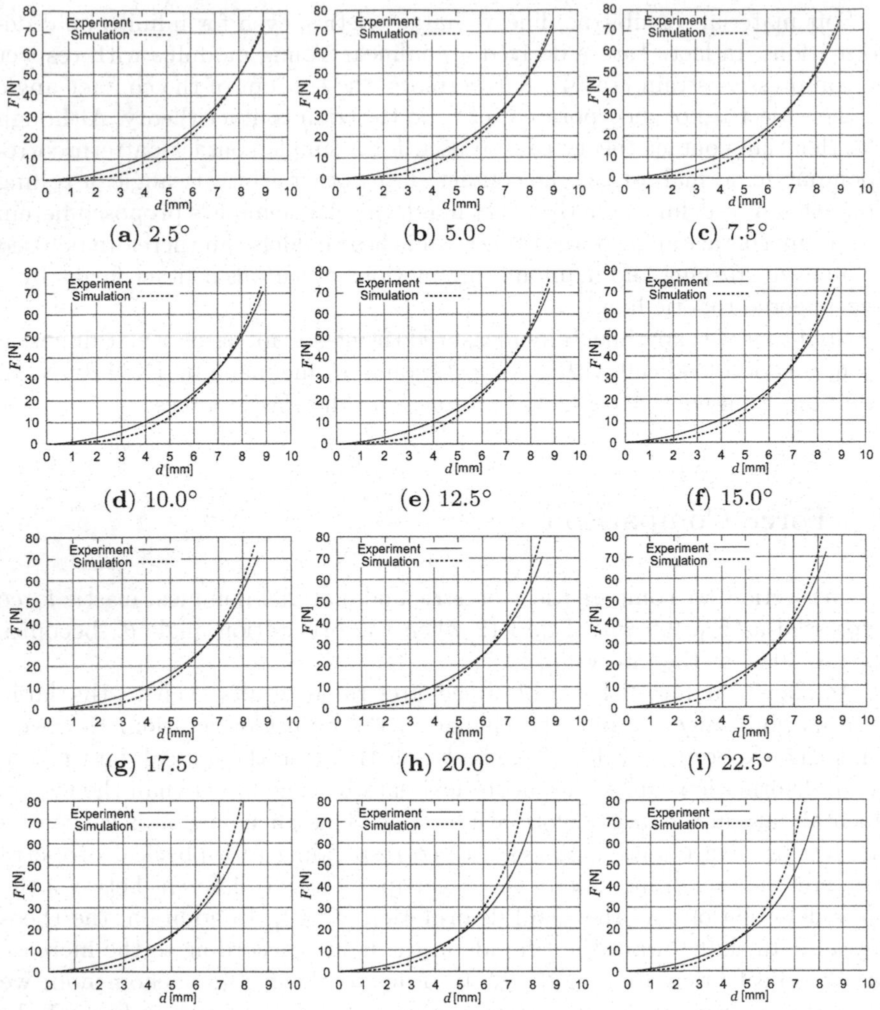

(a) 2.5° (b) 5.0° (c) 7.5°

(d) 10.0° (e) 12.5° (f) 15.0°

(g) 17.5° (h) 20.0° (i) 22.5°

(j) 25.0° (k) 27.5° (l) 30.0°

Fig. 9.4 Comparison between the elastic forces in the geometric and material nonlinear models

simulation. Therefore, the force applied to the fingertip is directly transmitted to the bottom surface in simulations and thereby increases the force in Fig. 9.6a.

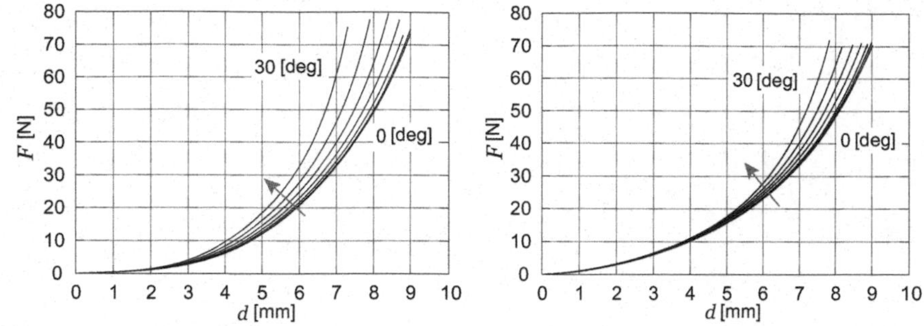

(**a**) Simulations (**b**) Experiments

Fig. 9.5 Comparison between simulations and experiments

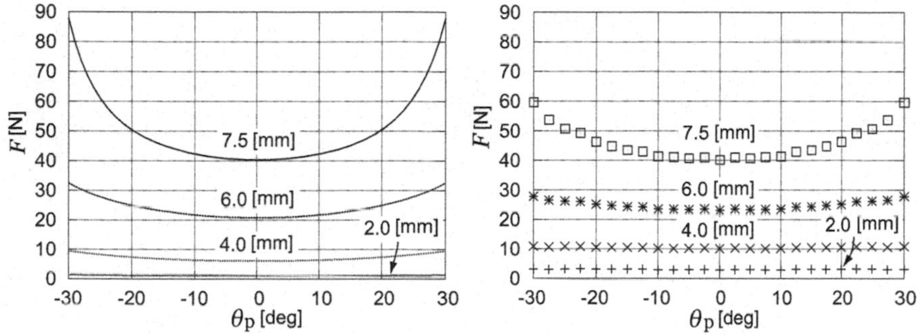

(**a**) Simulations (**b**) Experiments

Fig. 9.6 Local minimum of elastic forces

9.6 Concluding Remarks

In this chapter, we have defined the nonlinear Young's modulus for a soft fingertip, which is directly computed from stress-strain data measured by a compression test. Using the nonlinear Young's modulus, we have successfully modified the discrepancy between the linear spring fingertip model Eq. 3.12 and a more realistic nonlinear model Eq. 9.5.

Chapter 10
Non-Jacobian Control of Robotic Pinch Tasks

10.1 Introduction

The human thumb and index finger are a superior combination for performing dexterous manipulation and secure pinching movements. Although many researchers in the field of robotic hands have concentrated on sophisticated control designs and their theoretical verifications, complicated computations containing system kinematics, dynamics, and a Jacobian matrix in the brain have not been clarified yet and remain a matter of debate. This chapter provides an extremely straightforward control scheme for achieving object orientation control in robotic pinch tasks, in which 1-DOF and 2-DOF robotic hands having a thumb and an index finger manipulate a target object. Through simulation of the pinching movements based on the proposed control design, we show that non-Jacobian control can be accomplished by employing the characteristics of anatomical structure of the hand and placing a pair of soft fingertips at the distal ends of both fingers.

Most of the mechanical structures of robotic manipulators are composed of finite revolute joints. At present, however, such manipulators predominate only in research and are generally not used in practical applications. This results in the fact that the Jacobian matrix can describe the relationship between the task space and the joint space variables of the manipulators. Eventually, the description associated with the revolute joint angles for robots is preferable to prismatic joints. In addition, the Jacobian and its family (including transpose and inverse matrices) are usually used for computing joint torques required for a force control strategy on the tip of the manipulator. As described above, the Jacobian has been key in both the dynamics and kinematics of robotic manipulators ever since Dr. Carl Gustav Jacob Jacobi (1804–1851) developed the functional matrix (Jacobian matrix). Is it possible to completely eliminate the Jacobian family from robot control problems?

Generally, it was believed that the human brain did not perform complicated computations of trajectory planning for redundant arms or produce sophisticated control laws for dexterous manipulation or any other fine tasks. These observations are clearly verified because humans are able to simultaneously accomplish multiple tasks very easily. As a result, the traditional robot control scheme is far from the ultimate objective, whereby robotic manipulation approaches human manipulation. Refined control designs that have been used thus far for robots and their performance should simply be a supplementary tool for humanlike robots. We assume that the physical movement function and anatomical bodily characteristics other than higher-level brain functions contribute greatly to the natural and high performances of humans. The relatively low-level mechanical structure of the human body (*e.g.*, musculotendon-skeletal mechanics) plays a key role in achieving precise and delicate manipulation tasks through the tip of the index finger and the thumb. The elimination of the requirement to compute the Jacobian family in a series of control processes for manipulators and conventional robotic hands would be groundbreaking for the field of robotics.

The present paper proposes a new control scheme that is able to complete precise orientation control of a target object grasped by an articulated multifingered robotic hand, on which a set of deformable soft fingertips is mounted, in order to mimic the human fingertips and leverage their flexible mechanisms that affect stable pinching motions. This control method is constructed on the basis of a sensory feedback design, which forms a *serial two-stage* structure in terms of the object orientation and joint angles of the hand. In particular, the proposed control law has no Jacobian matrix in its closed-loop. In this case, it is no longer necessary for each joint in the robotic system to be a revolute joint. This means that a mixed mechanical structure including both revolute and prismatic joints is feasible in robot design.

10.2 Kinematic Thumb Models in Previous Studies

The present study limits the operating range of the hand robot to a vertical 2D plane (Fig. 10.1) in which the influence of gravity is considered. From an anatomical viewpoint of finger movements, the mechanical structure of the thumb has not yet been established. That is, the kinematic model of the thumb has not been uniquely standardized in the biomechanical or robotic research fields. For instance, Cooney *et al.* [CLC+81] postulated that during active motion such as pinching and grasping, muscle forces and the compressive force across adjacent joints constrain the axial rotation of the finger, limiting it to such an extent that the axial rotation is not considered as a degree of freedom of the thumb. As a result, a set of universal joints on both the metacarpophalangeal (MP) [Kap07] and the trapeziometacarpal (TM) [Kap07] joints of the thumb has been adopted, resulting in a 5-DOF kinematic model to describe pinching motions. Giurintano *et al.* [GHB+95] also proposed a five-link kinematic model of the thumb, which was differ-

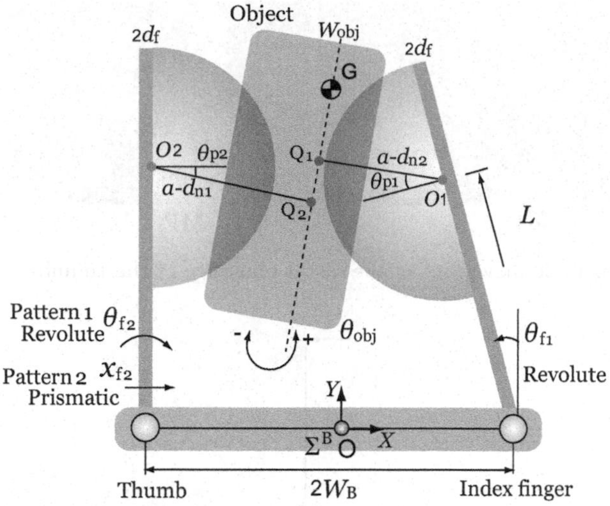

Fig. 10.1 Soft-fingered manipulation

ent from Cooney's model in that the universal joint structure was not employed because the nonnegligible geometric offsets on each joint are present in terms of their flexion-extension and abduction-adduction movements. Note that the TM joint is often called the carpometacarpal joint in other studies [GHB+95, CM06, HBM+92]. On the other hand, in recent years, it has been reported that the 5-DOF kinematic model is inherently inappropriate for describing the biomechanical thumb structure rigorously. Valero-Cuevas *et al.* [VJT03] mentioned that the 5-DOF thumb model doubled the magnitude of the thumb-tip forces and produced unrealistically large thumb-tip torques in experimental pinching motions. In addition, Clewley *et al.* [CGV08] postulated that the *effective number* of DOFs of each task is different and that knowing this number is essential in order to mimic subtle and agile movements of human fingers in terms of the musculoskeletal redundancy of the biomechanical finger structure. As described above, generally acceptable full-link mechanisms of the thumb based on anatomical knowledge have not yet been formulated.

In this chapter, as shown in Fig. 10.1, we consider an extremely simple robotic hand structure (2 DOFs) that consists of an opposed pair of an index finger and a thumb, on which a set of deformable soft fingertips are mounted for stable manipulation. This opposed mechanism was assumed in order to determine whether expected feasible tasks using the individual mechanism of revolute and prismatic joints differ from each other during pinching motions. In fact, the translational movement of the distal link is readily confirmable from the human thumb motion shown in Fig. 10.2. This can also be shown theoretically using the 3-DOF manipulator mechanism shown in Fig. 10.3,

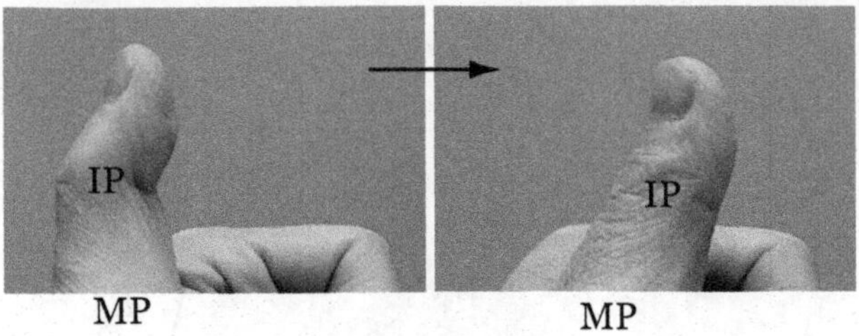

Fig. 10.2 Prismatic movement of the distal phalange of the thumb

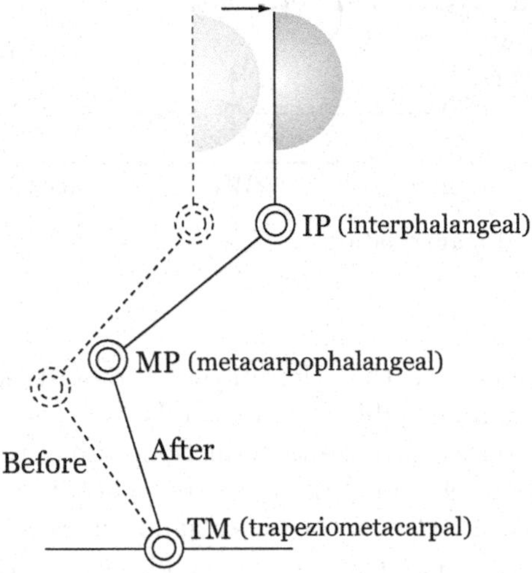

Fig. 10.3 A 3-DOF finger mechanism in the 2D plane and the prismatic movement of the distal link

which is the minimum number of DOFs required to achieve translational movement at the distal link. Thus, as a preliminary step, the present study confines the model to a simple 3-DOF revolute joint mechanism for the thumb in the 2D plane. In particular, this study concentrates on a minimal-DOF configuration of the opposed formation, which can be composed by treating a pair of distal links located at each index finger and at the thumb, in order to properly evaluate the relationship between the joint structure and the feasible tasks and to exclude the influence of mechanical redundancy and its associated control issues.

10.3 Equations of Motion

Equations of motion of two-fingered manipulation expressed as a dynamic constrained system having a pair of revolute joints (RR joints for short) were presented in Chaps. 7 and 8. Hence, in this section we formulate a set of dynamic equations of motion for a pair of revolute and prismatic joints (RP joints for short) on the basis of Lagrange's method in the presence of dynamically changeable constraints during manipulation tasks.

As shown in Fig. 10.1, let $G(x_{obj}, y_{obj})$ be the center of gravity of a grasped object, and let θ_{obj} be the orientation angle of the object with respect to the base coordinate system $\sum^{B}$. In addition, let W_{obj} be the object width, and let $2W_B$ be the distance between both joints of the hand. Let L be the length of both fingers, and let $2d_f$ be the width of each finger. Finally, let O_i be the origin of the ith fingertip. The present study classifies a couple of patterns of joint motions individually (Fig. 10.1) and attempts to discuss the different manipulation capabilities of each of the RR and RP joint mechanisms. Therefore, a set of system variables for the robotic hand system is defined as $(x_{obj}, y_{obj}, \theta_{obj}, \theta_{f1}, \theta_{f2})$ or $(x_{obj}, y_{obj}, \theta_{obj}, \theta_{f1}, x_{f2})$, respectively. In particular, in the case of the latter structure of the hand, the equations of motion of the total system are described in what follows.

The origin of both fingertips is represented as

$$\begin{aligned} O_{1x} &= W_B - L\sin\theta_{f1} - d_f\cos\theta_{f1}, & O_{1y} &= L\cos\theta_{f1} - d_f\sin\theta_{f1}, \\ O_{2x} &= -W_B + d_f + x_{f2}, & O_{2y} &= L. \end{aligned} \tag{10.1}$$

Recalling the geometric constraint that appears along the normal direction of the grasped object, this is rewritten substituting $w = 0$ into Eq. 7.1 as

$$\begin{aligned} C_{ni} = {} & (-1)^i(x_{obj} - O_{ix})C_{obj} + (-1)^i(y_{obj} - O_{iy})S_{obj} \\ & -(a - d_{ni}) - \frac{W_{obj}}{2} = 0, \quad (i = 1, 2), \end{aligned} \tag{10.2}$$

where the simplified forms, S_{obj} and C_{obj}, denote $\sin\theta_{obj}$ and $\cos\theta_{obj}$, respectively. In Eq. 10.2, the first and second terms on the right-hand side represent the distance between the center of gravity of the object and point O_i, which remains normal to the object contact surface.

On the other hand, the *Pfaffian constraints* [MLS94, SHV05] governing the tangential motions of the object and the soft fingertip can be expressed by considering the velocities of both displacements due to object rolling motions. Note that, in the present study, the effective rolling radii of both fingertips are assumed to be a, which, based on Chang's study [CC95], is equivalent to the mechanical radius of the fingertips. As shown in Fig. 10.4, let GQ_i be the intersection between the straight line O_iA_i and a line extending from point G parallel to the y-axis of the object coordinate system $\sum^{obj}$. The distance GQ_i is then represented as

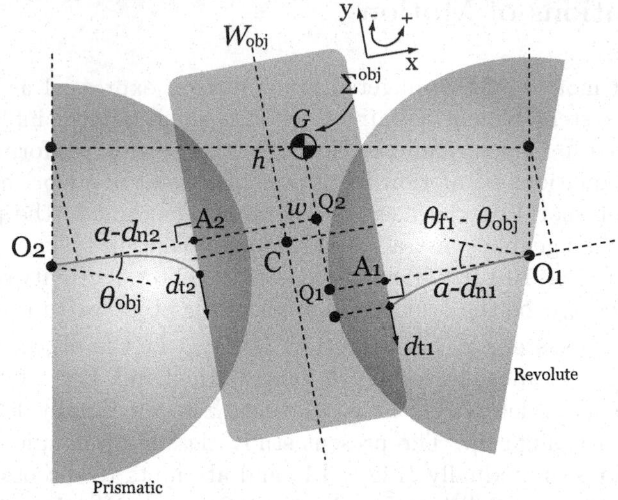

Fig. 10.4 The geometry of soft-fingered grasping is illustrated, in which the object rolling motion on the hemispherical surface and the tangential movement with the lateral deformation of the fingertips occur simultaneously during the manipulation. The object movement for the tangential direction is defined as a set of Pfaffian constraints

$$\mathrm{GQ}_i = -(x_{\mathrm{obj}} - \mathrm{O}_{ix})\sin\theta_{\mathrm{obj}} + (y_{\mathrm{obj}} - \mathrm{O}_{iy})\cos\theta_{\mathrm{obj}}. \tag{10.3}$$

When the grasped object rotates toward the positive direction (counterclockwise) on the right fingertip while maintaining the effective rolling radius a, then GQ_1 increases gradually. However, even if the object simultaneously slides along the y-direction with a certain deflection d_{t1} of the fingertip (Fig. 10.4), point Q_1 itself does not change from the extended line of $\mathrm{O}_1\mathrm{A}_1$ as long as the definition of Q_1 is preserved, resulting in shortening GQ_1. As a result, GQ_1 lengthens due to the positive rolling motion but shortens due to the sliding of the object. Hence, this geometrically combined relationship can be expressed using a velocity-formed description as

$$\dot{C}_{\mathrm{t}i} = \dot{\mathrm{GQ}}_i + a\dot{\theta}_{\mathrm{p}i} + \dot{d}_{\mathrm{t}i} = 0, \tag{10.4}$$

where

$$\theta_{\mathrm{p1}} = \theta_{\mathrm{f1}} - \theta_{\mathrm{obj}}, \quad \theta_{\mathrm{p2}} = \theta_{\mathrm{obj}}. \tag{10.5}$$

Note that Eq. 10.4 corresponds to a set of Pfaffian constraints in a minimal-DOF soft-fingered manipulation.

Generally, Lagrange equations of motion of dynamically constrained systems are given as follows:

$$\boldsymbol{M}\ddot{\boldsymbol{q}} + \boldsymbol{D}\dot{\boldsymbol{q}} - \boldsymbol{f}_{\mathrm{p}} - \boldsymbol{\Phi}^{\mathrm{T}}\boldsymbol{\lambda} = \boldsymbol{f}_{\mathrm{ext}} + \boldsymbol{u}, \tag{10.6}$$

where $\boldsymbol{M}$ is an inertia matrix containing rotational and translational movements of the present system, and $\boldsymbol{D}$ denotes a viscous damping matrix that is the intrinsic parameter of soft-finger materials. In addition, $\boldsymbol{q}$ stands for a set of system variables $(x_{\rm obj}, y_{\rm obj}, \theta_{\rm obj}, \theta_{\rm f1}, x_{\rm f2})$ in the case of the RP joint mechanism, and $\boldsymbol{f}_{\rm p}$ is a vector of generalized potential forces/moments, which is expressed by differentiating entire potential energy (Eq. 7.3) with respect to the generalized coordinate. Symbol $\boldsymbol{\Phi}$ is a constraint matrix [PH86] obtained by differentiating each constraint equation with respect to $\boldsymbol{q}$ or $\dot{\boldsymbol{q}}$. In the present system, it can be described by the following calculations:

$$\boldsymbol{\Phi} = \begin{bmatrix} \dfrac{\partial C_{\rm n}}{\partial \boldsymbol{q}} \\ \dfrac{\partial C_{\rm t}}{\partial \dot{\boldsymbol{q}}} \end{bmatrix}. \tag{10.7}$$

Vector $\boldsymbol{\lambda}$ means undetermined multipliers that further correspond to constraint forces caused by the mechanical contact between the grasped object and both fingertips. Vector $\boldsymbol{f}_{\rm ext}$ denotes external forces and moments applied to the system and $\boldsymbol{u}$ indicates control input signals for the robotic hand.

10.4 Simulations

10.4.1 A Serial Two-phased Controller

Recall the following fundamental serial two-phased controller capable of achieving robust convergence of the orientation of a grasped object:

$$\theta_{{\rm f}i}^{\rm d} = -(-1)^i K_{\rm I}^{\theta} \int_0^t (\theta_{\rm obj} - \theta_{\rm obj}^{\rm d}) {\rm d}\tau, \tag{10.8}$$

$$u_i = -K_{\rm P}^{\theta}(\theta_{{\rm f}i} - \theta_{{\rm f}i}^{\rm d}) - K_{\rm D}^{\theta}\dot{\theta}_{{\rm f}i} + \tau_{\rm b}, \tag{10.9}$$

where $K_{\rm P}^{\theta}$, $K_{\rm D}^{\theta}$, and $K_{\rm I}^{\theta}$ denote, respectively, the proportional, derivative, and integral gains for the controller in which the superscript θ means that these gains are specially used for the object orientation control. This controller architecture adopts a discriminative design in appearance such that desired joint angles are dynamically and continuously produced during the manipulation control and are serially cascaded between the first and second stages, represented as Eqs. 10.8 and 10.9, as shown in Fig. 10.5. Note that this controller has no Jacobian matrix, and the biased torque $\tau_{\rm b}$ in the second phase is an arbitrary constant, rather than being obtained by multiplying the transpose of Jacobian matrix into certain grasping forces. This method can be employed in the case of the revolute joint mechanism illustrated as pattern 1 in Fig. 10.1. By utilizing the proposed controller and evaluating its control

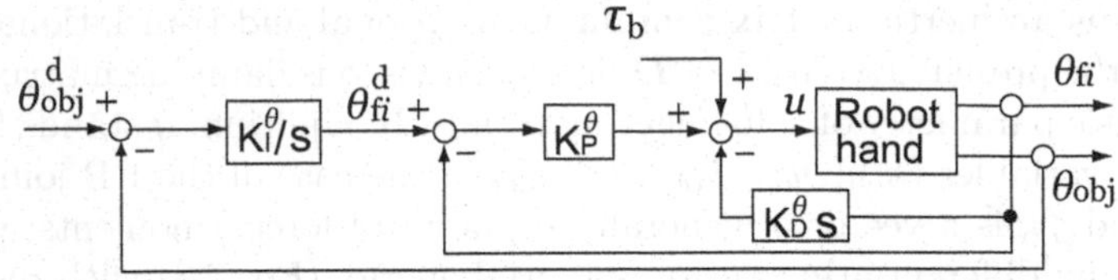

Fig. 10.5 Block diagram of a serial two-phased controller capable of achieving robust convergence of the orientation of a grasped object

Table 10.1 Simulation parameters

parameters	values (orientation control)
$\theta^{\mathrm{d}}_{\mathrm{obj}}$	$-2° \rightarrow -5° \rightarrow -8°$
K^{θ}_{P}	300 Nm
K^{θ}_{D}	14 Nm·sec
K^{θ}_{I}	0.003
τ_{b}	10 Nm
	values (position control)
$x^{\mathrm{d}}_{\mathrm{obj}}$	4 mm $\rightarrow$ −6 mm $\rightarrow$ −6 mm
K^{x}_{P}	300 N/m
K^{x}_{D}	14 N·sec/m
K^{x}_{I}	0.1
τ_{b}	10 N

performances and using numerical simulations, we verify that non-Jacobian control can be achieved in soft-fingered robotic manipulation. In addition, we indicate for the first time that a minimal degree-of-freedom anthropomorphic robotic hand (2 DOFs) is able to perform either independent tasks selectively among the position control and orientation control of a target object to be manipulated. Finally, we organize the new findings associated with the relationship between the mechanical degrees of freedom and the number of feasible handling tasks by the robotic hand by comparing each mechanism in several combinations, including the revolute and prismatic joints. Note that in the proposed control structure, as shown in Fig. 10.5 and Eqs. 10.8 and 10.9, we do not apply a compensator for the gravitational force, although the robotic hand system is considered in a vertical 3D space with gravity.

10.4.2 Revolute Joint vs. Prismatic Joint (RP Joints)

First, the introduction of another set of serial two-phased controller designed for object position control (Fig. 10.1) can be achieved by modifying the previous orientation controller as

$$x_{\mathrm{f2}}^{\mathrm{d}} = -K_{\mathrm{I}}^{x} \int_{0}^{t} (x_{\mathrm{obj}} - x_{\mathrm{obj}}^{\mathrm{d}}) \mathrm{d}\tau, \tag{10.10}$$

$$u_2 = -K_{\mathrm{P}}^{x}(x_{\mathrm{f2}} - x_{\mathrm{f2}}^{\mathrm{d}}) - K_{\mathrm{D}}^{x}\dot{x}_{\mathrm{f2}} + \tau_{\mathrm{b}}. \tag{10.11}$$

In the above equations, K_{P}^{x}, K_{D}^{x}, and K_{I}^{x} denote the proportional, derivative, and integral gains, respectively. Taken together, Eqs. 10.10 and 10.11 are used to control the object position and the prismatic joint, and Eqs. 10.8 and 10.9, into which $i = 1$ is substituted, are used for the object orientation and the rotational index finger. All of the parameters used in both of these controllers are listed in Table 10.1. Parameters of the hand system are shown in Table 7.1. A straightforward task given to the robotic hand in this simulation is simultaneous control of the object position and orientation, the desired values $(\theta_{\mathrm{obj}}^{\mathrm{d}}, x_{\mathrm{obj}}^{\mathrm{d}})$ of which are also listed in Table 10.1. The designated pattern is set to vary at 1-s intervals. Finally, we can examine the dynamic behavior of the total robotic hand system by carrying out numerical computations for Eq. 10.6 together with the control inputs (u_1, u_2). In addition, in this case, the elastic potential energy can be modified because the joint angle of the left finger, θ_{f2}, is omitted from Eq. 4.5 as follows:

$$\begin{aligned} P_1 + P_2 = \pi E \left\{ \frac{d_{\mathrm{n1}}^3}{3\cos^2(\theta_{\mathrm{f1}} - \theta_{\mathrm{obj}})} + d_{\mathrm{n1}}^2 d_{\mathrm{t1}} \tan(\theta_{\mathrm{f1}} - \theta_{\mathrm{obj}}) + d_{\mathrm{n1}} d_{\mathrm{t1}}^2 \right\} \\ + \pi E \left\{ \frac{d_{\mathrm{n2}}^3}{3\cos^2\theta_{\mathrm{obj}}} + d_{\mathrm{n2}}^2 d_{\mathrm{t2}} \tan\theta_{\mathrm{obj}} + d_{\mathrm{n2}} d_{\mathrm{t2}}^2 \right\}. \end{aligned} \tag{10.12}$$

In the initial setup of the robot configuration related to the base coordinate $\sum^{\mathrm{B}}$, it is assumed that $x_{\mathrm{obj}} = 0$ mm, $y_{\mathrm{obj}} = L$, $\theta_{\mathrm{obj}} = 0°$, $\theta_{\mathrm{f1}} = 0°$, $x_{\mathrm{f2}} = -W_{\mathrm{B}}$ (Fig. 10.1). This means that both fingers grasp a planar object maintaining a geometric parallel between the object and both fingers in the initial configuration. From the initial condition, a set of desired trajectories $(x_{\mathrm{obj}}^{\mathrm{d}}, \theta_{\mathrm{obj}}^{\mathrm{d}})$ is given to the system equations. Figure 10.6 shows the simulation results in the RP joint mechanism. As shown in Figs. 10.6a and b, x_{obj} and θ_{obj} both converge precisely to the desired trajectory. We further find that θ_{obj} can vary with no change in the x_{obj} trajectory. This fact means that the position and orientation controls can be separated and performed independently, resulting in the revolute joint contributing only to the orientation control of the grasped object, and on the other hand, the prismatic joint acts as a function only for controlling object position. That is, there exists a certain type of *role sharing* for pinching tasks between the index finger and thumb when employing different types of joints (RP joints) with which the adjoining target object is in contact. These new findings can also be explained by the concept of the LMEE of soft fingertips, as demonstrated in Chap. 3 [IH06].

In addition, Figs. 10.6c and d show that the revolute and prismatic joints both converge to a certain angle and position, respectively. However, large errors exist in each time step. In particular, the position trajectory of the pris-

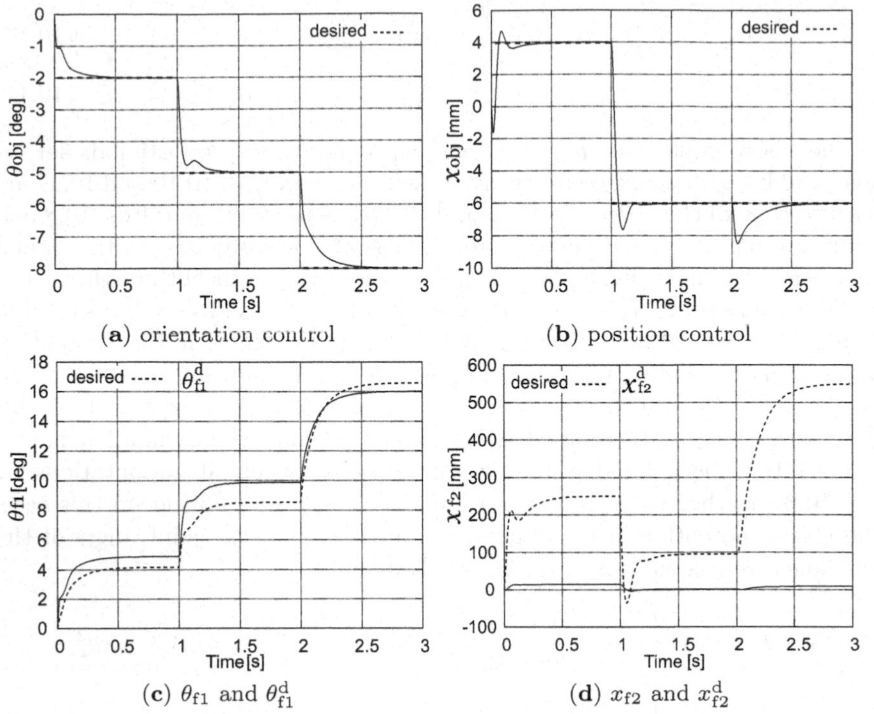

Fig. 10.6 Simulation results of independent-simultaneous control of the position and orientation of a grasped object when adopting the revolute and prismatic joint mechanism (RP joints)

matic joint is completely out of control throughout the simulation in that the discrepancy becomes extremely large. Therefore, the control design method proposed in the present paper likely leads to a failure. These discrepancies, however, can be defined as admissible errors as long as the object position and orientation converge to desired values. The cause of this phenomenon is intimately linked with the LMEE.

The LMEE has been found based on a physically spontaneous idea that elastic deformations of soft fingertips have some sort of minimum level of elastic energy [IH06]. In particular, in the case of actual pinching motions, since multiple geometric and kinematic constraints exist during dynamic manipulation tasks, this extended perspective of elastic energy has been defined as *LMEE with constraints* (LMEEwC for short) [IH07a]. The LMEEwC corresponds to an equilibrium point in terms of physical meaning. Therefore, in this case, a set of system variables that ought to be the equilibrium point is uniquely determined as, *e.g.*, $(x^{\star}_{obj}, y^{\star}_{obj}, \theta^{\star}_{obj}, \theta^{\star}_{f1}, x^{\star}_{f2})$. In order to simultaneously control two or more system variables among the set of the unique solution, *i.e.*, LMEEwC, the chosen variables must be known in advance in order to avoid such inconsistencies whereby one variable is not a member of

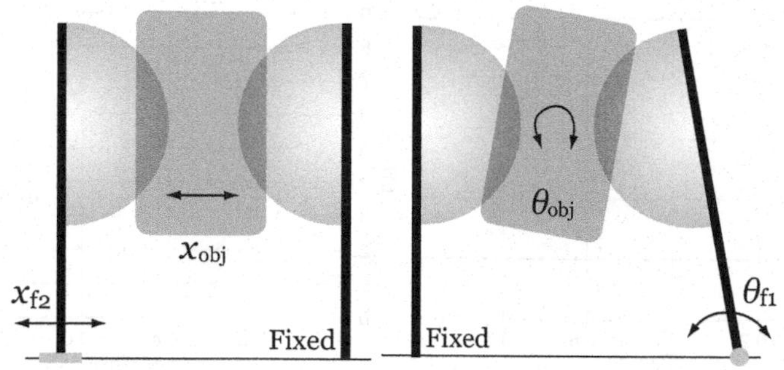

(**a**) Translational movement (**b**) Rotational movement

Fig. 10.7 Diagram of a 1-DOF robotic hand

the LMEEwC. This equilibrium point, however, cannot be analytically computed due to the strong nonlinearity of the deformation model of the elastic soft fingertips.

In the present study, when an arbitrary desired object position $x^{\mathrm{d}}_{\mathrm{obj}}$ is assumed to be an element among the LMEEwC, the desired joint displacement $x^{\mathrm{d}}_{\mathrm{f2}}$ of Eq. 10.10, which is artificially produced and depends on the integral gain K^x_{I}, unlike $x^{\mathrm{d}}_{\mathrm{obj}}$, never becomes an element of the LMEEwC. Therefore, the joint displacement x_{f2} indicated in Eq. 10.11 must not converge to $x^{\mathrm{d}}_{\mathrm{f2}}$ with no error. If it does, the convergence will rather disturb the physical stabilization of the LMEEwC. In general, the PD control for the joints expressed in Eqs. 10.9 and 10.11 may remain with some errors constitutionally because of the lack of an integral controller, whereas such errors are not preferable in traditional control mechanics. In other words, the joint errors of Figs. 10.6c and d correspond to *essential avoidance* of convergence to the desired joints. As a result, the existing discrepancies of the figures must be preserved. We can therefore conclude that the converged value (continuous lines) in each time step coincides with each member of the LMEEwC.

From the viewpoint of the LMEEwC, the problem of why both of the desired values of $x^{\mathrm{d}}_{\mathrm{obj}}$ and $\theta^{\mathrm{d}}_{\mathrm{obj}}$ can be independently determined remains unsolved. Therefore, we consider extremely simple robotic hands having only a single 1-DOF prismatic or revolute joint, that is, we simulate two cases in which either the thumb or the index finger is fixed (Fig. 10.7). In this case, what movements and tasks can each remaining finger perform?

In the case of a pair of prismatic and fixed joints, position control of a grasped object along the translational direction, which is equivalent to joint motion, is possible (Fig. 10.7a). On the other hand, in the case of the revolute joint (Fig. 10.7b), Fig. 10.8 shows that the orientation control of the object can be realized despite the considerably large deviation of the revolute joint.

These overall results can be interpreted such that the index finger of the revolute joint assumes an orientation control and the thumb of the prismatic joint fulfills the role of position control of a grasped object. Thus, individual

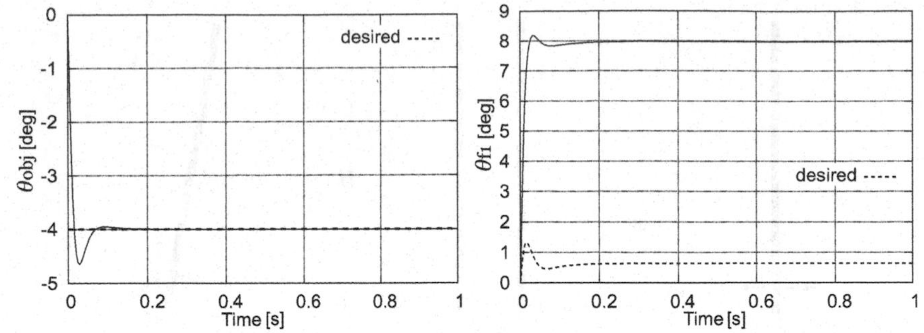

(**a**) Orientation control (**b**) θ_{f1} and $\theta_{\mathrm{f1}}^{\mathrm{d}}$

Fig. 10.8 Simulation results of object orientation control on the condition that the left finger is fixed and the right finger only rotates, as shown in Fig. 10.7b

control of robotic joints configured with a pair of different mechanisms is possible. In other words, the functional role sharing of the index finger and the thumb for complicated pinching motions simplifies even the traditional control architecture of multifingered robotic hands. This implies that a control scheme without the Jacobian matrix family is a feasible methodology, as long as the deformable soft fingers mounted on the distal links of the hand grasp and manipulate the target object. We can thus conclude that the LMEEwC is able to ensure stable and robust pinching motions and relates directly to the possibility of non-Jacobian control newly proposed herein. These examples of the 1 DOF hand will be described in greater detail in Sects. 10.4.4 and 10.4.5.

Next, we present an additional case study that examines the dynamic trajectory of the entire system when the desired trajectory of a target object to be manipulated varies discontinuously as a step function. As in the case of the previous results shown in Fig. 10.6, Eqs. 10.8 and 10.9 and Eqs. 10.10 and 10.11 are distinctly employed for controlling the right revolute and left prismatic fingers, respectively. A pair of the desired variables is therefore comprised of the object position and orientation $(x_{\mathrm{obj}}, \theta_{\mathrm{obj}})$, and the biased force and torque τ_{b} remain constant and are both set equal to the value prepared in Table 10.1. Here, K_{I}^{x} was designed to be $K_{\mathrm{I}}^{x} = 0.06$ for comprehensible subsequent comparisons. Figures 10.9a and b show the dynamic trajectories of $(x_{\mathrm{obj}}, \theta_{\mathrm{obj}})$ together with the desired object trajectory assigned in advance, which is depicted in both figures as dotted lines. We can easily see that both variables converge robustly to each desired value, whereas there exist relatively large overshoots on the x_{obj} trajectory. The reason for this is that the response of θ_{obj} depicted in Fig. 10.9b varied drastically and widely at time steps other than those from 2 s to 3 s. That is, the influence of the inertia of the entire system due to the swinging motion of the object and fingers generated explicit overshoots. In addition, large errors can be seen in the revolute joint angle shown in Fig. 10.9d, and extremely large discrepancies appear in

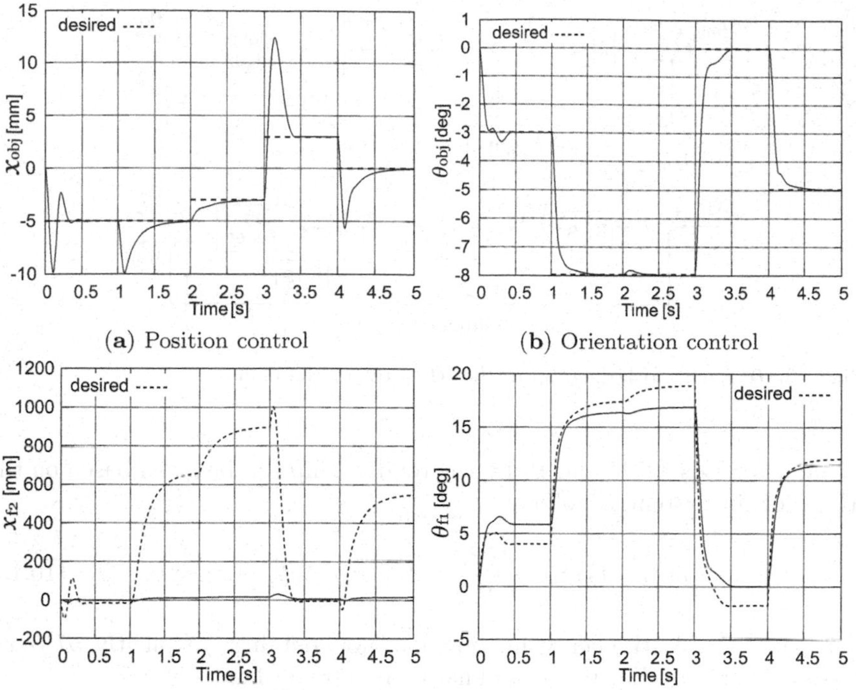

(**a**) Position control (**b**) Orientation control

(**c**) x_{f2} and x^{d}_{f2} (**d**) θ_{f1} and θ^{d}_{f1}

Fig. 10.9 These simulation results show that simultaneous control of (x_{obj}, θ_{obj}) can be achieved by means of 2 DOF robotic fingers that consist of a pair comprised of a revolute joint and a prismatic joint. In this simulation, we set the integral gain for position control such that $K^{x}_{I} = 0.06$, and the other parameters obey Table 10.1. Note that τ_b represents not only the torque but also the biased force required for the prismatic joint movement

the results for the prismatic joint position shown in Figs. 10.9c and 10.6d. However, since the objective of this task was completed, these discrepancies are considered to be admissible deviations. Thus, these errors need not vanish during the manipulation tasks.

Let us next verify the above successful convergence of (x_{obj}, θ_{obj}) from the viewpoint of statics during the manipulation motion. Let us formulate a static relationship around the axis of the revolute joint in the steady state. First, since the joint mechanism is transformed into a prismatic joint, E_{n2} and F_{t2}, respectively, used in Eqs. 7.17 and 7.22 for the normal and tangential constraints are then modified as

$$E_{n2} = -\cos\theta_{obj}, \tag{10.13}$$

$$F_{t2} = \sin\theta_{obj}. \tag{10.14}$$

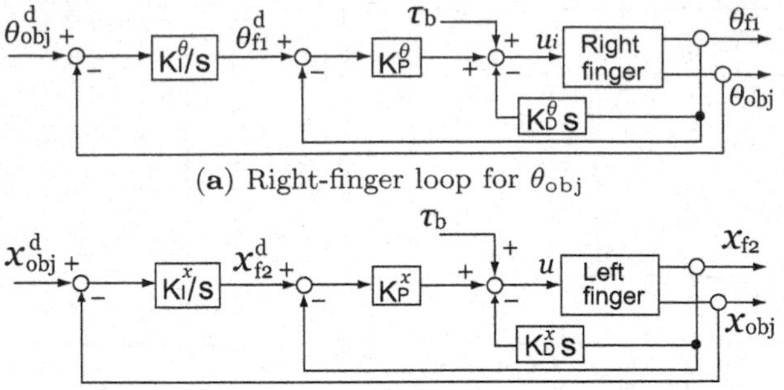

(a) Right-finger loop for θ_{obj}

(b) Left-finger loop for x_{obj}

Fig. 10.10 Each block diagram in the RP joint mechanism

Then, using Eqs. 10.11 and 10.12, the relationship can be given based on the statics of the system, as follows:

$$E_{\mathrm{n}2}\lambda_{\mathrm{n}2} + F_{\mathrm{t}2}\lambda_{\mathrm{t}2} - \frac{\partial P}{\partial x_{\mathrm{f}2}} - K_{\mathrm{P}}^{x}(x_{\mathrm{f}2} - x_{\mathrm{f}2}^{\mathrm{d}}) + \tau_{\mathrm{b}} = 0. \tag{10.15}$$

Here, since the elastic energy function does not contain $x_{\mathrm{f}2}$ explicitly, $\partial P/\partial x_{\mathrm{f}2}$ is zero. Therefore, Eq. 10.15 can finally be expressed as

$$E_{\mathrm{n}2}\lambda_{\mathrm{n}2} + F_{\mathrm{t}2}\lambda_{\mathrm{t}2} - K_{\mathrm{P}}^{x}(x_{\mathrm{f}2} - x_{\mathrm{f}2}^{\mathrm{d}}) + \tau_{\mathrm{b}} = 0. \tag{10.16}$$

On the other hand, another static relation in the steady state including the potential moment, the control input, and the moment generated by the two types of constraint must be given by Eqs. 10.9 and 10.12 and refer to Eqs. 7.17 and 7.22 as

$$E_{\mathrm{n}1}\lambda_{\mathrm{n}1} + F_{\mathrm{t}1}\lambda_{\mathrm{t}1} - \frac{\partial P}{\partial \theta_{\mathrm{f}1}} - K_{\mathrm{P}}^{\theta}(\theta_{\mathrm{f}1} - \theta_{\mathrm{f}1}^{\mathrm{d}}) + \tau_{\mathrm{b}} = 0. \tag{10.17}$$

Note that the fourth term of the above equation denotes a potential moment that appears due to the differentiation of gravitational potential for the right finger with respect to the rotational angle of the finger. We can see that Eq. 10.16 has the dimension of force physical quantity, while Eq. 10.17 has the dimension of moment physical quantity. Therefore, these equations can independently satisfy each steady-state configuration during the dynamic system behavior. Thus, x_{obj} is controlled by only the prismatic joint, whereas θ_{obj} is controlled by only the revolute joint. In particular, these movements can be achieved, satisfying complete independence in terms of the mechanical structure, resulting in a noteworthy pinch performance such that a straightforward 2 DOF robotic hand can simultaneously attain position and orientation controls of the grasped object. That is, the individual control of $(x_{\mathrm{obj}}, \theta_{\mathrm{obj}})$ can be achieved by independently constructing decoupled control loops between

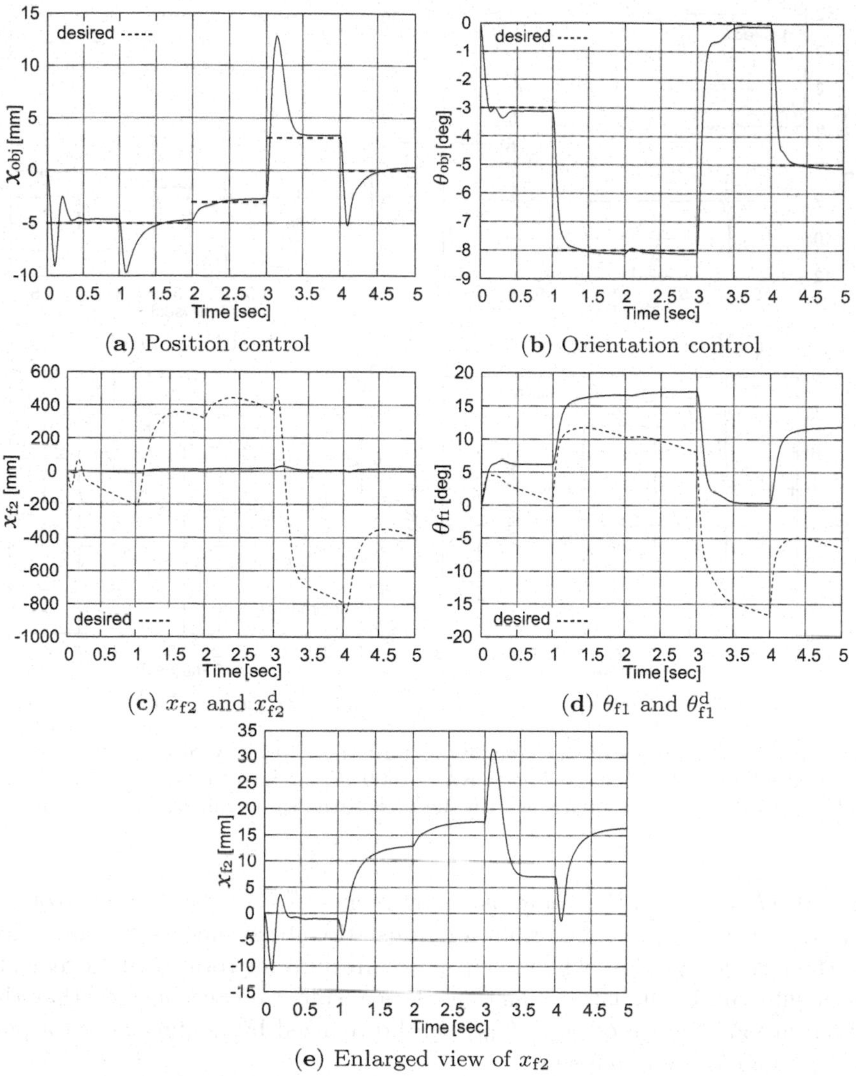

Fig. 10.11 In this simulation, we also examined the trajectory of (x_{obj}, θ_{obj}). Unlike Fig. 10.9, continuously increasing biased torque/force is added to the system at a rate of 60 N/s and 20 Nm/s every 1 ms from the initial value listed in Table 10.1. This result shows stable convergence of (x_{obj}, θ_{obj}), but slight errors can be seen in every time step

these two variables, as shown in Fig. 10.10. Because of the preferable inconsistency of mechanical structure between the revolute and prismatic joints, which have different physical dimensions, the kinematics associated with x_{obj} can be described as Eq. 10.16, in which translational movements of the system are involved, while the kinematics related to θ_{obj} can be expressed as

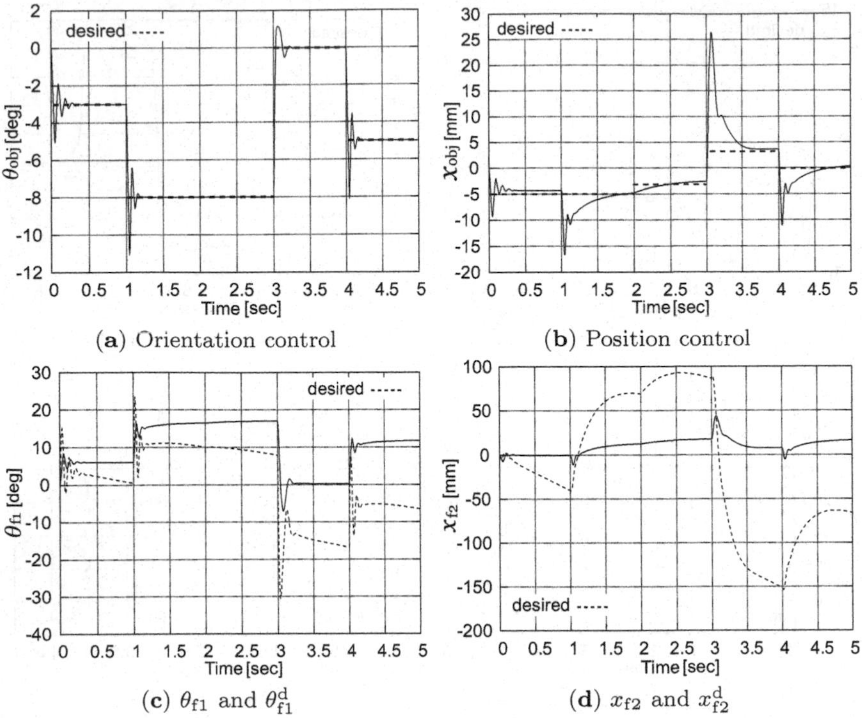

Fig. 10.12 The results of this simulation show the trajectory of $(x_{\mathrm{obj}}, \theta_{\mathrm{obj}})$ where we used $K_{\mathrm{I}}^{\theta} = 0.03$, $K_{\mathrm{I}}^{x} = 0.006$, and $K_{\mathrm{P}}^{x} = 1500$. The relative ratio between K_{I}^{θ} and K_{I}^{x} is set to 100:1, in comparison with the rate of the same gains used in Fig. 10.11

Eq. 10.17, in which the rotational movements of the system are involved. In other words, force equilibrium for translational movements is maintained in the former equation, while moment equilibrium for rotational movements is maintained in the latter equation. As a result, we can conclude that the simultaneous control of $(x_{\mathrm{obj}}, \theta_{\mathrm{obj}})$ can be realized by making use of a pair of *contradictory mechanisms*.

Next, we show another case where the biased torque and force increase continuously at the rate of 60 N/s and 20 Nm/s per millisecond from the initial value listed in Table 10.1. Accordingly, the last values of τ_{b} become 310 N and 110 Nm, respectively, for each input signal of (u_1, u_2). Figure 10.11 shows the results of $(x_{\mathrm{obj}}, \theta_{\mathrm{obj}})$, and both trajectories have small deflections, but nonnegligible steady-state errors in every time step. In accordance with the continuous increase of τ_{b}, both joint angles of the right finger and the prismatic motion of the left finger continue to increase/decrease along with the change in the desired trajectories that are virtually produced in the individual first stage expressed as Eqs. 10.8 and 10.10, as shown in Figs. 10.11c-e. In other words, the object position and orientation retain each steady-state

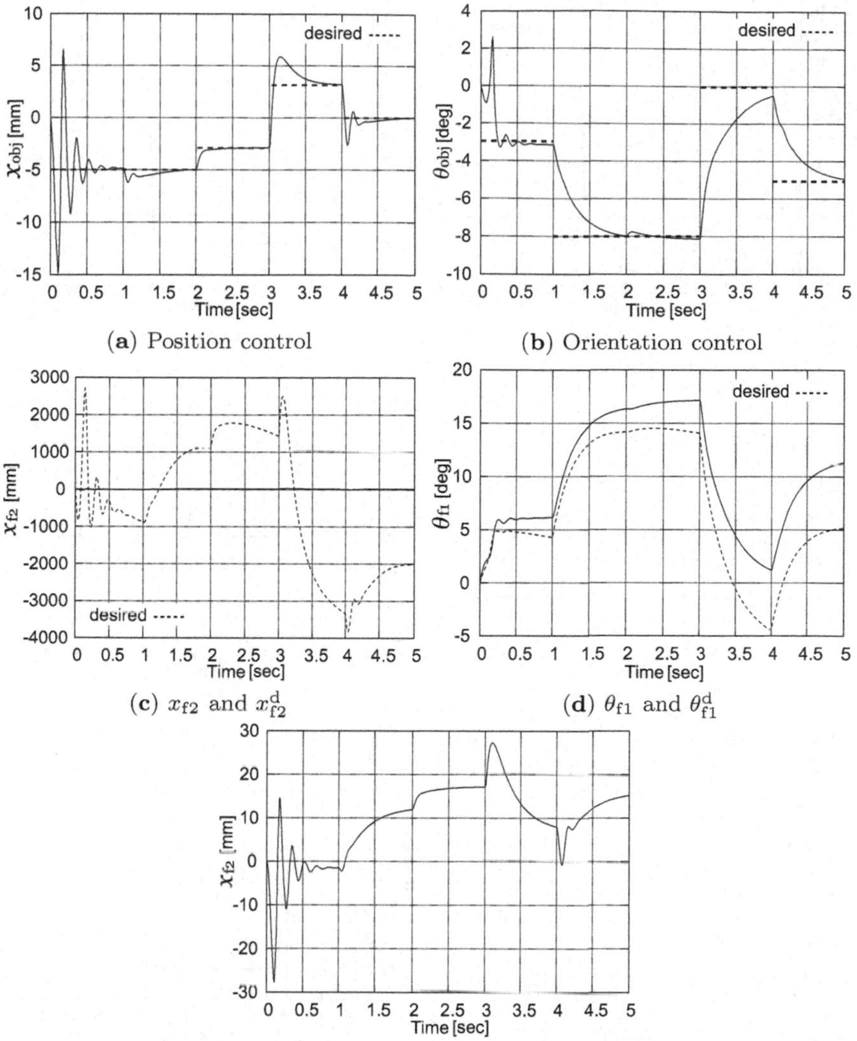

(**a**) Position control

(**b**) Orientation control

(**c**) x_{f2} and $x_{\mathrm{f2}}^{\mathrm{d}}$

(**d**) θ_{f1} and $\theta_{\mathrm{f1}}^{\mathrm{d}}$

(**e**) Enlarged view of x_{f2}

Fig. 10.13 The results of this simulation show the trajectory of $(x_{\mathrm{obj}}, \theta_{\mathrm{obj}})$ where we used $K_{\mathrm{I}}^{\theta} = 0.0008$, $K_{\mathrm{I}}^{x} = 0.6$, $K_{\mathrm{P}}^{\theta} = 900$, and $K_{\mathrm{P}}^{x} = 70$. The relative ratio between K_{I}^{θ} and K_{I}^{x} is set to 1:100, in comparison with the rate of the same gains used in Fig. 10.11

value, even though both of joints move continually. This means that the strength of the pinching motion, which is called the *pinching force*, can be altered without changing the object information. As a result, in the case of the RP joint mechanism, the three objectives of pinching force control, position control, and orientation control of the target object cannot be satisfied

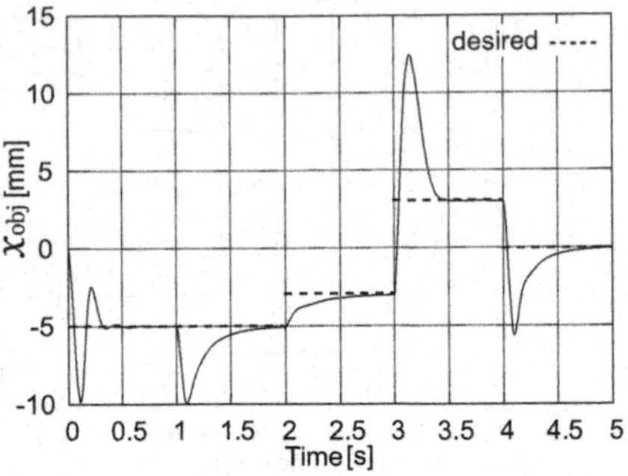

(**a**) Position control (**b**) Orientation control

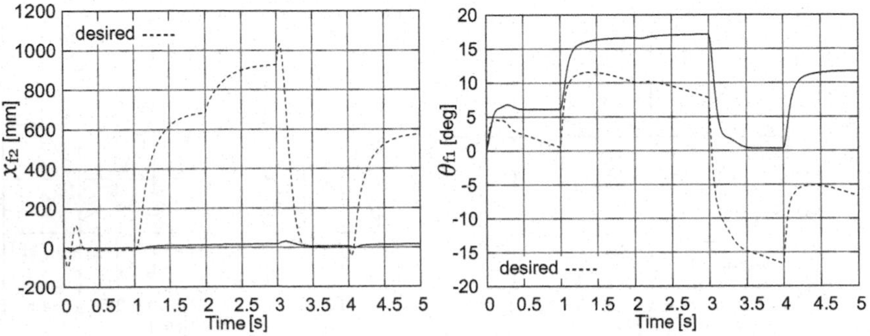

(**c**) x_{f2} and x_{f2}^{d} (**d**) θ_{f1} and θ_{f1}^{d}

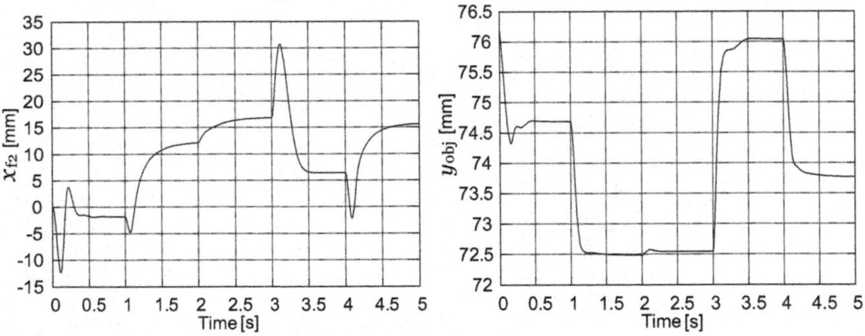

(**e**) Enlarged view of x_{f2} (**f**) y_{obj}

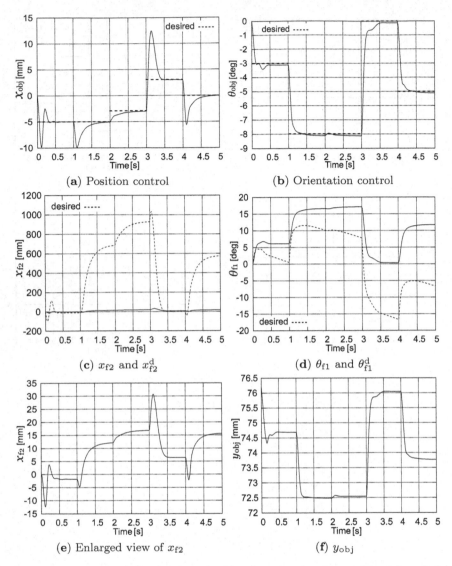

Fig. 10.14 Simulation result when the biased torque for the revolute joint of the right finger increased and the opposite biased force for the prismatic joint of the left finger was set to be constant. Simulation parameters obey Table 10.1, but $K_{\rm I}^{x} = 0.06$

simultaneously. However, even though slight deflections of $x_{\rm obj}$ and $\theta_{\rm obj}$ exist, the true influence of the deflections on a task may probably be small. This is because whether the task is successfully done is rather important, and these slight deflections may not affect the success of the experiment. Therefore, whether the deflections correspond to a fatal error for the control system is unclear as long as we only consider the results of Figs. 10.11a and b.

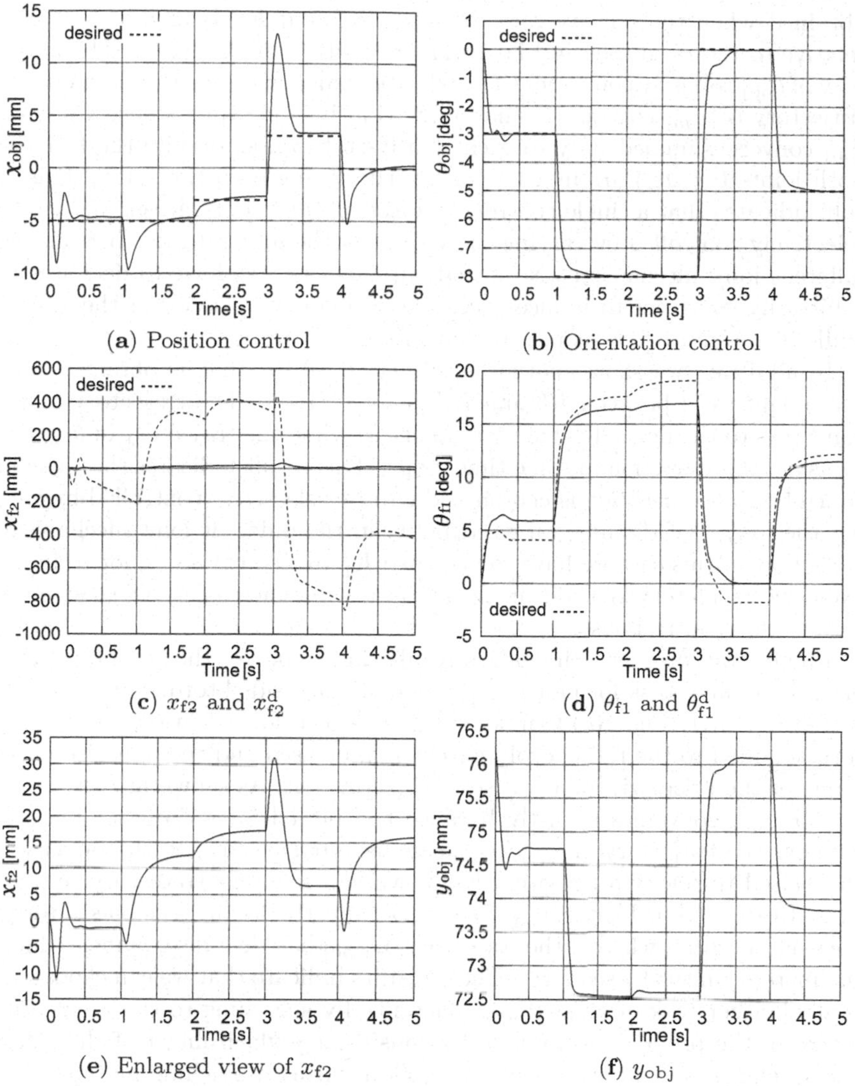

Fig. 10.15 Simulation result when the biased force for the prismatic joint of the left finger increased and the opposite biased torque for the revolute joint of the right finger was set to be constant. Simulation parameters obey Table 10.1, but $K_{\mathrm{I}}^{x} = 0.06$

As stated above, the simultaneous control of three control tasks, *i.e.*, pinching force, object position, and object orientation, using fewer than 3 DOF of the robot, is generally difficult. However, we found a control method to solve this problem. This method involves choosing either the position control task or the orientation control task on a priority basis by changing the weight of the control law between x_{obj} and θ_{obj}. For instance, the simulation results obtained when the relative ratio between K_{I}^{θ} and K_{I}^{x} is modified to 100:1 relative to the ratio used in Fig. 10.11 are plotted in Fig. 10.12. The trajectory of θ_{obj} robustly converges to the desired value with no error, whereas the trajectory of x_{obj} contains a small deviation in some time steps. Moreover, θ_{obj} converges immediately to each steady-state value, despite the fact that both joints $(\theta_{\mathrm{f1}}, x_{\mathrm{f2}})$ are moving constantly during each time step. This result indicates that a single control of *object orientation* is chosen and can be selectively controlled by varying the weight of the integral gains between the different joint mechanisms. Note that, in this case, we have to decrease K_{P}^{θ} arbitrarily so as not to induce excessive oscillatory behavior in the system, while increasing K_{P}^{x} so as not to incur a delay in response.

In addition, we show another result (Fig. 10.13) that is obtained when the weight ratio between K_{I}^{θ} and K_{I}^{x} is set to 1:100. The trajectory of x_{obj} converges to the desired value with no error, while the trajectory of θ_{obj} contains a small deviation in some time steps. This result indicates that a single control of *object position* is chosen and can be selectively controlled by varying the weight of the integral gains between the different joint mechanisms. Note that in this case we have to decrease K_{P}^{x} arbitrarily so as not to induce excessive oscillatory behavior in the system, while increasing K_{P}^{θ} so as not to incur a delay in response.

Finally, we consider a simulation result that indicates the accomplishment of an individual task for $(x_{\mathrm{obj}}, \theta_{\mathrm{obj}})$ when the one-sided torque of the fingers increases. First, Fig. 10.14 shows the task execution when only the biased torque added to the right revolute joint is increased and the opposite biased force for the prismatic joint was set to be constant. As shown in Figs. 10.14a and b, x_{obj} converges robustly because the biased force does not increase, whereas the trajectory of θ_{obj} has clear discrepancies due to the increase in the biased torque. On the other hand, we can see a contradictory result in Figs. 10.15a and b. That is, θ_{obj} converges robustly because the biased torque does not increase, whereas the trajectory of x_{obj} has clear discrepancies due to the increase in the biased force. These results indicate that we can choose and control either task of $(x_{\mathrm{obj}}, \theta_{\mathrm{obj}})$ preferentially according to the momentary status of the progress of the task by considering which finger of the thumb and forefinger should be actively controlled during a pinching motion.

10.4.3 Revolute Joint vs. Revolute Joint (RR Joints)

From the above results, we can expect that the object orientation can at least be controlled in the case of the RR joint mechanism, as suggested in Fig. 10.16a. This figure shows that the object orientation robustly converges to each desired value that varies every second.

Next, reviewing the results of RP joints, we know that the position and orientation of the object can be controlled precisely, indicating that the double

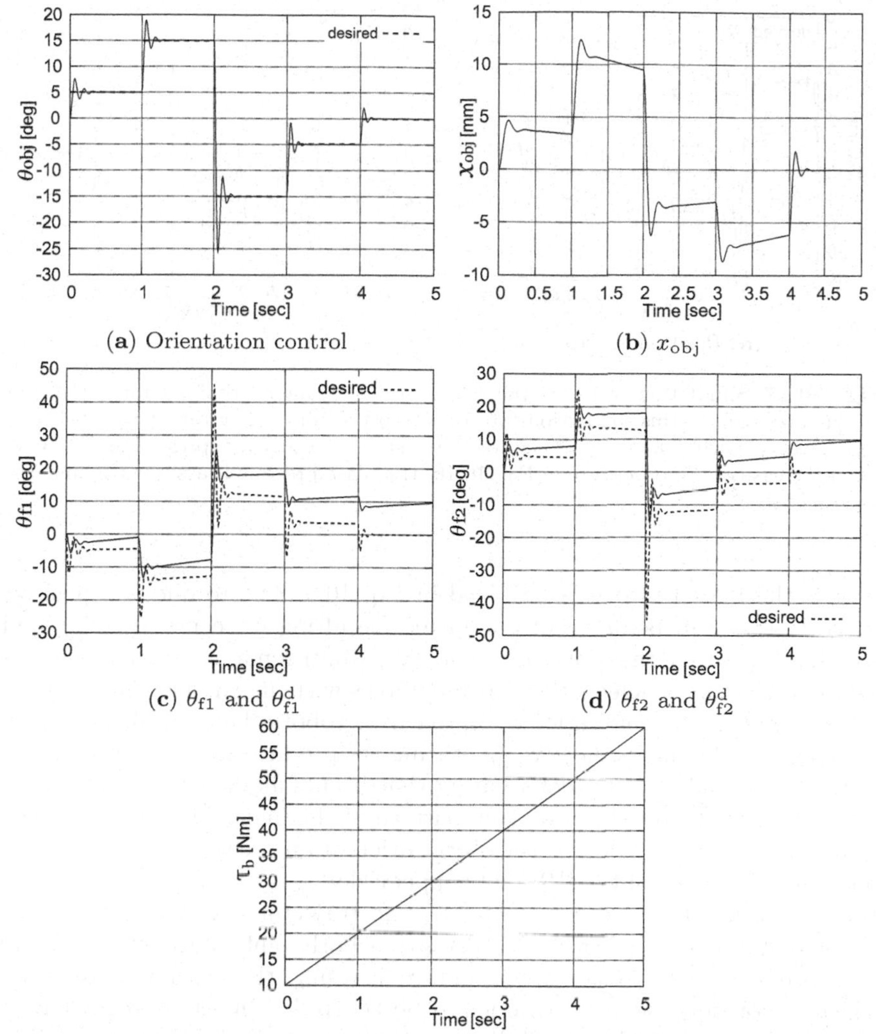

(**a**) Orientation control

(**b**) $x_{\rm obj}$

(**c**) $\theta_{\rm f1}$ and $\theta_{\rm f1}^{\rm d}$

(**d**) $\theta_{\rm f2}$ and $\theta_{\rm f2}^{\rm d}$

(**e**) Increment of $\tau_{\rm b}$

Fig. 10.16 Simulation results of the object orientation control with continuous increase of the biased torque in the RR joint mechanism: The applied biased torque was determined on the basis of the continuous increasing rate of 10 Nm/sec. In this simulation, the apparent errors of both the revolute joints never induce the failure of the orientation control. In addition, robotic experiments of the orientation control task using the RR joint mechanism were already presented in Fig. 8.13 and Fig. 8.15

outputs $(x_{\rm obj}, \theta_{\rm obj})$ are regulated by the double inputs $(x_{\rm f2}, \theta_{\rm f1})$, respectively. Then, in the case of the RR joints, one question may arise. What is another achievable remaining task (output) other than the object orientation in terms of the double inputs $(\theta_{\rm f1}, \theta_{\rm f2})$? Figure 10.16 indicates that the an-

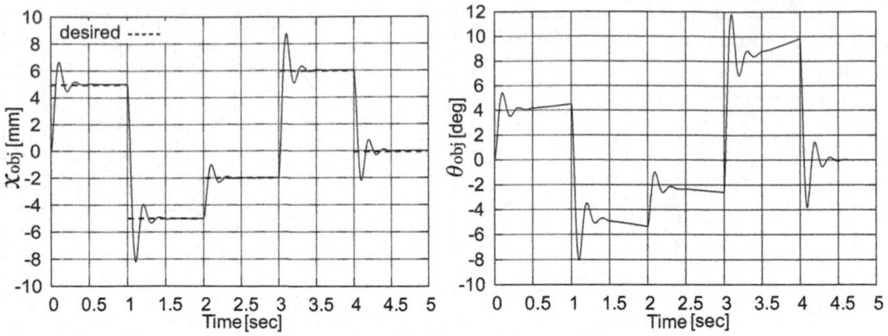

(**a**) Position control (**b**) θ_{obj}

Fig. 10.17 Simulation result of the RR joint mechanism obtained for the object position control. In this simulation, we converted the controlling target variable from θ_{obj} to x_{obj} within the integral sign of the first stage control expressed as Eq. 10.8. As in the case of the procedure in Fig. 10.16, the biased torque is assumed to increase continuously

swer is the biased torque τ_b involved in Eq. 10.9. The torque is a positive constant that can be arbitrarily adjusted, resulting in secure grasping and an effect such that the input torque u_i (Eq. 10.9) remains positive. The increase in the torque affects the grasping forces without any movement of the finger joints in the conventional approach to robotic hand applications. In the case of soft fingers, however, both fingers produce inward rotation due to the additional fingertip deformation that occurs along with the increased torque. Figure 10.16e shows a given pattern of the linear increase in τ_b. Despite the continuous increase, the object orientation precisely converges, and the allowable errors of the RR joints appear clearly, as shown in Figs. 10.16c and d. During the convergence of θ_{obj}, the trajectory of x_{obj} tends to increase or decrease. The reason for this is that both joints of the fingers rotate inward due to the increase in τ_b, and, as a result, the other variable x_{obj}, which is not designated as a target to be controlled, must be regulated so that θ_{obj}, which should be controlled, converges to a specified value smoothly without being affected by the changes in other variables. That is, all of the variables, including $(x_{obj}, y_{obj}, \theta_{obj}, \theta_{f1}, \theta_{f2})$, satisfy an LMEEwC, even during the dynamic behavior of the system. In other words, we should create an idle variable capable of moving freely in order to satisfy the LMEEwC. The individual *admissible deviation* of the joint angles arises for the same reason. As mentioned in Chap. 8, rather than a conformation of PID control, a PD control for the joints is absolutely imperative for the control strategy in soft-fingered manipulation, because, in general, the PD control scheme has the potential to remain some sort of steady-state deviations in comparison with the PID control. This result is reflected on Figs. 10.16c and d.

Next, we consider a special case of position control of a grasped object using Eqs. 10.8 and 10.9, in which the control equations are transformed into equations that are capable of being implemented as below:

$$\theta_{fi}^{d} = -(-1)^{i} K_{I}^{x} \int_{0}^{t} (x_{obj} - x_{obj}^{d}) d\tau, \tag{10.18}$$

$$u_i = -K_{P}^{\theta}(\theta_{fi} - \theta_{fi}^{d}) - K_{D}^{\theta} \dot{\theta}_{fi} + \tau_b. \tag{10.19}$$

In this transformation of Eq. 10.18, the control variable is only replaced from θ_{obj} to x_{obj}, and K_I^θ changes to K_I^x. The simulation result of the position control using Eqs. 10.18 and 10.19 is shown in Fig. 10.17. This result indicates that the object position converges precisely to the desired value x_{obj}^d, which means that if the control variable is successfully chosen during the manipulating movement in a dynamic sense, either the object position or the orientation control can be selectively achieved under the inherent secure pinching performance by means of only a 2 DOF robotic hand structure. We refer to this potential capability as *task-selection control.* In addition, the object orientation θ_{obj}, which is not controlled in this case, increases gradually, as shown in Fig. 10.17b. This occurs because the task of x_{obj} can be skillfully controlled when attempting to control the object position on a priority basis.

As a result, the RR joint mechanism is able to independently accomplish either object orientation control or position control by a feedback structure and the biased-torque control by an open-loop structure. In particular, we have shown for the first time that a task-selection control can be readily attained during dynamic pinching of biomechanical robotic hands.

Next, we consider the case of a failure when attempting to control two independent variables (x_{obj}, θ_{obj}) simultaneously using a robot with a RR joint mechanism having no prismatic joint. The orientation control is assigned to the right finger joint, and the left finger controls the object position. A corresponding control law to be utilized in the first-stage controller of the control scheme is then expressed as

$$\theta_{f1}^{d} = K_{I}^{\theta} \int_{0}^{t} (\theta_{obj} - \theta_{obj}^{d}) d\tau, \tag{10.20}$$

$$\theta_{f2}^{d} = -K_{I}^{x} \int_{0}^{t} (x_{obj} - x_{obj}^{d}) d\tau. \tag{10.21}$$

In this simulation, each desired value is set so as to satisfy $x_{obj} = 5$ mm and $\theta_{obj} = 3°$. Here, we used the same controller for the second stage, represented as Eq. 10.19, for both fingers. Note that the biased torque applied to both fingers is assumed to be constant, such that $\tau_b = 10$ Nm. As shown in Figs. 10.18a and b, both variables deviate greatly from each of the preliminarily designated desired values, finally resulting in a failure to manipulate the target object. The object trajectory becomes unstable and does not converge to the desired value, and the virtual desired joint angles $(\theta_{f1}^d, \theta_{f2}^d)$ vary unstably in one direction because these values are computed by the accumu-

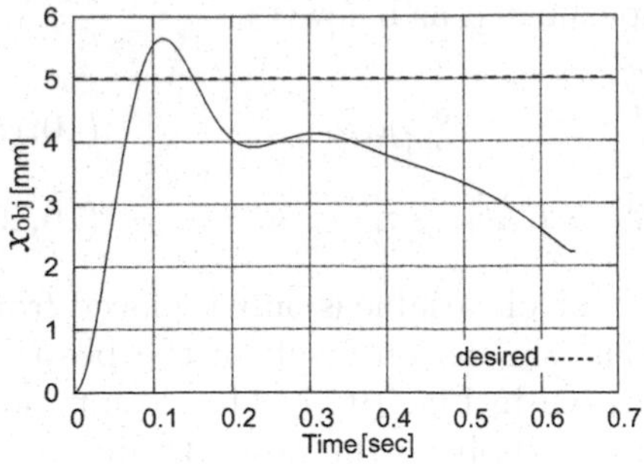

(**a**) Position control

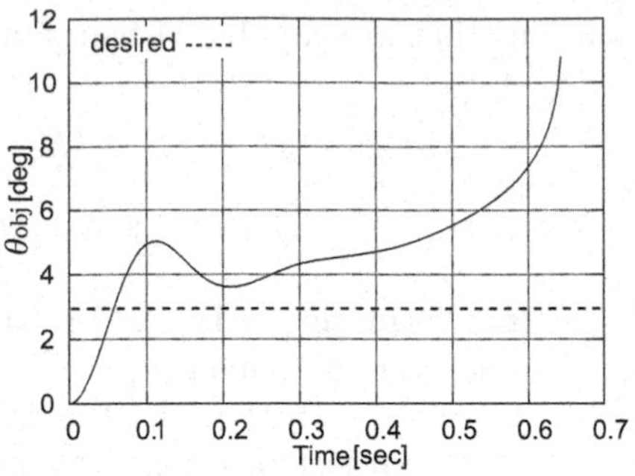

(**b**) Orientation control

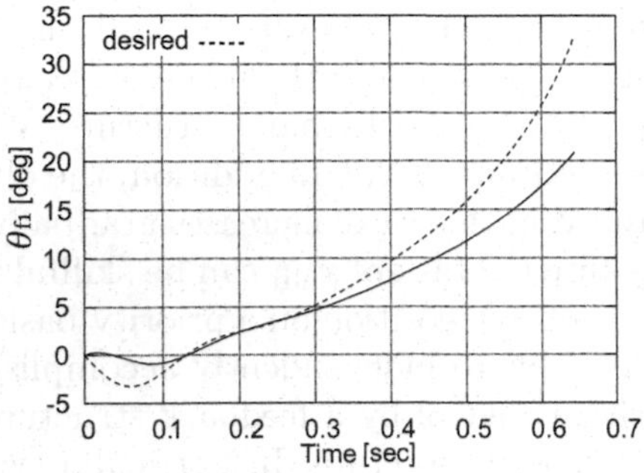

(**c**) θ_{f1} and θ_{f1}^{d}

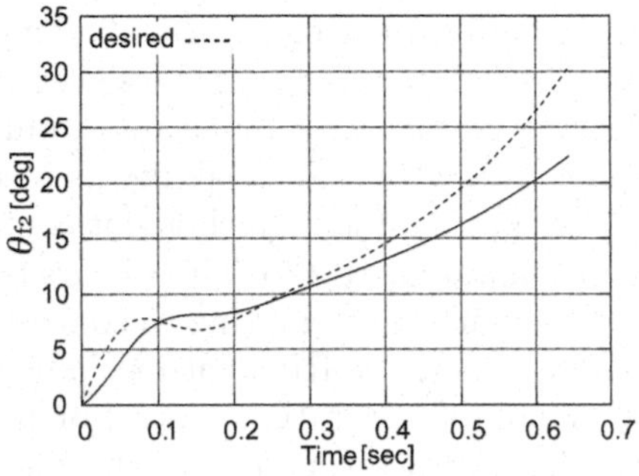

(**d**) θ_{f2} and θ_{f2}^{d}

Fig. 10.18 This simulation shows a case of failure when attempting to control (x_{obj}, θ_{obj}) simultaneously by means of a robotic hand having an RR joint mechanism. Unlike the previous procedure of Figs. 10.16 and 10.17, the biased torque is assumed to be constant, such that $\tau_b = 10$ Nm

lation of the error in the integral controllers at the first stage, as shown in Figs. 10.18c and d.

Through this simulation, we investigate why two-variable control for (x_{obj}, θ_{obj}) cannot be achieved, even though the relationship between the number of mechanical DOFs, $(\theta_{f1}, \theta_{f2})$, and the number of target tasks, (x_{obj}, θ_{obj}), is satisfied due to the use of constant biased torque. We perform a static analysis of the system and the RP joint mechanism. Let us recall the static relation Eq. 10.17 and, in this case, formulate this relation for the revolute joint of the left finger by using the elastic energy of both fingers P derived in Eq. 4.5, as follows:

$$E_{n1}\lambda_{n1} + F_{t1}\lambda_{t1} - \frac{\partial P}{\partial \theta_{f1}} - K_P^{\theta}(\theta_{f1} - \theta_{f1}^{d}) + \tau_b = 0, \tag{10.22}$$

$$E_{n2}\lambda_{n2} + F_{t2}\lambda_{t2} - \frac{\partial P}{\partial \theta_{f2}} - K_P^{\theta}(\theta_{f2} - \theta_{f2}^{d}) + \tau_b = 0. \tag{10.23}$$

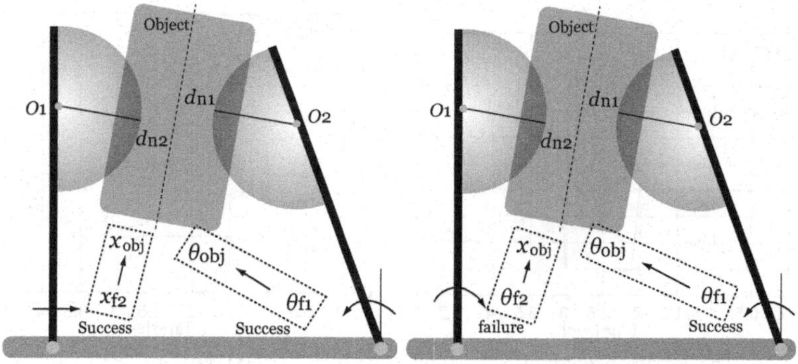

(**a**) Success in RP joints (**b**) Failure in RR joints

Fig. 10.19 Conceptual diagram of success and failure for simultaneous control of x_{obj} and θ_{obj}. This diagram shows that the RP joint mechanism can perform independent and simultaneous control of $(x_{\mathrm{obj}}, \theta_{\mathrm{obj}})$ if the biased torque remains constant. On the other hand, the RR joint mechanism can perform independent and simultaneous control of either set of $(\theta_{\mathrm{obj}}, \tau_{\mathrm{b}})$ or $(x_{\mathrm{obj}}, \tau_{\mathrm{b}})$, whereas this mechanism cannot achieve the control of x_{obj} and θ_{obj}, even when τ_{b} becomes constant

We know at a glance from Eqs. 10.22 and 10.23 that both equations have the dimension of the physical quantity of moment and correspond to a static relationship in the steady state of the system. Even when attempting position control of a grasped object using the rotational mechanism of the left finger, the object position may provisionally approach a desired value. At the same time, the object orientation varies inevitably according to the rotation of the left finger because of a structural consistency related to the rotational movement. In the case of the RP joint mechanism, x_{obj} and θ_{obj} are both successfully controlled by the prismatic joint and the revolute joint respectively, as shown in Fig. 10.19a. The most important point in this case is that the control loop for each variable is completely different from that of the other variable. However, in this case, Eq. 10.23 corresponds to an equilibrium equation in the steady state of the revolute joint of the left finger. Both Eqs. 10.22 and 10.23, therefore, govern rotational movements for achieving object orientation control. As a result, this RR joint cannot produce translational movements that are required for controlling the position of the pinched object when attempting to perform simultaneous control of $(x_{\mathrm{obj}}, \theta_{\mathrm{obj}})$, as shown in Fig. 10.19b.

10.4.4 Prismatic 1-DOF Hand (P Joint)

For the prismatic 1-DOF hand that consists of one fixed finger and one prismatic joint, a very simple robot configuration is described as

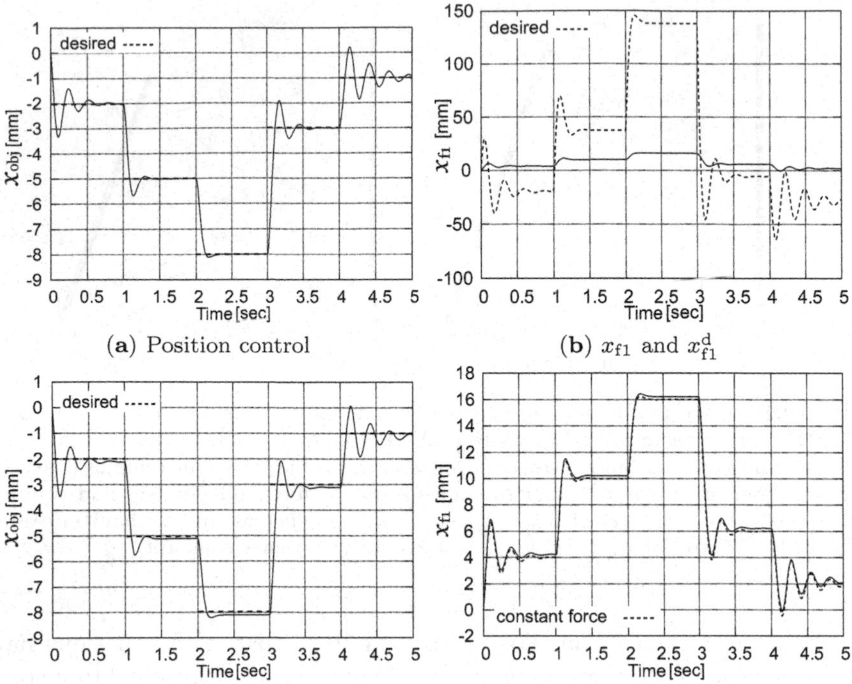

(**a**) Position control (**b**) x_{f1} and x_{f1}^{d}

(**c**) Position control under variable τ_b (**d**) Discrepancy of x_{f1}

Fig. 10.20 Simulation results of the prismatic 1-DOF Hand, in which the object position control is executed under constant biased force $\tau_b = 10$ N and under continuously-increasing biased force at the rate of 20 N/sec during the simulation

$$\begin{aligned} O_{1x} &= W_B - d_f - x_{f1}, \quad & O_{1y} &= L, \\ O_{2x} &= -W_B + d_f, \quad & O_{2y} &= L. \end{aligned} \tag{10.24}$$

Note that this robot configuration is a left-right reversal relative to that of Fig. 10.7a. Moreover, the elastic potential energy in the case of the prismatic joint is then modified as

$$\begin{aligned} P_1 + P_2 = \pi E &\left\{ \frac{d_{n1}^3}{3\cos^2\theta_{obj}} + d_{n1}^2 d_{t1}\tan\theta_{obj} + d_{n1} d_{t1}^2 \right\} \\ &+ \pi E \left\{ \frac{d_{n2}^3}{3\cos^2\theta_{obj}} + d_{n2}^2 d_{t2}\tan\theta_{obj} + d_{n2} d_{t2}^2 \right\}. \end{aligned} \tag{10.25}$$

The right finger therefore moves by the corresponding control inputs expressed in Eqs. 10.10 and 10.11. In this simulation, we set the initial variables to $(x_{obj}, y_{obj}, \theta_{obj}) = (0, 0, 0)$ and the biased force τ_b in the prismatic joint to be constant.

Figure 10.20 shows the simulation result in which the position control of the object was added and the desired position x_{obj}^{d} was given as a series of step functions, which is depicted in Fig. 10.20a as a dotted line. As shown in the figures, the object position converges robustly to the desired step trajectory. Furthermore, relatively large unexpected oscillations at the beginning and end of this simulation are observed, as shown in Fig. 10.20a. This results from an inherent physical phenomenon of the soft fingers whereby the fingertip stiffness becomes small when the fingertip deformation remains fractional while the magnitude of the stiffness rapidly increases as the deformation further expands [IH06]. In addition, Fig. 10.20b shows the trajectory of the right finger together with the virtual desired trajectory of the finger that is dynamically produced according to the proposed control method, expressed as Eq. 10.10, as x_{obj} approaches the desired value. Although a considerably large error of the prismatic joint is observed, the control target here is x_{obj}, and when considering the successful convergence of x_{obj}, the error does not significantly affect the controlling system. That is, this type of error can be referred to as an *admissible deviation.* As a result, this section reveals that the object position as a single output can be controlled by a single input of the prismatic finger joint.

Next, we consider the case in which the biased force varies at a certain rate during the task and verify the system dynamics in this case. In this simulation, we set the initial biased force at $\tau_{\mathrm{b}} = 10$ N and set this force to change at a rate of 20 N/s in all steps. Therefore, τ_{b} finally reaches 110 N at 5 s. Since the time interval is 1 ms, the increase is considerd to be almost continuous. Figure 10.20c shows the trajectory of the object position, and Fig. 10.20d presents the prismatic joint trajectory along with the same result (dotted line) for comparison obtained when τ_{b} is constant, which is depicted in Fig. 10.20b as the continuous line. The figures show that the system stability is maintained continuously, although there exist slight errors in both x_{obj} and x_{f1}. This reason for this is that the restoring forces and moments determine the object position uniquely due to the elastic deformation of the soft fingertips. That is, these small errors occur because the prismatic joint is further compressed by the increase of τ_{b}, and, as a result, the previous force/moment equilibrium point, including x_{f1}, shifts to a neighboring stable equilibrium condition. This observation can be further explained by the existence of the LMEEwC, descrbed in Chaps. 4 and 6 [IH07a].

10.4.5 Rotational 1-DOF Hand (R Joint)

Let us consider the case in which one finger rotates around its joint axis and another is fixed on a base, as shown in Fig. 10.21. Recalling Eq. 10.1, the coordinates of the fingertip origin O is easily rewritten as

$$\begin{aligned} \mathrm{O}_{1x} &= W_{\mathrm{B}} - L\sin\theta_{\mathrm{f1}} - d_{\mathrm{f}}\cos\theta_{\mathrm{f1}}, \quad & \mathrm{O}_{1y} &= L\cos\theta_{\mathrm{f1}} - d_{\mathrm{f}}\sin\theta_{\mathrm{f1}}, \\ \mathrm{O}_{2x} &= -W_{\mathrm{B}} + d_{\mathrm{f}}, & \mathrm{O}_{2y} &= L. \end{aligned} \tag{10.26}$$

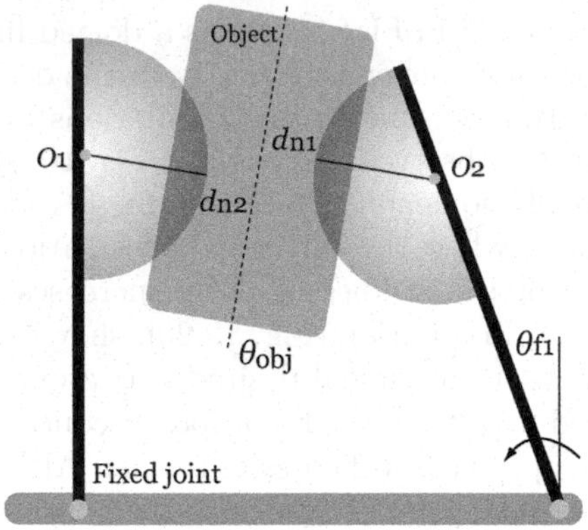

Fig. 10.21 Rotational 1-DOF robotic hand

By implementing numerical computation for Eq. 10.6 with the Runge–Kutta method on the basis of the CSM, the dynamic behavior of the rotational 1-DOF mechanical hand, including a grasped rectangular object, is revealed. In this robotic configuration, the elastic potential energy function is equivalent to Eq. 10.12. In addition, for the verification of subsequent simulations we formulate preliminarily the moment due to the deformation of the right fingertip around the corresponding joint angle as follows:

$$\frac{\partial(P_1+P_2)}{\partial\theta_{\mathrm{f1}}}=\pi E\left\{\frac{2d_{\mathrm{n1}}^3\sin(\theta_{\mathrm{f1}}-\theta_{\mathrm{obj}})}{3\cos^3(\theta_{\mathrm{f1}}-\theta_{\mathrm{obj}})}+\frac{d_{\mathrm{n1}}^2d_{\mathrm{t1}}}{\cos^2(\theta_{\mathrm{f1}}-\theta_{\mathrm{obj}})}\right\}. \quad (10.27)$$

The above computation can be performed by differentiating Eq. 10.12 with respect to the revolute joint. Another moment generated by the gravitational force is also given by $-M_{\mathrm{f}i}gL\sin\theta_{\mathrm{f1}}$, which is obtained by the same differentiation of the last term of Eq. 7.3. In total, the generalized force exerted on the right finger joint is therefore represented by summing the moments. This generalized force acts as a restoring moment around the joint angle of the right finger. According to Eq. 10.27, we should design an optimal input torque to maintain stable manipulation in the control system.

Figure 10.22 shows the simulation results in which the object orientation, θ_{obj}, is controlled so as to be $-6°$ throughout this simulation, and a control law is examined such that

$$\theta_{\mathrm{f1}}^{\mathrm{d}}=K_{\mathrm{I}}^{\theta}\int_0^t(\theta_{\mathrm{obj}}-\theta_{\mathrm{obj}}^{\mathrm{d}})\mathrm{d}\tau, \quad (10.28)$$

$$u_1=-K_{\mathrm{P}}^{\theta}(\theta_{\mathrm{f1}}-\theta_{\mathrm{f1}}^{\mathrm{d}})-K_{\mathrm{D}}^{\theta}\dot{\theta}_{\mathrm{f1}}+\tau_{\mathrm{b}}. \quad (10.29)$$

In this simulation, a discontinuous step torque function maintaining an interval of 200 ms from 1 s from 2 s, as shown in Fig. 10.22e, is added. The object orientation converges to the desired angle consistently, as shown in

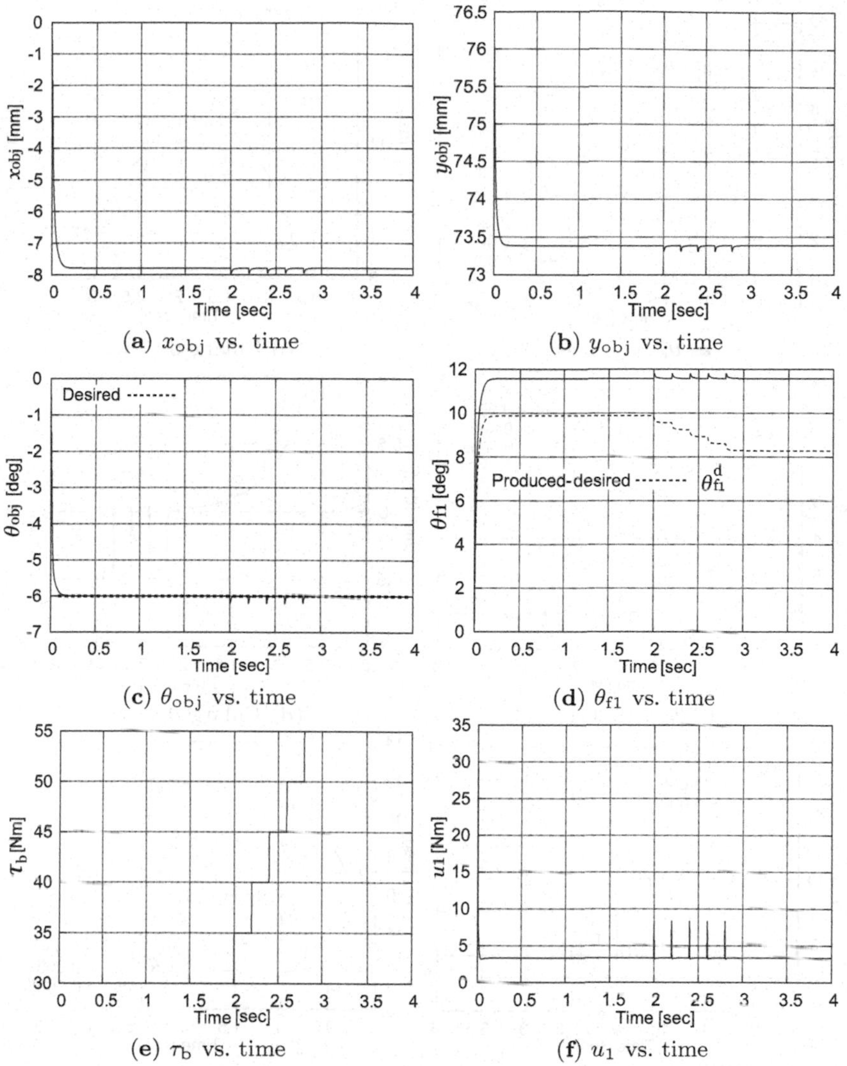

(a) x_{obj} vs. time (b) y_{obj} vs. time

(c) θ_{obj} vs. time (d) θ_{f1} vs. time

(e) τ_{b} vs. time (f) u_1 vs. time

Fig. 10.22 Simulation results of the rotational 1-DOF robotic hand. The controlled target is the object orientation and its desired value is set to be $-6°$ throughout this simulation. The integral and proportional gains satisfy $K_{\mathrm{I}}^{\theta} = 0.01$ and $K_{\mathrm{P}}^{\theta} = 900$, respectively, and the other parameters obey Table 10.1

Fig. 10.22c. During robust manipulation, the joint angle of the right finger has a large error and increases to less than approx. 4°, as shown in Fig. 10.22d. That is, the error of θ_{f1} does not influence whether θ_{obj} comes to rest at the equilibrium point in the steady state, although the object information $(x_{\mathrm{obj}}, y_{\mathrm{obj}}, \theta_{\mathrm{obj}})$ instantaneously deviates from the stable point during the discontinuous increase in the torque.

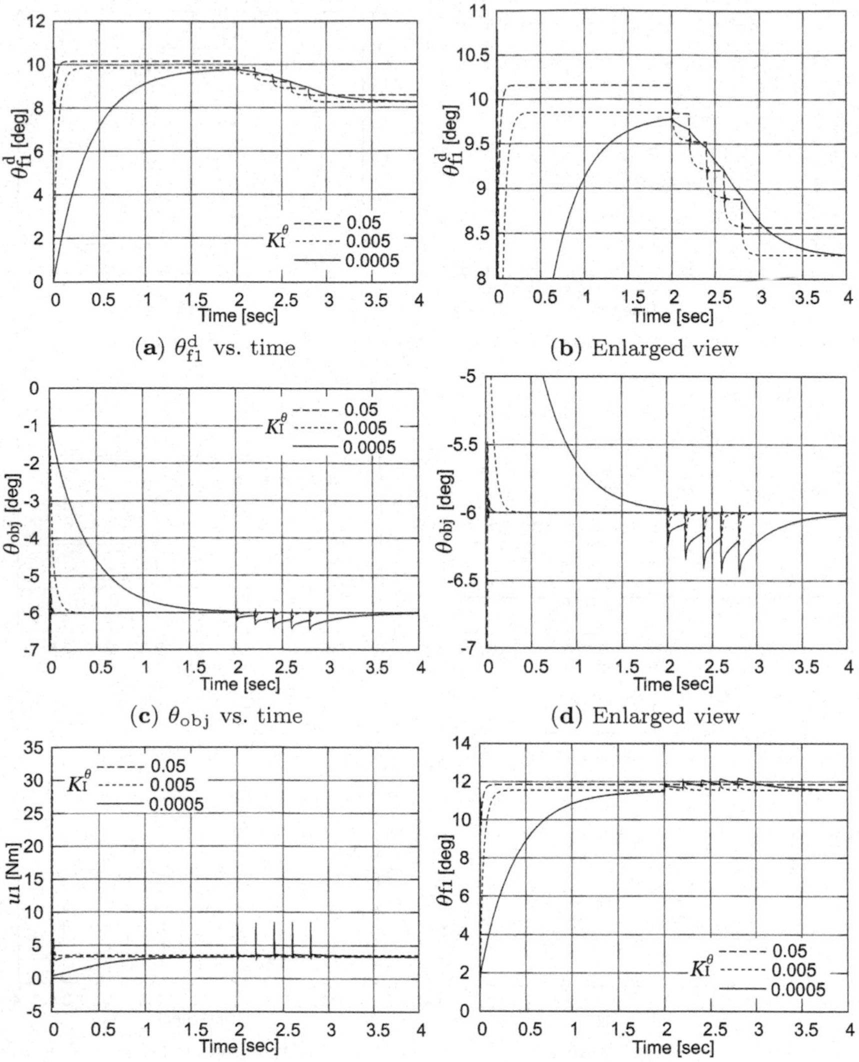

(**a**) $\theta_{\rm f1}^{\rm d}$ vs. time (**b**) Enlarged view

(**c**) $\theta_{\rm obj}$ vs. time (**d**) Enlarged view

(**e**) u_1 vs. time (**f**) $\theta_{\rm f1}$ vs. time

Fig. 10.23 Comparison of the responses of $\theta_{\rm f1}^{\rm d}$ and $\theta_{\rm obj}$ when $K_{\rm I}^{\theta}$=0.05, 0.005, and 0.0005, and $K_{\rm P}^{\theta}$=900. In this simulation, the increasing torque also follows the pattern of Fig. 10.22e. As well as Fig. 10.22, the only target variable to be controlled is the object orientation $\theta_{\rm obj}$, and its desired value is set to $-6°$ in this simulation

Here, the joint angle, $\theta_{\rm f1}$, does not change, even though the biased torque increases greatly. This means that the right finger is not pushed ahead and is able to maintain the desired posture $\theta_{\rm obj}$ even if the grasping torque increases rapidly. The reason for this is that the static force/moment relationship on the fixed left finger is locally balanced between the fingertip and the grasped object. In other words, since there exists a natural balance on the fixed side due to the restoring moment for the object, a relationship such that $\boldsymbol{f}_p + \boldsymbol{u} = \boldsymbol{0}$ on the opposite side is only required to be locally satisfied in the steady state, which is described in part in Eq. 10.6. That is, we can recall Eq. 10.17 for static analysis of this case:

$$E_{\mathrm{n1}}\lambda_{\mathrm{n1}} + F_{\mathrm{t1}}\lambda_{\mathrm{t1}} - \frac{\partial P}{\partial \theta_{\mathrm{f1}}} - K_{\mathrm{P}}^{\theta}(\theta_{\mathrm{f1}} - \theta_{\mathrm{f1}}^{\mathrm{d}}) + \tau_{\mathrm{b}} = 0. \tag{10.30}$$

To ascertain the physical mechanism of the convergence from the viewpoint of statics, Eq. 10.30 should be analyzed by comparing all of the figures of Fig. 10.22. First, as shown in Fig. 10.22f, u_1 remains constant in the steady state. Therefore, considering Eq. 10.29, u_1 can be represented as

$$u_1 = -K_{\mathrm{P}}^{\theta}(\theta_{\mathrm{f1}} - \theta_{\mathrm{f1}}^{\mathrm{d}}) + \tau_{\mathrm{b}} \triangleq \xi \quad \text{(const.)} \tag{10.31}$$

in the steady state. Furthermore, the revolute joint angle θ_{f1} also remains constant, as shown in Fig. 10.22d. Thus, when the biased torque τ_{b} increases, it is balanced by $K_{\mathrm{P}}^{\theta}\theta_{\mathrm{f1}}^{\mathrm{d}}$. In other words, the procedure of the desired rotational joint that is dynamically produced in the first stage of the control law skillfully absorbs the variation of the applied biased torque. Equation 10.30 can therefore be rewritten as

$$E_{\mathrm{n1}}\lambda_{\mathrm{n1}} + F_{\mathrm{t1}}\lambda_{\mathrm{t1}} - \frac{\partial P}{\partial \theta_{\mathrm{f1}}} + \xi = 0. \tag{10.32}$$

Recalling the convergence of θ_{obj} and the fact that θ_{f1} remains constant in the steady state, the first and second terms of Eq. 10.32 also become constant. As a result, we can conclude that the LMEEwC is a unique solution in the case of a 1-DOF rotational hand mechanism and plays an important role in naturally reducing the magnitude of the elastic energy, resulting in a simple description of the control law for robotic manipulation. In contrast, since the system configuration containing the grasped object and both fingers does not change, even though the applied torque is augmented significantly, in the future, we should investigate what the increasing torque actually corresponds to in the biological motor control system of the human hand.

Next, let us validate the function of the integral gain K_{I}^{θ} and elucidate the dynamic influence on robotic manipulation motions. Fig. 10.23 shows the simulation results when K_{I}^{θ} varies in three patterns such that $K_{\mathrm{I}}^{\theta} = 0.05, 0.005, 0.0005$ and the increasing manner of the torque follows Fig. 10.22e. On the whole, the response of θ_{obj}, θ_{f1}, and $\theta_{\mathrm{f1}}^{\mathrm{d}}$ improved, and a relatively large value of K_{I}^{θ} caused the speed to increase dramatically. Since the accumulated amount of the error of the integral controller expressed as Eq. 10.28 depends directly on the integral gain K_{I}^{θ}, the virtual desired trajectories of $\theta_{\mathrm{f1}}^{\mathrm{d}}$ differ in the three cases, as shown in Figs. 10.23a and b. Furthermore, the joint angle θ_{f1} slightly rotates in the direction for increased compression of the fingertip

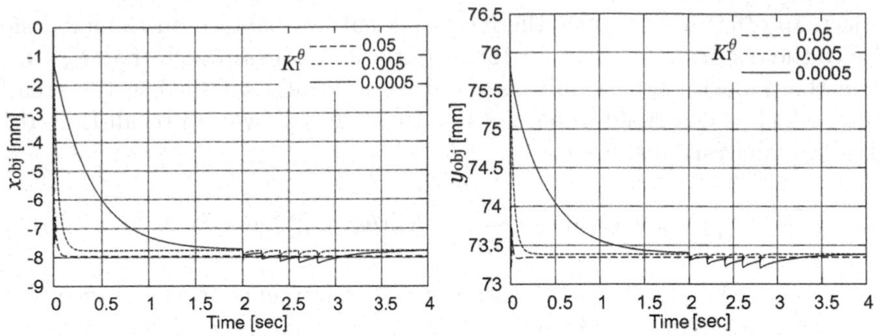

(**a**) x_{obj} vs. time (**b**) y_{obj} vs. time

Fig. 10.24 Trajectories of x_{obj} and y_{obj}, which are based on the Fig. 10.23

as the gain increases, as shown in Fig. 10.23f. On the other hand, the entire torque u_1 input to the robot varies only slightly in the three cases.

Based on these observations, it is possible to adjust θ_{f1} freely to some extent by varying K_{I}^{θ} while maintaining the controlling variable, θ_{obj}, to converge to a predetermined desired value. Accordingly, we may wonder why the joint angle is able to vary even in this extremely simple 1-DOF finger despite the fact that the object information remains constant. As shown in Fig. 10.24, each of the steady-state trajectories of x_{obj} and y_{obj} depicts different lines when K_{I}^{θ} varies. Nevertheless, θ_{obj} never changes, as shown in Fig. 10.23c. Hence, it is somewhat preferable to change the integral gain rather than the biased torque because K_{I}^{θ} can be used to improve the response speed of the total system and contributes to achieving a powerful pinch movement induced by further rotation of the finger. Thus, if we can construct a robotic hand and implement the above control scheme for the robot, the magnitude or strength of the robotic grasps can be readily regulated by users through certain electrical components such as variable resistors placed on an electronic circuit. In addition, it is important that these two tasks can be performed simultaneously with respect to the sole parameter K_{I}^{θ}. As in Eq. 10.18, since the rotational joint mechanism, regardless of the number of degrees of freedom, is capable of realizing the task-selection control, we can recognize that the rotational joint structure of the hands is considerably superior to that of other robotic hands and even the human hand.

Next, let us validate the function of the proportional gain K_{P}^{θ} and elucidate its dynamic influence on robotic manipulation motions. Figure 10.25 shows the simulation results, in which we first find that only the virtually produced desired joint angle $\theta_{\mathrm{f1}}^{\mathrm{d}}$ has changed significantly according to the variation of K_{P}^{θ} in the steady state, while the actual joint angle θ_{f1} changed only slightly, as shown in Fig. 10.25f. Therefore, the relationship of Eq. 10.31 is also satisfied in this case. In other words, the virtual desired joint angle appears to vary flexibly and effectively so that the input signal u_1 remains

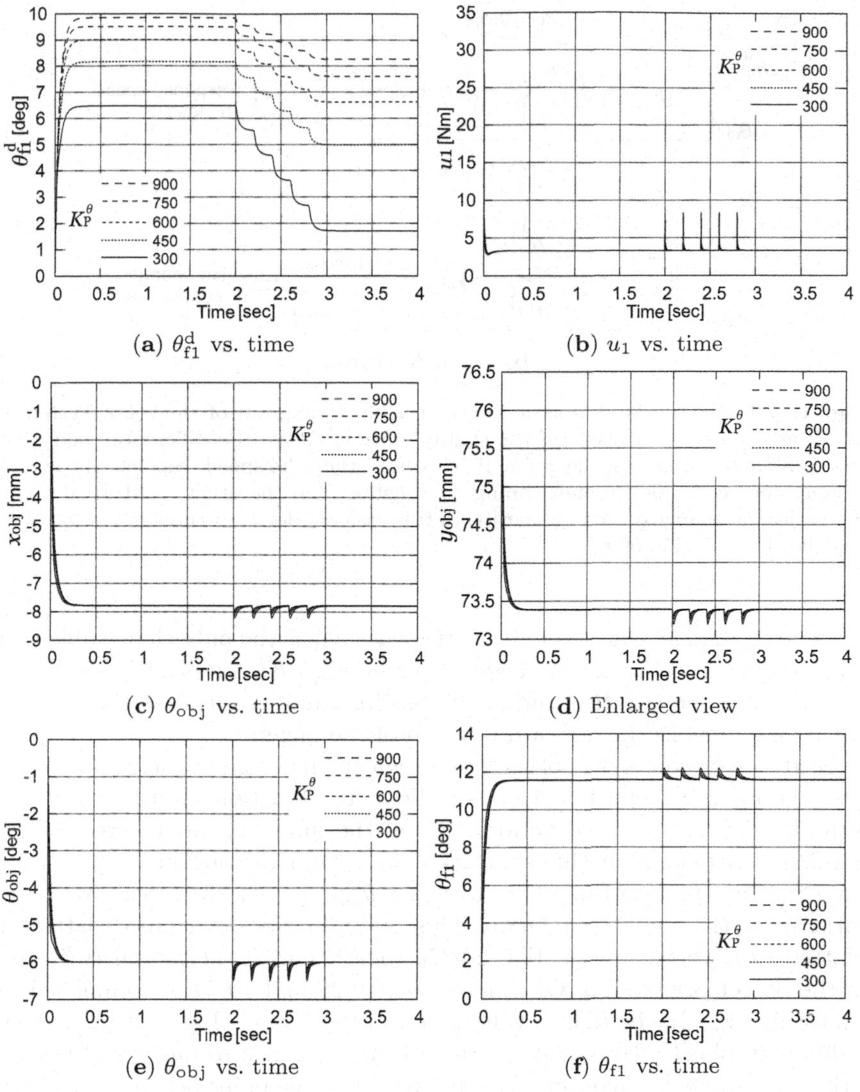

(**a**) θ_{f1}^d vs. time
(**b**) u_1 vs. time
(**c**) θ_{obj} vs. time
(**d**) Enlarged view
(**e**) θ_{obj} vs. time
(**f**) θ_{f1} vs. time

Fig. 10.25 Comparison of the trajectory responses when K_P^θ varies such that $K_P^\theta = 300, 450, 600, 750, 900$, and $K_I^\theta = 0.005$ is constant. As in Fig. 10.22, the only target variable to be controlled is the object orientation θ_{obj}, and its desired value is set to be $-6°$

constant. These results indicate that the variance of K_P^θ does not directly contribute to the dynamic behavior of the system, but makes θ_{f1}^d follow different trajectories, as long as K_I^θ remains constant. This intrinsic fact occurs internally in the control structure and never appears outwardly, as shown in Fig. 10.26b. As a result, in the former case, four variables $(x_{obj}, y_{obj}, \theta_{f1}, \theta_{f1}^d)$

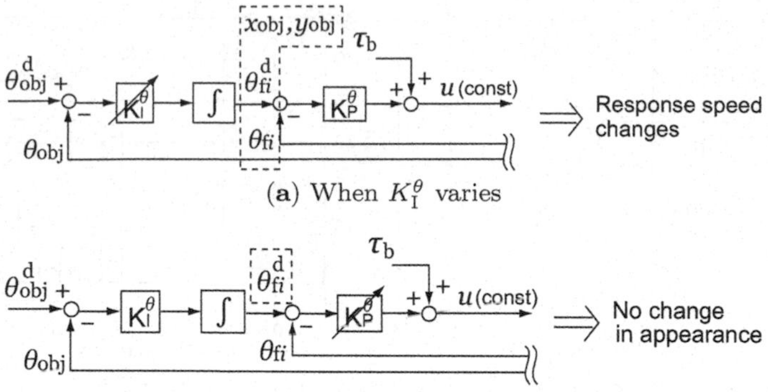

(**a**) When K_I^θ varies

(**b**) When K_P^θ varies

Fig. 10.26 Partial diagrammatic views of a block diagram of the robot system in the steady state. As a result of the steady state analysis, the differential controller was eliminated from these figures. **a** If K_I^θ varies, then the speed response of the total system changes explicitly maintaining u_1 constant. On the other hand, **b** although K_P^θ varies, no change occurs outwardly. In this case, u_1 does not change, but $\theta_{\mathrm{f1}}^{\mathrm{d}}$ alone absorbs the variation of τ_b

are able to change, whereas in the latter case, $\theta_{\mathrm{f1}}^{\mathrm{d}}$ is the only changeable variable, as shown in Fig. 10.26. Thus, we can design K_P^θ extensively so that u_1 does not become negative during the pinching motion because the negative value may provoke the outward rotation of the fingers.

Next, let us consider another case in which the desired orientation of the grasped object is determined so as to be a step function in order to further evaluate the task-oriented performance of the minimal one-fingered robotic hand. In particular, in this simulation we give a discontinuous pattern of the $\theta_{\mathrm{obj}}^{\mathrm{d}}$ trajectory to Eq. 10.28 such that $\theta_{\mathrm{obj}}^{\mathrm{d}} = -4°, -6°, -8°, -3°, -1°$ is satisfied at the same 1-s interval. This preliminarily determined pattern is plotted as a dotted line in Fig. 10.27a together with the actual trajectory of the object orientation. Of course, Eqs. 10.28 and 10.29 are employed for controlling θ_{obj} in the right revolute joint of the robot. Figure 10.27a shows individual robust convergences of the orientation. As a result, the object position simultaneously converges to an unpredictable point because the object position is not the target variable to be controlled, as shown in Figs. 10.27c and d. At the same time, Fig. 10.27b exhibits relatively large discrepancies of θ_{f1} between the desired joint angles, which are dynamically and virtually produced at Eq. 10.28, and their actual data. However, taking into account that the controlled target is θ_{obj}, which consistently converges to the desired value, we can define the individual discrepancy of θ_{f1} on each time step as an *admissible deviation*. As a result, the minimal robot mechanism having a 1-DOF revolute joint is capable of controlling a single output such as object orientation. Note that the secure pinching motion of the object is invariably performed during the tasks. Conventionally, it has been said that another de-

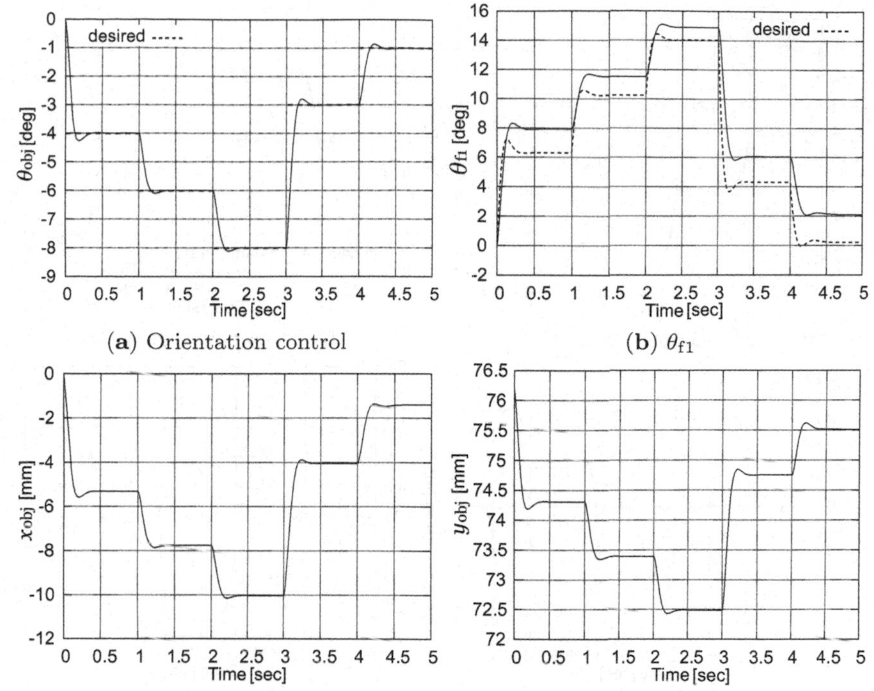

(**a**) Orientation control (**b**) θ_{f1} (**c**) x_{obj} (**d**) y_{obj}

Fig. 10.27 The results of this simulation show the robust convergence of the object orientation, which is controlled such that θ^{d}_{obj} changes $-4°, -6°, -8°, -3°, -1°$ in order. Here, the constant τ_b and three gains $(K^{\theta}_{I}, K^{\theta}_{P}, K^{\theta}_{D})$ are all listed in Table 10.1. Existing discrepancies of the joint angle are defined as a certain type of *admissible deviations*

gree of freedom would be absolutely imperative to achieve the secure pinching motion [HAK$^+$01].

In addition, we show the results of dynamic responses when the biased torque varies from the initial value up to 110 Nm at the specified rate of 20 Nm/s, which rises continuously at intervals of 1 ms. In this case, Fig. 10.28a shows significant discrepancies of the θ_{obj} trajectory in all time steps. Furthermore, by depicting the trajectory of θ_{f1} along with the same variable previously obtained in Fig. 10.27 when τ_b remains constant, we can see slight errors in each time step, as shown in Fig. 10.28b. All trajectories of the object position (x_{obj}, y_{obj}), which are not to be controlled directly, appear to converge accurately according to the stability of the system. Hence, any system variable should function as an absorber of sorts in order to cancel out the continuous change in τ_b. This absorber corresponds to the desired joint angle that is virtually produced in Eq. 10.28. This phenomenon can be clearly seen in Fig. 10.28e based on Eq. 10.31.

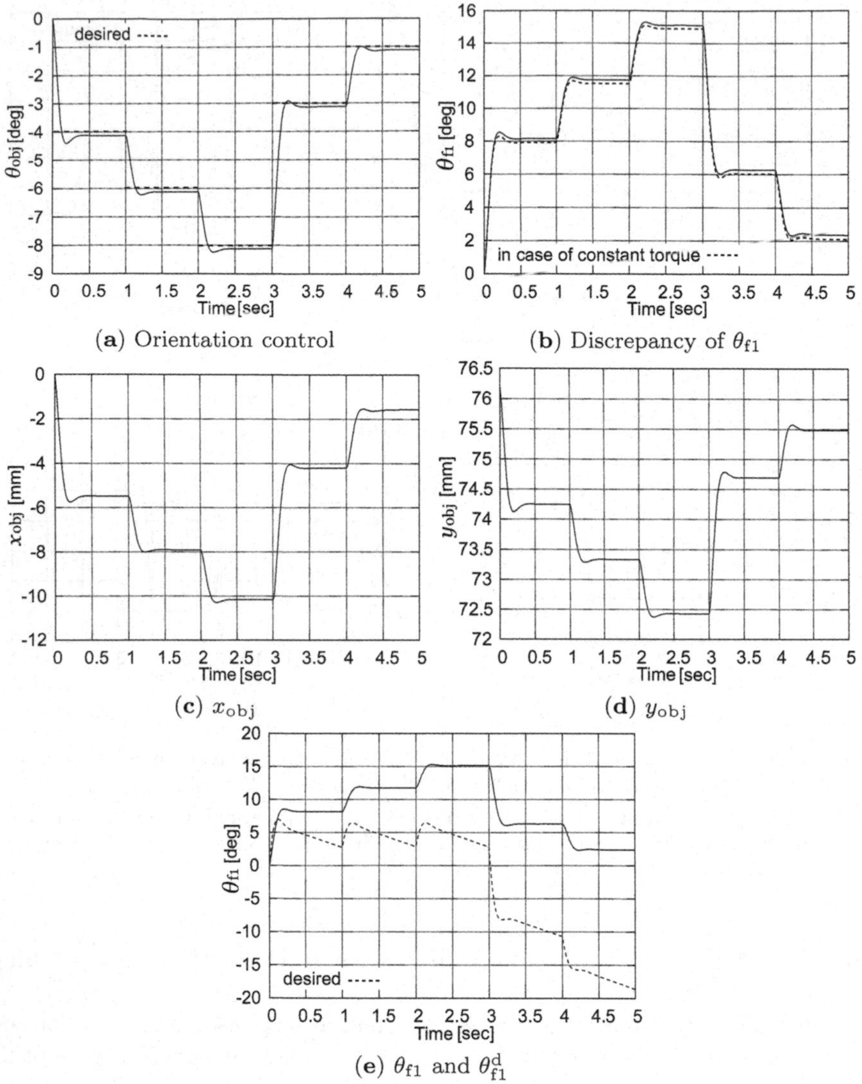

Fig. 10.28 The results of this simulation show significant steady-state deviations in θ_{obj}. Here, the biased torque τ_b increases continuously at the specified rate of 20 Nm/sec from 10 Nm to 110 Nm

Next, we consider a special case that examines position control using a robot having a 1 DOF revolute joint as well as the simulation procedure of Fig. 10.17. Recalling the original control strategy expressed as Eqs. 10.18 and 10.19, a modified control law in this case is then rewritten as

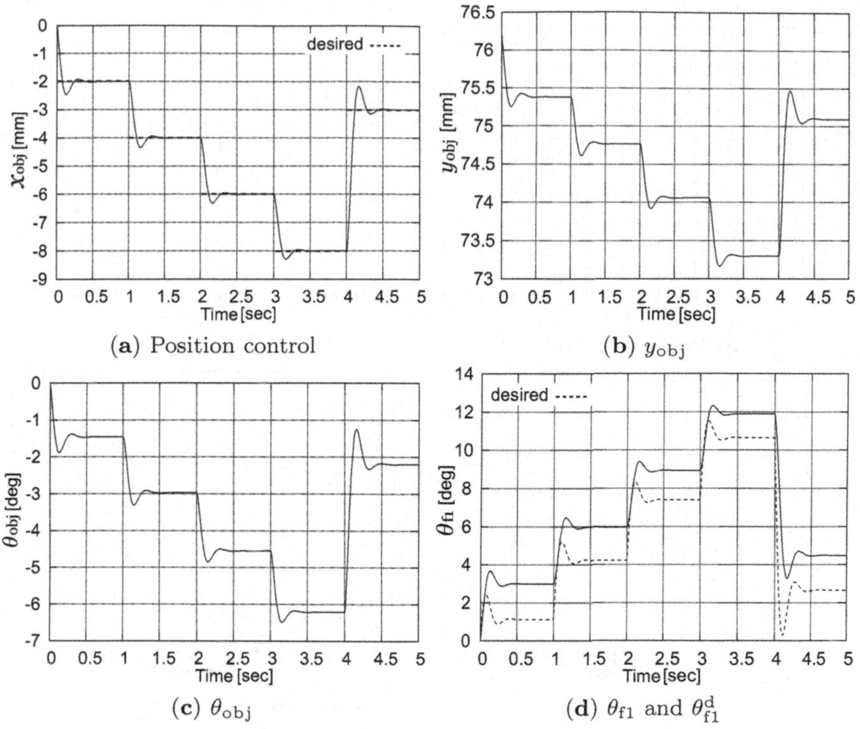

Fig. 10.29 The trajectory of x_{obj} is shown when the target variable to be controlled is substituted from θ_{obj} to x_{obj} of Eq. 10.18. This simulation is examined when constant biased torque is applied, $\tau_b = 10$ Nm

$$\theta_{f1}^{d} = K_I^x \int_0^t (x_{obj} - x_{obj}^{d}) d\tau, \tag{10.33}$$

$$u_1 = -K_P^{\theta}(\theta_{f1} - \theta_{f1}^{d}) - K_D^{\theta}\dot{\theta}_{f1} + \tau_b. \tag{10.34}$$

When a constant biased torque is applied to the system, the object position x_{obj} accurately converges to the desired value, as shown in Fig. 10.29a, and other object information (y_{obj}, θ_{obj}) is sent to individual equilibrium points, as shown in Figs. 10.29b and c. As stated previously, there exist *stable admissible deviations* on the trajectory of θ_{f1} shown in Fig. 10.29d. Note that we have determined a pattern of the desired object position that becomes consistently negative because the contact between the soft fingertip and the grasped object cannot be satisfied if the right finger rotates in the clockwise direction across 0° in the case of a 1-DOF rotational mechanism. In addition, we show the case in Fig. 10.30 where the biased torque varies. As in Fig. 10.28, the trajectory of x_{obj} contains a small but significant steady-state error in all time steps, as shown in Fig. 10.30a. At the same time, y_{obj} and θ_{obj} go to an LMEEwC in this case along with the movement of x_{obj}, as

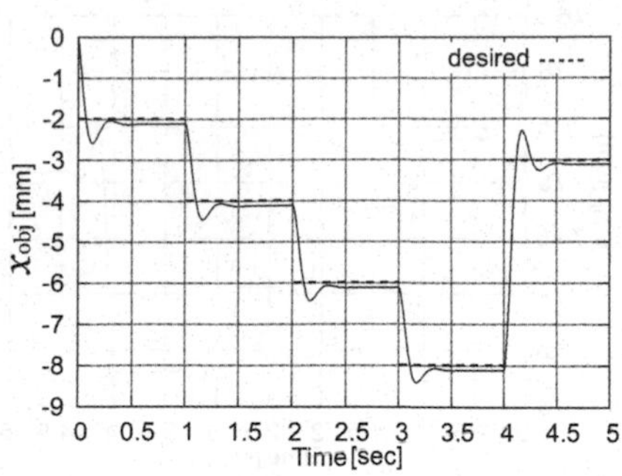

(**a**) Position control

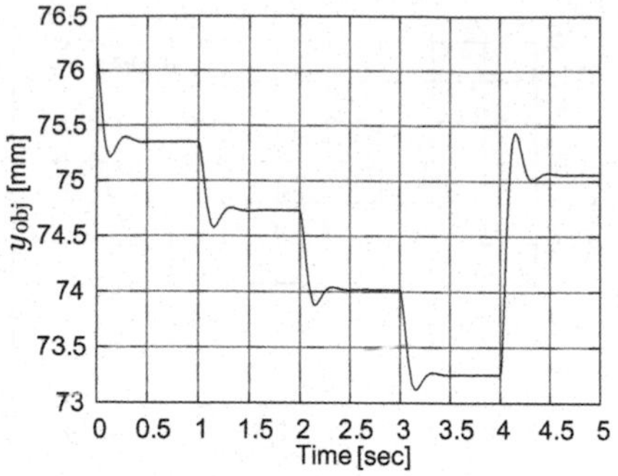

(**b**) y_{obj}

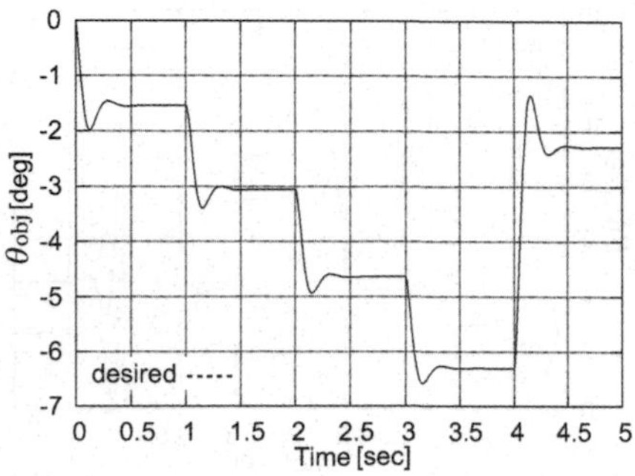

(**c**) θ_{obj}

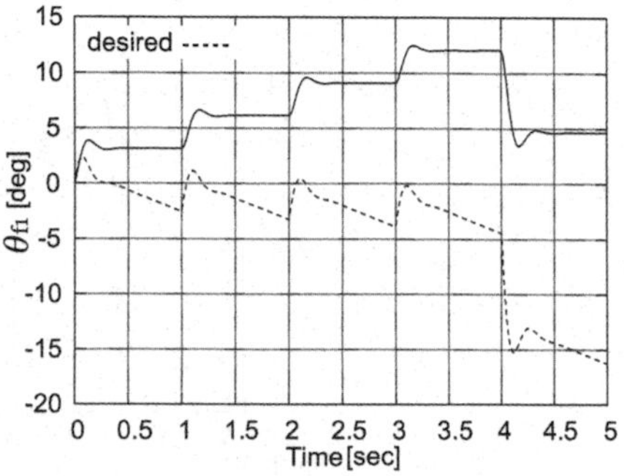

(**d**) θ_{f1} and θ_{f1}^{d}

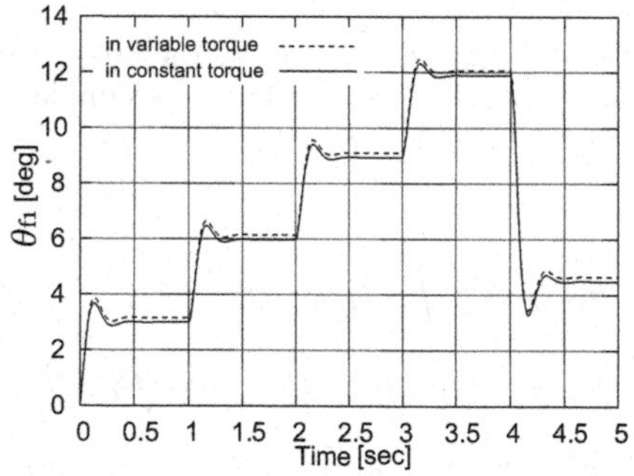

(**e**) Comparison of θ_{f1}

Fig. 10.30 The trajectory of x_{obj} is shown when the target variable to be controlled is substituted from θ_{obj} to x_{obj} of Eq. 10.18. This simulation is examined when τ_b increased at a rate of 20 Nm/sec every millisecond from the initial value listed in Table 10.1

shown in Figs. 10.30b and c. The virtually produced desired trajectory of θ_{f1} descends through the dynamic movement, as shown in Fig. 10.30d. Despite this, the stability of the object is not affected by the extremely large error becsue the continuous increase in τ_b is regulated and eliminated in an orderly fashion by the increase in discrepancy of θ_{f1}.

Table 10.2 Results and new findings for the number of degrees of freedom: ○ (possible), △ (possible but including slight error), × (impossible), P (prismatic), R (revolute)

No.	Structure	Stable pinch	Position	Orientation	Joint force	Joint torque
1	1-DOF P	(○)	○	–	Const	–
2		(○)	△	–	Increase	–
3	1-DOF R	(○)	–	○	–	Const
4		(○)	–	△	–	Increase
5		(○)	○	–	–	Const
6		(○)	△	–	–	Increase
7	2-DOF P+P	(○)	○	–	Const	–
8		(○)	○	–	Increase	–
9	2-DOF R+R	(○)	–	○	–	Const
10		(○)	–	○	–	Increase
11		(○)	○	–	–	Const
12		(○)	○	–	–	Increase
13		(○)	×	×	–	Const
14	2-DOF R+P	(○)	○	○	Const	Const
15		(○)	△	○	Increase	Const
16		(○)	○	△	Const	Increase
17		(○)	○*	○*	Increase	Increase

10.5 Observations and Discussions

The relationship between the mechanical degrees of freedom of a two-fingered robotic hand and the manipulation ability of that mechanism are summarized in Table 10.2. Soft-fingered manipulation enables a pinched object to remain stable due to the inherent stability induced by a certain physical equilibrium, which is defined here as LMEEwC. That is, since the stable pinching is always preserved in this case, we added brackets for the category of *stable pinching* in Table 10.2. The second and third categories of *position* and *orientation* indicate whether each task is achievable with no explicit error. In this table, the symbol △ indicates that the controlled variable converges but a significant steady-state error occurs. The symbol $*$ denotes that either the position or orientation of a grasped object can be selectively controlled by changing the relative rate of each of the integral and proportional gains between position control and orientation control. Therefore, the result described at the bottom of the table indicates that if either x_{obj} or θ_{obj} is successfully controlled, the remaining variable is not completely controlled and has some steady-state error that corresponds to △.

Table 10.2 shows that the rotational joint mechanism has advantages in the sense that both the position and orientation of the pinched object can be selectively controlled by choosing the target variable appropriately during manipulation movements. We defined this skillful ability in soft finger operations as *task-selection control.*

In both cases of the 1-DOF prismatic and revolute joint mechanisms, we notice that each target task is improved from a situation that has significant but small errors when stopping the increase of biased torque/force (Nos. 1–6). The human hand is usually not able to continue to increase pinching forces for individual tasks because the applied forces reach upper bound due to the performance limitation of the motor control system and the musculoskeletal structure. The biased torque that appears in the proposed control method therefore converges and becomes constant over time. As a result, the significant error of the target variable to be controlled vanishes, and then the corresponding task has been completed. In other words, in these mechanisms we are not able to precisely control information of the grasped object when implementing the continuous increase of pinching force.

On the other hand, in the case of the 2-DOF RR mechanism, each task of position and orientation control is completed despite whether the biased torque varies or remains constant (Nos. 9–12). The reason for this is that the rotational movement due to the revolute joint corresponds to a 2D motion on a Cartesian coordinate. That is, the motion of the 1-DOF revolute joint is constrained to a circular movement, resulting in a 1D motion on the polar coordinate because the radial motion of the joint does not occur. By contrast, when describing this motion on the Cartesian coordinate, both movements of x- and y-coordinates arise during the rotation of the joint. Therefore, by selecting either the prismatic movement or the rotational movement of a target object in the first-stage, the structure of revolute joint is capable of achieving each command such as the position and orientation of a grasped object. Eventually, both fingers actually display an inward-rotation movement while maintaining the object in a certain orientation when the biased torque increases. As a result, in this case a powerful pinching motion with large deformation of the soft fingers can be achieved while maintaining the target variable at a desired value through the manipulation process.

As in the case of the 1-DOF prismatic and revolute joint mechanisms, in the case of 2-DOF RP mechanism, each target task is improved from a situation including significant errors when stopping the increase of biased torque/force (Nos. 14–16). Eventually, the performance of this mechanism results in the simultaneous control of two target variables, which is described in the top row of this mechanism listed in Table 10.2 (No. 14). In particular, this mechanism is superior to any other mechanism because it is able to independently accomplish three individual tasks: *stable pinching*, *position control*, and *orientation control*. Soft-fingered manipulation has the ability to control more variables than the number of mechanical degrees of freedom in the case of a mechanism that consists of different types of joint structures.

In addition, the achievement of each task when the biased torque increases is only seen in the results of Nos. 10 and 12. This result indicates that a strong pinching motion can be achieved, that is, the magnitude of the pinching force can be arbitrarily changed by the increase/decrease in the biased torque. Thus, we can selectively choose either pinching force control or fine object control by converting the movement of the left finger between revolute and prismatic joints. Human fingers construct a multi-DOF structure that realizes flexion-extension, abduction-adduction, and pronation-supination movements, resulting in the production of complex motions and dexterous pinching performances. Therefore, the translational movement of the distal phalanx of the thumb, which faces normal to the finger cushion, can be generated by combining the complex movements of the musculoskeletal structure, as shown in Fig. 10.2. Likewise, the rotational movement of the distal phalanx can be readily produced by the structure. Hence, we are able to perform some sort of *task-selection control* in pinching tasks by the thumb and forefinger by switching the role of thumb movements.

10.6 Concluding Remarks

The present study has demonstrated that the biomimicry of the anatomical hand structure has the potential to drastically simplify the control architecture of robotic hands. Jacobian matrix, which is commonly accepted knowledge of manipulator control, had been utilized in describing robot kinematics, dynamics, and also the field of robot control because the matrix had been useful for the analysis for the static movement of robots over the past 30 years. However, only because of that extreme convenience, it has long been thought that the Jacobian matrix is computed within the brain and used even in human movement and motor control. But the compelling evidence of that proposition has not been given yet. The distinctive biomechanical configuration provides a certain role-sharing function that contains different mechanical joints. The present study provides some interesting perspectives based on several simulational observations. In particular, this chapter revealed that the softness and flexibility of robotic fingers are important and take on several roles for maintaining stable pinching motions. The present study concentrates on a very simple robotic finger structure having a hemispherical fingertip constructed of soft materials and consisting of a minimal-degree-of-freedom mechanism, which achieved high-performance humanlike pinching motions. In addition, for the first time we have shown the potential to eliminate the Jacobian matrix from control strategies and the dynamic systems of soft-fingered robotic hands.

In the future, we must formulate the dynamics of three- or four-link finger structures that internalize redundancy based on the musculoskeletal system. Finally, we should develop a more appropriate control scheme for fulfilling refined tasks and advanced performance by means of a multilinked two-fingered robotic hand.

Chapter 11
Three-dimensional Grasping and Manipulation

11.1 Introduction

This chapter focuses on 3D soft-fingered grasping and manipulation. We extend the 2D fingertip model described in Chap. 4 to the 2D fingertip model, taking 3D rolling contact into consideration.

11.2 Quaternions

This section reviews *quaternions*, which enable us to describe the rotation of rigid bodies without trigonometric functions [Kui02].

Let us derive the rotation around a unit vector $\boldsymbol{a} = [\,a_x,\, a_y,\, a_z\,]^{\mathrm{T}}$ by angle α. The unit vector $\boldsymbol{a}$ denotes the axis of rotation. Let $R(\boldsymbol{a}, \alpha)$ be the corresponding rotation matrix. An arbitrary vector $\boldsymbol{x}$ can be decomposed into two components $\boldsymbol{x}^{\parallel}$ and $\boldsymbol{x}^{\perp}$; the former is parallel to the unit vector $\boldsymbol{a}$, and the latter is perpendicular to the unit vector $\boldsymbol{a}$. Since $\boldsymbol{x}^{\parallel}$ is the projection of vector $\boldsymbol{x}$ to a line specified by unit vector $\boldsymbol{a}$, we have

$$\begin{aligned}\boldsymbol{x}^{\parallel} &= (\boldsymbol{a}^{\mathrm{T}}\boldsymbol{x})\boldsymbol{a} = (\boldsymbol{a}\boldsymbol{a}^{\mathrm{T}})\boldsymbol{x},\\ \boldsymbol{x}^{\perp} &= \boldsymbol{x} - \boldsymbol{x}^{\parallel} = (I - \boldsymbol{a}\boldsymbol{a}^{\mathrm{T}})\boldsymbol{x}.\end{aligned}$$

Note that vectors $\boldsymbol{a}$, $\boldsymbol{x}^{\perp}$, and $\boldsymbol{a} \times \boldsymbol{x}^{\perp}$ are orthogonal to one another and form a right-handed system. Since $\boldsymbol{x}^{\parallel}$ is parallel to $\boldsymbol{a}$, rotation $R(\boldsymbol{a}, \alpha)$ transforms $\boldsymbol{x}^{\parallel}$ into itself:

$$R\boldsymbol{x}^{\parallel} = \boldsymbol{x}^{\parallel}.$$

Any vector involved in a plane composed of $\boldsymbol{x}^{\perp}$ and $\boldsymbol{a} \times \boldsymbol{x}^{\perp}$ rotates in this plane. In particular, $R(\boldsymbol{a}, \alpha)$ transforms $\boldsymbol{x}^{\perp}$ into $C_\alpha \boldsymbol{x}^{\perp} + S_\alpha \boldsymbol{a} \times \boldsymbol{x}^{\perp}$:

$$R\boldsymbol{x}^{\perp} = C_{\alpha}\boldsymbol{x}^{\perp} + S_{\alpha}\boldsymbol{a} \times \boldsymbol{x}^{\perp},$$

where $C_{\alpha} = \cos\alpha$ and $S_{\alpha} = \sin\alpha$. As a result, vector $\boldsymbol{x}$ is transformed as follows:

$$R\boldsymbol{x} = R(\boldsymbol{x}^{\parallel} + \boldsymbol{x}^{\perp}) = \boldsymbol{x}^{\parallel} + C_{\alpha}\boldsymbol{x}^{\perp} + S_{\alpha}\boldsymbol{a} \times \boldsymbol{x}^{\perp}.$$

Since $\boldsymbol{a} \times \boldsymbol{x}^{\parallel} = 0$, we have

$$\boldsymbol{a} \times \boldsymbol{x}^{\perp} = \boldsymbol{a} \times (\boldsymbol{x}^{\parallel} + \boldsymbol{x}^{\perp}) = \boldsymbol{a} \times \boldsymbol{x} = \left[\,\boldsymbol{a}\times\,\right]\boldsymbol{x},$$

where skew-symmetric matrix $[\,\boldsymbol{a}\times\,]$ describes the outer product:

$$\left[\,\boldsymbol{a}\times\,\right] = \begin{bmatrix} 0 & -a_z & a_y \\ a_z & 0 & -a_x \\ -a_y & a_x & 0 \end{bmatrix}.$$

From the above equations we have

$$\begin{aligned} R\boldsymbol{x} &= (\boldsymbol{a}\boldsymbol{a}^{\mathrm{T}})\boldsymbol{x} + C_{\alpha}(I - \boldsymbol{a}\boldsymbol{a}^{\mathrm{T}})\boldsymbol{x} + S_{\alpha}\left[\,\boldsymbol{a}\times\,\right]\boldsymbol{x} \\ &= \left\{C_{\alpha}I + (1 - C_{\alpha})\boldsymbol{a}\boldsymbol{a}^{\mathrm{T}} + S_{\alpha}\left[\,\boldsymbol{a}\times\,\right]\right\}\boldsymbol{x}. \end{aligned}$$

Thus,

$$R(\boldsymbol{a}, \alpha) = C_{\alpha}I + (1 - C_{\alpha})\boldsymbol{a}\boldsymbol{a}^{\mathrm{T}} + S_{\alpha}\left[\,\boldsymbol{a}\times\,\right].$$

Consequently, we have

$$R(\boldsymbol{a}, \alpha) = \begin{bmatrix} C_{\alpha} + \overline{C}_{\alpha}a_x^2 & \overline{C}_{\alpha}a_x a_y - S_{\alpha}a_z & \overline{C}_{\alpha}a_x a_z + S_{\alpha}a_y \\ \overline{C}_{\alpha}a_y a_x + S_{\alpha}a_z & C_{\alpha} + \overline{C}_{\alpha}a_y^2 & \overline{C}_{\alpha}a_y a_z - S_{\alpha}a_x \\ \overline{C}_{\alpha}a_z a_x - S_{\alpha}a_y & \overline{C}_{\alpha}a_z a_y + S_{\alpha}a_x & C_{\alpha} + \overline{C}_{\alpha}a_z^2 \end{bmatrix}, \quad (11.1)$$

where $\overline{C}_{\alpha} = 1 - C_{\alpha}$. The above matrix describes the rotation around a specified unit vector. The time derivative of matrix $R(\boldsymbol{a}, \alpha)$ is given by

$$\frac{\mathrm{d}R}{\mathrm{d}t}(\boldsymbol{a}, \alpha) = \frac{\partial R}{\partial \alpha}\dot{\alpha} + \frac{\partial R}{\partial a_x}\dot{a}_x + \frac{\partial R}{\partial a_y}\dot{a}_y + \frac{\partial R}{\partial a_z}\dot{a}_z,$$

where

$$\frac{\partial R}{\partial \alpha} = \begin{bmatrix} (a_x^2 - 1)S_{\alpha} & S_{\alpha}a_x a_y - C_{\alpha}a_z & S_{\alpha}a_x a_z + C_{\alpha}a_y \\ S_{\alpha}a_y a_x + C_{\alpha}a_z & (a_y^2 - 1)S_{\alpha} & S_{\alpha}a_y a_z - C_{\alpha}a_x \\ S_{\alpha}a_z a_x - C_{\alpha}a_y & S_{\alpha}a_z a_y + C_{\alpha}a_x & (a_z^2 - 1)S_{\alpha} \end{bmatrix}$$

and

$$\frac{\partial R}{\partial a_x} = \begin{bmatrix} 2\overline{C}_\alpha a_x & \overline{C}_\alpha a_y & \overline{C}_\alpha a_z \\ \overline{C}_\alpha a_y & 0 & -S_\alpha \\ \overline{C}_\alpha a_z & S_\alpha & 0 \end{bmatrix},$$

$$\frac{\partial R}{\partial a_y} = \begin{bmatrix} 0 & \overline{C}_\alpha a_x & S_\alpha \\ \overline{C}_\alpha a_x & 2\overline{C}_\alpha a_y & \overline{C}_\alpha a_z \\ -S_\alpha & \overline{C}_\alpha a_z & 0 \end{bmatrix},$$

$$\frac{\partial R}{\partial a_z} = \begin{bmatrix} 0 & -S_\alpha & \overline{C}_\alpha a_x \\ S_\alpha & 0 & \overline{C}_\alpha a_y \\ \overline{C}_\alpha a_x & \overline{C}_\alpha a_y & 2\overline{C}_\alpha a_z \end{bmatrix}.$$

Note that when vector $\boldsymbol{a}$ is constant, that is, the axis of rotation is fixed, $\dot{R} = (\partial R/\partial \alpha)\dot{\alpha}$.

The above matrix includes trigonometric functions. Let us simplify the above description. First, let us define $q_0 = \cos(\alpha/2)$, which yields

$$C_\alpha = 2q_0^2 - 1, \quad \overline{C}_\alpha = 2\sin^2\frac{\alpha}{2}, \quad S_\alpha = 2q_0 \sin\frac{\alpha}{2}.$$

Let us then introduce $[\, q_1,\, q_2,\, q_3 \,]^{\mathrm{T}} = -\sin(\alpha/2)[\, a_x,\, a_y,\, a_z \,]^{\mathrm{T}}$, which yields

$$R(\boldsymbol{q}) = \begin{bmatrix} 2(q_0^2+q_1^2)-1 & 2(q_1q_2+q_0q_3) & 2(q_1q_3-q_0q_2) \\ 2(q_1q_2-q_0q_3) & 2(q_0^2+q_2^2)-1 & 2(q_2q_3+q_0q_1) \\ 2(q_1q_3+q_0q_2) & 2(q_2q_3-q_0q_1) & 2(q_0^2+q_3^2)-1 \end{bmatrix}, \tag{11.2}$$

where $\boldsymbol{q} = [\, q_0,\, q_1,\, q_2,\, q_3 \,]^{\mathrm{T}}$. Note that the four parameters q_0 through q_3 must satisfy the following constraint:

$$Q \stackrel{\triangle}{=} q_0^2 + q_1^2 + q_2^2 + q_3^2 - 1 = 0. \tag{11.3}$$

A set of the four parameters is referred to as a *quaternion*. This description excludes trigonometric functions. All elements in rotation matrix $R(\boldsymbol{q})$ and constraint Q are quadratic forms of quaternion elements q_0 through q_3.

Let us compute the angular velocity described in quaternions form. Computing the time derivative of constraint $Q = 0$, we have

$$2(q_0\dot{q}_0 + q_1\dot{q}_1 + q_2\dot{q}_2 + q_3\dot{q}_3) = 0.$$

Let $C{-}\xi\eta\zeta$ be a coordinate system attached to a rigid body. Let $\boldsymbol{\xi}$, $\boldsymbol{\eta}$, and $\boldsymbol{\zeta}$ be unit vectors along the ξ-, η-, and ζ-axes. The first, second, and third columns of rotation matrix R coincide with the three unit vectors: $R = [\, \boldsymbol{\xi} \,|\, \boldsymbol{\eta} \,|\, \boldsymbol{\zeta} \,]$. Let $\boldsymbol{\omega} = [\, \omega_\xi,\, \omega_\eta,\, \omega_\zeta \,]^{\mathrm{T}}$ be the angular velocity vector. Since the angular velocity vector is defined as $[\, \boldsymbol{\omega}\times \,] = \dot{R}R^{\mathrm{T}}$, we have the following equations:

$$\omega_\xi = 2(q_1\dot{q}_0 - q_0\dot{q}_1 + q_3\dot{q}_2 - q_2\dot{q}_3), \tag{11.4}$$
$$\omega_\eta = 2(q_2\dot{q}_0 - q_3\dot{q}_1 - q_0\dot{q}_2 + q_1\dot{q}_3), \tag{11.5}$$
$$\omega_\zeta = 2(q_3\dot{q}_0 + q_2\dot{q}_1 - q_1\dot{q}_2 - q_0\dot{q}_3). \tag{11.6}$$

The above four equations are described in vector form as follows:

$$2\begin{bmatrix} q_0 & q_1 & q_2 & q_3 \\ q_1 & -q_0 & q_3 & -q_2 \\ q_2 & -q_3 & -q_0 & q_1 \\ q_3 & q_2 & -q_1 & -q_0 \end{bmatrix} \begin{bmatrix} \dot{q}_0 \\ \dot{q}_1 \\ \dot{q}_2 \\ \dot{q}_3 \end{bmatrix} = \begin{bmatrix} 0 \\ \omega_\xi \\ \omega_\eta \\ \omega_\zeta \end{bmatrix}. \tag{11.7}$$

Let $\boldsymbol{h}_0^{\mathrm{T}}$ through $\boldsymbol{h}_3^{\mathrm{T}}$ be the 0th through 3rd row vectors of the coefficient matrix in Eq. 11.7. The relationship between $\boldsymbol{\omega}$ and $\dot{\boldsymbol{q}}$ is given by $\boldsymbol{\omega} = H\dot{\boldsymbol{q}}$, where

$$H \triangleq \begin{bmatrix} \boldsymbol{h}_1^{\mathrm{T}} \\ \boldsymbol{h}_2^{\mathrm{T}} \\ \boldsymbol{h}_3^{\mathrm{T}} \end{bmatrix} = 2\begin{bmatrix} q_1 & -q_0 & q_3 & -q_2 \\ q_2 & -q_3 & -q_0 & q_1 \\ q_3 & q_2 & -q_1 & -q_0 \end{bmatrix}.$$

Let I be the inertia matrix of a rigid body. The rotational kinetic energy of the rigid body is then described as follows:

$$K_{\mathrm{rot}} = \frac{1}{2}\boldsymbol{\omega}^{\mathrm{T}} I \boldsymbol{\omega} = \frac{1}{2}\dot{\boldsymbol{q}}^{\mathrm{T}} J\, \dot{\boldsymbol{q}},$$

where

$$J \triangleq H^{\mathrm{T}} I H = \begin{bmatrix} \boldsymbol{h}_1 \ \boldsymbol{h}_2 \ \boldsymbol{h}_3 \end{bmatrix} \begin{bmatrix} I_\xi & I_{\xi\eta} & I_{\xi\zeta} \\ I_{\eta\xi} & I_\eta & I_{\eta\zeta} \\ I_{\zeta\xi} & I_{\zeta\eta} & I_\zeta \end{bmatrix} \begin{bmatrix} \boldsymbol{h}_1^{\mathrm{T}} \\ \boldsymbol{h}_2^{\mathrm{T}} \\ \boldsymbol{h}_3^{\mathrm{T}} \end{bmatrix} \tag{11.8}$$

is a 4×4 matrix. Letting $J_{m,n}$ be the (m,n)th element of matrix J, we have

$$K_{\mathrm{rot}} = \frac{1}{2}\sum_{m=0}^{3}\sum_{n=0}^{3} J_{m,n}\, \dot{q}_m \dot{q}_n.$$

Since matrix J depends on quaternion elements q_0 through q_3 but not on their derivatives $\dot{q}_0$ through $\dot{q}_3$, we have

$$\frac{\partial K_{\mathrm{rot}}}{\partial q_i} = \frac{1}{2}\sum_{m=0}^{3}\sum_{n=0}^{3} \frac{\partial J}{\partial q_i}\dot{q}_m \dot{q}_n,$$
$$\frac{\partial K_{\mathrm{rot}}}{\partial \dot{q}_i} = \sum_{n=0}^{3} J_{i,n}\, \dot{q}_n, \qquad (i = 0, 1, 2, 3).$$

The contribution of kinetic energy K_{rot} to a set of Lagrange equations of motion with respect to q_0 through q_3 is thus given by

$$\frac{\partial K_{\rm rot}}{\partial q_i} - \frac{\rm d}{{\rm d}t}\frac{\partial K_{\rm rot}}{\partial \dot{q}_i} = \frac{1}{2}\sum_{m=0}^{3}\sum_{n=0}^{3}\frac{\partial J_{m,n}}{\partial q_i}\dot{q}_m\dot{q}_n - \sum_{m=0}^{3}\sum_{n=0}^{3}\frac{\partial J_{i,n}}{\partial q_m}\dot{q}_m\dot{q}_n - \sum_{n=0}^{3} J_{i,n}\ddot{q}_n, \quad (i = 0, 1, 2, 3).$$

This contribution is described by

$$\sum_{n=0}^{3} W_{i,n}\dot{q}_n + \sum_{n=0}^{3} Y_{i,n}\dot{q}_n - \sum_{n=0}^{3} J_{i,n}\ddot{q}_n, \tag{11.9}$$

where

$$W_{i,n} = \frac{1}{2}\left\{\sum_{m=0}^{3}\left(\frac{\partial J_{m,n}}{\partial q_i} - \frac{\partial J_{i,m}}{\partial q_n}\right)\dot{q}_m\right\}, \tag{11.10}$$

$$Y_{i,n} = \frac{1}{2}\left\{-\sum_{m=0}^{3}\frac{\partial J_{i,n}}{\partial q_m}\dot{q}_m\right\}. \tag{11.11}$$

Since $W_{i,n} + W_{n,i}$ vanishes, the matrix $W_{\rm quat}$, the (i,n)th element of which is given by $W_{i,n}$, is skew-symmetric. Since $Y_{i,n}$ corresponds to $Y_{n,i}$, the matrix $Y_{\rm quat}$, the (i,n)th element of which is given by $Y_{i,n}$, is symmetric. The contribution of kinetic energy $K_{\rm rot}$ is then summarized in vector form as follows:

$$\frac{\partial K_{\rm rot}}{\partial \boldsymbol{q}} - \frac{\rm d}{{\rm d}t}\frac{\partial K_{\rm rot}}{\partial \dot{\boldsymbol{q}}} = W_{\rm quat}\dot{\boldsymbol{q}} + Y_{\rm quat}\dot{\boldsymbol{q}} - J\ddot{\boldsymbol{q}}. \tag{11.12}$$

The Lagrangian of the rotation of a rigid body is then formulated as

$$\mathcal{L} = K_{\rm rot} + \lambda_{\rm Q} Q. \tag{11.13}$$

The Lagrange equations of motion with respect to quaternion $\boldsymbol{q}$ are described as

$$\frac{\rm d}{{\rm d}t}\frac{\partial \mathcal{L}}{\partial \dot{\boldsymbol{q}}} - \frac{\partial \mathcal{L}}{\partial \boldsymbol{q}} = 0, \tag{11.14}$$

yielding

$$J\ddot{\boldsymbol{q}} - \boldsymbol{h}_0\lambda_{\rm Q} = W_{\rm quat}\dot{\boldsymbol{q}} + Y_{\rm quat}\dot{\boldsymbol{q}}. \tag{11.15}$$

Stabilization of the holonomic constraint Q is given by

$$\ddot{Q} + 2\alpha\dot{Q} + \alpha^2 Q = 0. \tag{11.16}$$

Since $\dot{Q} = \boldsymbol{h}_0^{\rm T}\dot{\boldsymbol{q}}$ and $\ddot{Q} = \boldsymbol{h}_0^{\rm T}\ddot{q} + 2\dot{\boldsymbol{q}}^{\rm T}\dot{\boldsymbol{q}}$, we have

Table 11.1 Partial derivatives of R with respect to q_i

	$\partial R/\partial q_i$
$i=0$	$\begin{bmatrix} 4q_0 & 2q_3 & -2q_2 \\ -2q_3 & 4q_0 & 2q_1 \\ 2q_2 & -2q_1 & 4q_0 \end{bmatrix}$
$i=1$	$\begin{bmatrix} 4q_1 & 2q_2 & 2q_3 \\ 2q_2 & 0 & 2q_0 \\ 2q_3 & -2q_0 & 0 \end{bmatrix}$
$i=2$	$\begin{bmatrix} 0 & 2q_1 & -2q_0 \\ 2q_1 & 4q_2 & 2q_3 \\ 2q_0 & 2q_3 & 0 \end{bmatrix}$
$i=3$	$\begin{bmatrix} 0 & 2q_0 & 2q_1 \\ -2q_0 & 0 & 2q_2 \\ 2q_1 & 2q_2 & 4q_3 \end{bmatrix}$

Table 11.2 Second-order partial derivatives of R with respect to q_i and q_j

	$j=0$	$j=1$	$j=2$	$j=3$
$i=0$	$\begin{bmatrix} 4 & 0 & 0 \\ 0 & 4 & 0 \\ 0 & 0 & 4 \end{bmatrix}$	$\begin{bmatrix} 0 & 0 & 0 \\ 0 & 0 & 2 \\ 0 & -2 & 0 \end{bmatrix}$	$\begin{bmatrix} 0 & 0 & -2 \\ 0 & 0 & 0 \\ 2 & 0 & 0 \end{bmatrix}$	$\begin{bmatrix} 0 & 2 & 0 \\ -2 & 0 & 0 \\ 0 & 0 & 0 \end{bmatrix}$
$i=1$	$\begin{bmatrix} 0 & 0 & 0 \\ 0 & 0 & 2 \\ 0 & -2 & 0 \end{bmatrix}$	$\begin{bmatrix} 4 & 0 & 0 \\ 0 & 0 & 0 \\ 0 & 0 & 0 \end{bmatrix}$	$\begin{bmatrix} 0 & 2 & 0 \\ 2 & 0 & 0 \\ 0 & 0 & 0 \end{bmatrix}$	$\begin{bmatrix} 0 & 0 & 2 \\ 0 & 0 & 0 \\ 2 & 0 & 0 \end{bmatrix}$
$i=2$	$\begin{bmatrix} 0 & 0 & -2 \\ 0 & 0 & 0 \\ 2 & 0 & 0 \end{bmatrix}$	$\begin{bmatrix} 0 & 2 & 0 \\ 2 & 0 & 0 \\ 0 & 0 & 0 \end{bmatrix}$	$\begin{bmatrix} 0 & 0 & 0 \\ 0 & 4 & 0 \\ 0 & 0 & 0 \end{bmatrix}$	$\begin{bmatrix} 0 & 0 & 0 \\ 0 & 0 & 2 \\ 0 & 2 & 0 \end{bmatrix}$
$i=3$	$\begin{bmatrix} 0 & 2 & 0 \\ -2 & 0 & 0 \\ 0 & 0 & 0 \end{bmatrix}$	$\begin{bmatrix} 0 & 0 & 2 \\ 0 & 0 & 0 \\ 2 & 0 & 0 \end{bmatrix}$	$\begin{bmatrix} 0 & 0 & 0 \\ 0 & 0 & 2 \\ 0 & 2 & 0 \end{bmatrix}$	$\begin{bmatrix} 0 & 0 & 0 \\ 0 & 0 & 0 \\ 0 & 0 & 4 \end{bmatrix}$

$$-\boldsymbol{h}_0^{\mathrm{T}}\ddot{\boldsymbol{q}} = 2\dot{\boldsymbol{q}}^{\mathrm{T}}\dot{\boldsymbol{q}} + 2\alpha\boldsymbol{h}_0^{\mathrm{T}}\dot{\boldsymbol{q}} + \alpha^2 Q. \tag{11.17}$$

Combining the above two equations, we have

$$\begin{bmatrix} J & -\boldsymbol{h}_0 \\ -\boldsymbol{h}_0^{\mathrm{T}} & \end{bmatrix}\begin{bmatrix} \ddot{\boldsymbol{q}} \\ \lambda_{\mathrm{Q}} \end{bmatrix} = \begin{bmatrix} W_{\mathrm{quat}}\dot{\boldsymbol{q}} + Y_{\mathrm{quat}}\dot{\boldsymbol{q}} \\ 2\dot{\boldsymbol{q}}^{\mathrm{T}}\dot{\boldsymbol{q}} + 2\alpha\boldsymbol{h}_0^{\mathrm{T}}\dot{\boldsymbol{q}} + \alpha^2 Q \end{bmatrix}. \tag{11.18}$$

Note that the coefficient matrix in the above equation is regular, even though matrix J is not regular.

Partial derivatives of rotation matrix $R(\boldsymbol{q})$ with respect to q_0 through q_3 are listed in Table 11.1. Second-order partial derivatives of rotation matrix $R(\boldsymbol{q})$ with respect to q_0 through q_3 are shown in Table 11.2. Partial derivatives

Table 11.3 Partial derivatives of $[\boldsymbol{\omega}\times]$ with respect to q_i and $\dot{q}_i$

	$\partial[\boldsymbol{\omega}\times]/\partial q_i$	$\partial[\boldsymbol{\omega}\times]/\partial \dot{q}_i$
$i=0$	$\begin{bmatrix} 0 & 2\dot{q}_3 & -2\dot{q}_2 \\ -2\dot{q}_3 & 0 & 2\dot{q}_1 \\ 2\dot{q}_2 & -2\dot{q}_1 & 0 \end{bmatrix}$	$\begin{bmatrix} 0 & -2q_3 & 2q_2 \\ 2q_3 & 0 & -2q_1 \\ -2q_2 & 2q_1 & 0 \end{bmatrix}$
$i=1$	$\begin{bmatrix} 0 & 2\dot{q}_2 & 2\dot{q}_3 \\ -2\dot{q}_2 & 0 & -2\dot{q}_0 \\ -2\dot{q}_3 & 2\dot{q}_0 & 0 \end{bmatrix}$	$\begin{bmatrix} 0 & -2q_2 & -2q_3 \\ 2q_2 & 0 & 2q_0 \\ 2q_3 & -2q_0 & 0 \end{bmatrix}$
$i=2$	$\begin{bmatrix} 0 & -2\dot{q}_1 & 2\dot{q}_0 \\ 2\dot{q}_1 & 0 & 2\dot{q}_3 \\ -2\dot{q}_0 & -2\dot{q}_3 & 0 \end{bmatrix}$	$\begin{bmatrix} 0 & 2q_1 & -2q_0 \\ -2q_1 & 0 & -2q_3 \\ 2q_0 & 2q_3 & 0 \end{bmatrix}$
$i=3$	$\begin{bmatrix} 0 & -2\dot{q}_0 & -2\dot{q}_1 \\ 2\dot{q}_0 & 0 & -2\dot{q}_2 \\ 2\dot{q}_1 & 2\dot{q}_2 & 0 \end{bmatrix}$	$\begin{bmatrix} 0 & 2q_0 & 2q_1 \\ -2q_0 & 0 & 2q_2 \\ -2q_1 & -2q_2 & 0 \end{bmatrix}$

of matrix $[\boldsymbol{\omega}\times]$ with respect to quaternion elements q_0 through q_3 and their time derivatives $\dot{q}_0$ through $\dot{q}_3$ are listed in Table 11.3. (See Appendix B.1.)

Assume that the ξ-, η-, and ζ-axes are the principal axes of a rigid object. Then, matrix I is diagonal, say, $I = \mathrm{diag}\{I_\xi, I_\eta, I_\zeta\}$. This yields

$$J = I_\xi H_\xi + I_\eta H_\eta + I_\zeta H_\zeta, \tag{11.19}$$

where

$$H_\xi = \boldsymbol{h}_1\boldsymbol{h}_1^{\mathrm{T}} = 4\begin{bmatrix} q_1q_1 & -q_0q_1 & q_1q_3 & -q_1q_2 \\ -q_0q_1 & q_0q_0 & -q_0q_3 & q_0q_2 \\ q_1q_3 & -q_0q_3 & q_3q_3 & -q_2q_3 \\ -q_1q_2 & q_0q_2 & -q_2q_3 & q_2q_2 \end{bmatrix},$$

$$H_\eta = \boldsymbol{h}_2\boldsymbol{h}_2^{\mathrm{T}} = 4\begin{bmatrix} q_2q_2 & -q_2q_3 & -q_0q_2 & q_1q_2 \\ -q_2q_3 & q_3q_3 & q_0q_3 & -q_1q_3 \\ -q_0q_2 & q_0q_3 & q_0q_0 & -q_0q_1 \\ q_1q_2 & -q_1q_3 & -q_0q_1 & q_1q_1 \end{bmatrix},$$

$$H_\zeta = \boldsymbol{h}_3\boldsymbol{h}_3^{\mathrm{T}} = 4\begin{bmatrix} q_3q_3 & q_2q_3 & -q_1q_3 & -q_0q_3 \\ q_2q_3 & q_2q_2 & -q_1q_2 & -q_0q_2 \\ -q_1q_3 & -q_1q_2 & q_1q_1 & q_0q_1 \\ -q_0q_3 & -q_0q_2 & q_0q_1 & q_0q_0 \end{bmatrix},$$

and

$$\frac{\partial J}{\partial q_0} = 4 \begin{bmatrix} 0 & -I_\xi q_1 & -I_\eta q_2 & -I_\zeta q_3 \\ -I_\xi q_1 & 2I_\xi q_0 & -I_\xi q_3 + I_\eta q_3 & I_\xi q_2 - I_\zeta q_2 \\ -I_\eta q_2 & I_\eta q_3 - I_\xi q_3 & 2I_\eta q_0 & -I_\eta q_1 + I_\zeta q_1 \\ -I_\zeta q_3 & -I_\zeta q_2 + I_\xi q_2 & I_\zeta q_1 - I_\eta q_1 & 2I_\zeta q_0 \end{bmatrix}, \quad (11.20)$$

$$\frac{\partial J}{\partial q_1} = 4 \begin{bmatrix} 2I_\xi q_1 & -I_\xi q_0 & I_\xi q_3 - I_\zeta q_3 & -I_\xi q_2 + I_\eta q_2 \\ -I_\xi q_0 & 0 & -I_\zeta q_2 & -I_\eta q_3 \\ -I_\zeta q_3 + I_\xi q_3 & -I_\zeta q_2 & 2I_\zeta q_1 & I_\zeta q_0 - I_\eta q_0 \\ I_\eta q_2 - I_\xi q_2 & -I_\eta q_3 & -I_\eta q_0 + I_\zeta q_0 & 2I_\eta q_1 \end{bmatrix}, \quad (11.21)$$

$$\frac{\partial J}{\partial q_2} = 4 \begin{bmatrix} 2I_\eta q_2 & -I_\eta q_3 + I_\zeta q_3 & -I_\eta q_0 & I_\eta q_1 - I_\xi q_1 \\ I_\zeta q_3 - I_\eta q_3 & 2I_\zeta q_2 & -I_\zeta q_1 & -I_\zeta q_0 + I_\xi q_0 \\ -I_\eta q_0 & -I_\zeta q_1 & 0 & -I_\xi q_3 \\ -I_\xi q_1 + I_\eta q_1 & I_\xi q_0 - I_\zeta q_0 & -I_\xi q_3 & 2I_\xi q_2 \end{bmatrix}, \quad (11.22)$$

$$\frac{\partial J}{\partial q_3} = 4 \begin{bmatrix} 2I_\zeta q_3 & I_\zeta q_2 - I_\eta q_2 & -I_\zeta q_1 + I_\xi q_1 & -I_\zeta q_0 \\ -I_\eta q_2 + I_\zeta q_2 & 2I_\eta q_3 & I_\eta q_0 - I_\xi q_0 & -I_\eta q_1 \\ I_\xi q_1 - I_\zeta q_1 & -I_\xi q_0 + I_\eta q_0 & 2I_\xi q_3 & -I_\xi q_2 \\ -I_\zeta q_0 & -I_\eta q_1 & -I_\xi q_2 & 0 \end{bmatrix}. \quad (11.23)$$

Based on the above equations, we can describe the equations of rotation of a rigid body by a quaternion.

11.3 Spatial Geometric Constraints Between an Object and a Fingertip

This section formulates normal and tangential constraints between a soft fingertip and a rigid object in 3D space, based on rolling contact kinematics in [MLS94] and rolling contact between a soft fingertip and a rigid body in [YAB$^+$06, YAB08]. A hemispherical soft fingertip of radius a is in contact with a planar surface S of a rigid object, as shown in Fig. 11.1. Let $\Sigma^{\rm obj}$ be a coordinate system attached to the object, and let $\Sigma^{\rm fin}$ be a coordinate system attached to the finger. Let $\Sigma^{\rm spc}$ be a system fixed in space. Let O be the origin of $\Sigma^{\rm spc}$, and let G be the origin of $\Sigma^{\rm obj}$. The position of the object is given by position vector $\boldsymbol{x}_{\rm obj}$ and its orientation is described by rotation matrix $R_{\rm obj}$. Let $\boldsymbol{n}^{\rm obj}$ be the outward normal vector of plane S described in the object coordinate system. The outward normal vector is then described in the spatial coordinate system as $\boldsymbol{n} = R_{\rm obj}\, \boldsymbol{n}^{\rm obj}$. Let $\boldsymbol{u}^{\rm obj}$ and $\boldsymbol{v}^{\rm obj}$ be tangential vectors along the surface described in the object coordinate system. Assume that $\boldsymbol{n}^{\rm obj}$, $\boldsymbol{u}^{\rm obj}$, and $\boldsymbol{v}^{\rm obj}$ form a right-handed coordinate system. That is, $\boldsymbol{n}^{\rm obj}$, $\boldsymbol{u}^{\rm obj}$, and $\boldsymbol{v}^{\rm obj}$ are unit vectors that are orthogonal to one another. The tangential vectors are then described in the spatial coordinate system as $\boldsymbol{u} = R_{\rm obj}\, \boldsymbol{u}^{\rm obj}$ and $\boldsymbol{v} = R_{\rm obj}\, \boldsymbol{v}^{\rm obj}$. Let Q be a point on planar surface S, and let $\boldsymbol{\xi}_{\rm Q}$ be its position in the object coordinate system. The spatial position of point Q is given by $\boldsymbol{x}_{\rm Q} = R_{\rm obj}\, \boldsymbol{\xi}_{\rm Q} + \boldsymbol{x}_{\rm obj}$. Any point on the planar surface

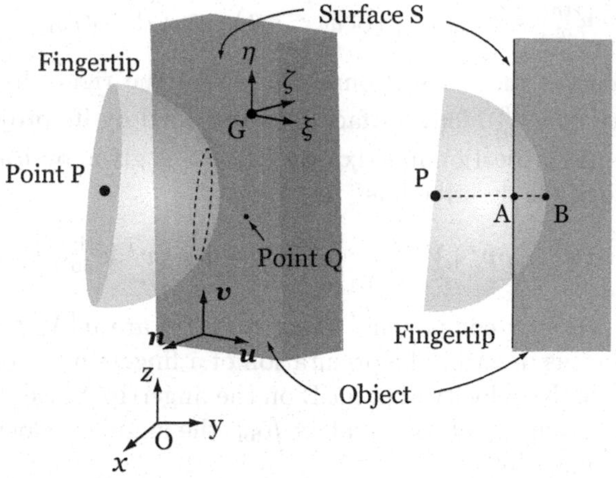

Fig. 11.1 Contact between a fingertip and a planar surface in 3D space

must then satisfy the following equation:

$$\boldsymbol{n}^{\mathrm{T}}(\boldsymbol{x} - \boldsymbol{x}_{\mathrm{Q}}) = 0.$$

Let P be the center of the hemispherical fingertip, and let A be its perpendicular projection on surface S. Let $\boldsymbol{x}_{\mathrm{fin}}$ be the position of point P. Let d_{n} be the maximum normal displacement of the hemispherical soft fingertip. The positional vector of point A is then given by $\boldsymbol{x}_{\mathrm{fin}} + (a - d_{\mathrm{n}})(-\boldsymbol{n})$. Since this point is on surface S, we have the following equation:

$$\boldsymbol{n}^{\mathrm{T}}\{\boldsymbol{x}_{\mathrm{fin}} + (a - d_{\mathrm{n}})(-\boldsymbol{n}) - (R_{\mathrm{obj}}\,\boldsymbol{\xi}_{\mathrm{Q}} + \boldsymbol{x}_{\mathrm{obj}})\} = 0.$$

This equation yields a normal constraint between a fingertip and an object:

$$C_{\mathrm{n}} \triangleq \boldsymbol{n}^{\mathrm{T}}(\boldsymbol{x}_{\mathrm{obj}} - \boldsymbol{x}_{\mathrm{fin}}) - d_{\mathrm{n}} + a + (\boldsymbol{n}^{\mathrm{obj}})^{\mathrm{T}}\boldsymbol{\xi}_{\mathrm{Q}} = 0. \tag{11.24}$$

This constraint is holonomic. Note that the fourth term $(\boldsymbol{n}^{\mathrm{obj}})^{\mathrm{T}}\boldsymbol{\xi}_{\mathrm{Q}}$ is a constant denoting either positive or negative distance between the origin of Σ^{obj} and surface S.

Assume that the deformation of a fingertip is the maximum at point B. The position of point B is given by $\boldsymbol{x}_{\mathrm{fin}} - a\boldsymbol{n}$. Let $\boldsymbol{V}_{\mathrm{obj}}^{\mathrm{tip}}$ be the velocity of point B on the object. Since the orientation of the object is given by rotation matrix R_{obj}, the angular velocity $\boldsymbol{\omega}_{\mathrm{obj}}$ of the object is determined by

$$[\boldsymbol{\omega}_{\mathrm{obj}}\times] = \dot{R}_{\mathrm{obj}}\,R_{\mathrm{obj}}^{\mathrm{T}}.$$

Since $\overrightarrow{GB} = (\boldsymbol{x}_{\mathrm{fin}} - a\boldsymbol{n}) - \boldsymbol{x}_{\mathrm{obj}}$, the velocity $\boldsymbol{V}_{\mathrm{obj}}^{\mathrm{tip}}$ is described as follows:

$$\boldsymbol{V}_{\rm obj}^{\rm tip} = \left[\boldsymbol{\omega}_{\rm obj}\times\right]\{(\boldsymbol{x}_{\rm fin} - a\boldsymbol{n}) - \boldsymbol{x}_{\rm obj}\} + \dot{\boldsymbol{x}}_{\rm obj}, \tag{11.25}$$

where $\dot{\boldsymbol{x}}_{\rm obj}$ denotes the translational velocity of the rigid object. The tangential velocity of $\boldsymbol{V}_{\rm obj}^{\rm tip}$ along surface S is described by its projection on the surface. Since the projection matrix on surface S is given by $\boldsymbol{u}\boldsymbol{u}^{\rm T} + \boldsymbol{v}\boldsymbol{v}^{\rm T}$, the tangential velocity is described as

$$(\boldsymbol{u}\boldsymbol{u}^{\rm T} + \boldsymbol{v}\boldsymbol{v}^{\rm T})\,\boldsymbol{V}_{\rm obj}^{\rm tip} = (\boldsymbol{u}^{\rm T}\boldsymbol{V}_{\rm obj}^{\rm tip})\,\boldsymbol{u} + (\boldsymbol{v}^{\rm T}\boldsymbol{V}_{\rm obj}^{\rm tip})\,\boldsymbol{v},$$

implying that velocity components along $\boldsymbol{u}$ and $\boldsymbol{v}$ are $\boldsymbol{u}^{\rm T}\boldsymbol{V}_{\rm obj}^{\rm tip}$ and $\boldsymbol{v}^{\rm T}\boldsymbol{V}_{\rm obj}^{\rm tip}$, respectively. Let us describe the orientation of a fingertip by rotation matrix $R_{\rm fin}$. Let $\boldsymbol{V}_{\rm fin}^{\rm tip}$ be the velocity of point B on the fingertip. Since the orientation of the finger is given by rotation matrix $R_{\rm fin}$, the angular velocity $\boldsymbol{\omega}_{\rm fin}$ of the finger is determined by

$$\left[\boldsymbol{\omega}_{\rm fin}\times\right] = \dot{R}_{\rm fin}\,R_{\rm fin}^{\rm T}.$$

Since $\overrightarrow{PB} = -a\boldsymbol{n}$, the velocity $\boldsymbol{V}_{\rm fin}^{\rm tip}$ is described as follows:

$$\boldsymbol{V}_{\rm fin}^{\rm tip} = \left[\boldsymbol{\omega}_{\rm fin}\times\right](-a\boldsymbol{n}) + \dot{\boldsymbol{x}}_{\rm fin}, \tag{11.26}$$

where $\dot{\boldsymbol{x}}_{\rm fin}$ denotes the translational velocity of the fingertip. The tangential velocity of $\boldsymbol{V}_{\rm fin}^{\rm tip}$ along surface S is described by its projection on the surface, which is described as

$$(\boldsymbol{u}\boldsymbol{u}^{\rm T} + \boldsymbol{v}\boldsymbol{v}^{\rm T})\,\boldsymbol{V}_{\rm fin}^{\rm tip} = (\boldsymbol{u}^{\rm T}\boldsymbol{V}_{\rm fin}^{\rm tip})\,\boldsymbol{u} + (\boldsymbol{v}^{\rm T}\boldsymbol{V}_{\rm fin}^{\rm tip})\,\boldsymbol{v},$$

implying that velocity components along $\boldsymbol{u}$ and $\boldsymbol{v}$ are $\boldsymbol{u}^{\rm T}\boldsymbol{V}_{\rm fin}^{\rm tip}$ and $\boldsymbol{v}^{\rm T}\boldsymbol{V}_{\rm fin}^{\rm tip}$, respectively. Let $d_{\rm u}$ and $d_{\rm v}$ be tangential deformations along $\boldsymbol{u}$ and $\boldsymbol{v}$. From the above discussion we find that the time rate of the tangential deformations are described by

$$\dot{d}_{\rm u} = \boldsymbol{u}^{\rm T}\boldsymbol{\Delta}^{\rm tip}, \tag{11.27}$$
$$\dot{d}_{\rm v} = \boldsymbol{v}^{\rm T}\boldsymbol{\Delta}^{\rm tip}, \tag{11.28}$$

where

$$\begin{aligned}\boldsymbol{\Delta}^{\rm tip} &\triangleq \boldsymbol{V}_{\rm obj}^{\rm tip} - \boldsymbol{V}_{\rm fin}^{\rm tip} \\ &= \left[\boldsymbol{\omega}_{\rm obj}\times\right](\boldsymbol{x}_{\rm fin} - \boldsymbol{x}_{\rm obj} - a\boldsymbol{n}) + \left[\boldsymbol{\omega}_{\rm fin}\times\right]a\boldsymbol{n} + \dot{\boldsymbol{x}}_{\rm obj} - \dot{\boldsymbol{x}}_{\rm fin}\end{aligned}$$

denotes the relative velocity at point B. Note that when $\dot{d}_{\rm u} = 0$ and $\dot{d}_{\rm v} = 0$, we have no tangential deformations. The above equations yield the tangential constraints between a fingertip and an object given by

$$\dot{C}_{\mathrm{u}} \triangleq \boldsymbol{u}^{\mathrm{T}} \boldsymbol{\Delta}^{\mathrm{tip}} - \dot{d}_{\mathrm{u}} = 0, \tag{11.29}$$

$$\dot{C}_{\mathrm{v}} \triangleq \boldsymbol{v}^{\mathrm{T}} \boldsymbol{\Delta}^{\mathrm{tip}} - \dot{d}_{\mathrm{v}} = 0. \tag{11.30}$$

The above two constraints are nonholonomic Pfaffian constraints.

Lagrange equations of motion require partial derivatives of holonomic constraints with respect to generalized coordinates. Note that the holonomic constraint C_{n} depends on the object position $\boldsymbol{x}_{\mathrm{obj}}$, quaternion elements q_0 through q_3 describing the object orientation, the fingertip position $\boldsymbol{x}_{\mathrm{fin}}$, and the normal displacement d_{n}. Computing partial derivatives of holonomic constraint C_{n} with respect to these variables, we have

$$\frac{\partial C_{\mathrm{n}}}{\partial \boldsymbol{x}_{\mathrm{obj}}} = \boldsymbol{n}, \tag{11.31}$$

$$\frac{\partial C_{\mathrm{n}}}{\partial q_i} = \left(\frac{\partial R_{\mathrm{obj}}}{\partial q_i} \boldsymbol{n}^{\mathrm{obj}} \right)^{\mathrm{T}} (\boldsymbol{x}_{\mathrm{obj}} - \boldsymbol{x}_{\mathrm{fin}}), \quad (i = 0, 1, 2, 3), \tag{11.32}$$

$$\frac{\partial C_{\mathrm{n}}}{\partial \boldsymbol{x}_{\mathrm{fin}}} = -\boldsymbol{n}, \tag{11.33}$$

$$\frac{\partial C_{\mathrm{n}}}{\partial d_{\mathrm{n}}} = -1. \tag{11.34}$$

Partial derivatives $\partial R_{\mathrm{obj}}/\partial q_0$ through $\partial R_{\mathrm{obj}}/\partial q_3$ are given in Table 11.1. An equation to stabilize the holonomic constraint C_{n} requires the second-order partial derivatives of the constraint:

$$\frac{\partial^2 C_{\mathrm{n}}}{\partial \boldsymbol{x}_{\mathrm{obj}} \partial q_i} = \frac{\partial R_{\mathrm{obj}}}{\partial q_i} \boldsymbol{n}^{\mathrm{obj}}, \quad (i = 0, 1, 2, 3), \tag{11.35}$$

$$\frac{\partial^2 C_{\mathrm{n}}}{\partial \boldsymbol{x}_{\mathrm{fin}} \partial q_i} = -\frac{\partial R_{\mathrm{obj}}}{\partial q_i} \boldsymbol{n}^{\mathrm{obj}}, \quad (i = 0, 1, 2, 3), \tag{11.36}$$

$$\frac{\partial^2 C_{\mathrm{n}}}{\partial q_i \partial q_j} = \left(\frac{\partial^2 R_{\mathrm{obj}}}{\partial q_i \partial q_j} \boldsymbol{n}^{\mathrm{obj}} \right)^{\mathrm{T}} (\boldsymbol{x}_{\mathrm{obj}} - \boldsymbol{x}_{\mathrm{fin}}), \quad (i, j = 0, 1, 2, 3). \tag{11.37}$$

Second-order partial derivatives $\partial^2 R_{\mathrm{obj}}/\partial q_i \partial q_j$ are given in Table 11.2.

Lagrange equations of motion require partial derivatives of Pfaffian constraints with respect to generalized velocities. Note that the Pfaffian constraint $\dot{C}_{\mathrm{u}}$ depends on $\dot{\boldsymbol{x}}_{\mathrm{obj}}$, $\dot{q}_0$ through $\dot{q}_3$, $\boldsymbol{\omega}_{\mathrm{fin}}$, and the time rate $\dot{d}_{\mathrm{u}}$. Computing partial derivatives of Pfaffian constraint $\dot{C}_{\mathrm{u}}$ with respect to these variables, we have

$$\frac{\partial \dot{C}_{\mathrm{u}}}{\partial \dot{\boldsymbol{x}}_{\mathrm{obj}}} = \boldsymbol{u}, \tag{11.38}$$

$$\frac{\partial \dot{C}_{\mathrm{u}}}{\partial \dot{q}_i} = \boldsymbol{u}^{\mathrm{T}} \frac{\partial \left[\boldsymbol{\omega}_{\mathrm{obj}} \times \right]}{\partial \dot{q}_i} (\boldsymbol{x}_{\mathrm{fin}} - \boldsymbol{x}_{\mathrm{obj}} - a\boldsymbol{n}), \quad (i = 0, 1, 2, 3), \tag{11.39}$$

$$\frac{\partial \dot{C}_{\mathrm{u}}}{\partial \boldsymbol{\omega}_{\mathrm{fin}}} = a\boldsymbol{n} \times \boldsymbol{u} = a\boldsymbol{v}, \tag{11.40}$$

$$\frac{\partial \dot{C}_{\mathrm{u}}}{\partial \dot{d}_{\mathrm{u}}} = -1. \tag{11.41}$$

See Appendix B.2 to derive Eq. 11.40. Partial derivatives $\partial \left[\boldsymbol{\omega}_{\mathrm{obj}} \times \right] / \partial \dot{q}_0$ through $\partial \left[\boldsymbol{\omega}_{\mathrm{obj}} \times \right] / \partial \dot{q}_3$ are given in Table 11.3. Replacing $\boldsymbol{u}$ in the above equations by $\boldsymbol{v}$, we obtain partial derivatives of $\dot{C}_{\mathrm{v}}$ with respect to $\dot{\boldsymbol{x}}_{\mathrm{obj}}$, $\dot{q}_0$ through $\dot{q}_3$, $\boldsymbol{\omega}_{\mathrm{fin}}$, and $\dot{d}_{\mathrm{v}}$.

$$\frac{\partial \dot{C}_{\mathrm{v}}}{\partial \dot{\boldsymbol{x}}_{\mathrm{obj}}} = \boldsymbol{v}, \tag{11.42}$$

$$\frac{\partial \dot{C}_{\mathrm{v}}}{\partial \dot{q}_i} = \boldsymbol{v}^{\mathrm{T}} \frac{\partial \left[\boldsymbol{\omega}_{\mathrm{obj}} \times \right]}{\partial \dot{q}_i} (\boldsymbol{x}_{\mathrm{fin}} - \boldsymbol{x}_{\mathrm{obj}} - a\boldsymbol{n}), \quad (i = 0, 1, 2, 3), \tag{11.43}$$

$$\frac{\partial \dot{C}_{\mathrm{v}}}{\partial \boldsymbol{\omega}_{\mathrm{fin}}} = a\boldsymbol{n} \times \boldsymbol{v} = -a\boldsymbol{u}, \tag{11.44}$$

$$\frac{\partial \dot{C}_{\mathrm{v}}}{\partial \dot{d}_{\mathrm{v}}} = -1. \tag{11.45}$$

An equation to stabilize the Pfaffian constraint $\dot{C}_{\mathrm{u}}$ requires partial derivatives of the constraint with respect to generalized coordinates:

$$\frac{\partial \dot{C}_{\mathrm{u}}}{\partial \boldsymbol{x}_{\mathrm{obj}}} = - \left[\boldsymbol{\omega}_{\mathrm{obj}} \times \right] \boldsymbol{u}, \tag{11.46}$$

$$\frac{\partial \dot{C}_{\mathrm{u}}}{\partial \boldsymbol{x}_{\mathrm{fin}}} = \left[\boldsymbol{\omega}_{\mathrm{obj}} \times \right] \boldsymbol{u}, \tag{11.47}$$

$$\begin{aligned} \frac{\partial \dot{C}_{\mathrm{u}}}{\partial q_i} &= \boldsymbol{u}^{\mathrm{T}} \left\{ \left[\boldsymbol{\omega}_{\mathrm{fin}} \times \right] - \left[\boldsymbol{\omega}_{\mathrm{obj}} \times \right] \right\} a \frac{\partial R_{\mathrm{obj}}}{\partial q_i} \boldsymbol{n}^{\mathrm{obj}} \\ &+ \boldsymbol{u}^{\mathrm{T}} \frac{\partial \left[\boldsymbol{\omega}_{\mathrm{obj}} \times \right]}{\partial q_i} (\boldsymbol{x}_{\mathrm{fin}} - \boldsymbol{x}_{\mathrm{obj}} - a\boldsymbol{n}) \\ &+ \left(\frac{\partial R_{\mathrm{obj}}}{\partial q_i} \boldsymbol{u}^{\mathrm{obj}} \right)^{\mathrm{T}} \boldsymbol{\Delta}^{\mathrm{tip}}, \quad (i = 0, 1, 2, 3). \end{aligned} \tag{11.48}$$

The partial derivatives of Pfaffian constraint $\dot{C}_{\mathrm{v}}$ with respect to generalized coordinates are as follows:

$$\frac{\partial \dot{C}_{\rm v}}{\partial \boldsymbol{x}_{\rm obj}} = - \left[\boldsymbol{\omega}_{\rm obj}\times\right] \boldsymbol{v}, \tag{11.49}$$

$$\frac{\partial \dot{C}_{\rm v}}{\partial \boldsymbol{x}_{\rm fin}} = \left[\boldsymbol{\omega}_{\rm obj}\times\right] \boldsymbol{v}, \tag{11.50}$$

$$\begin{aligned}\frac{\partial \dot{C}_{\rm v}}{\partial q_i} &= \boldsymbol{v}^{\rm T}\left\{\left[\boldsymbol{\omega}_{\rm fin}\times\right] - \left[\boldsymbol{\omega}_{\rm obj}\times\right]\right\} a \frac{\partial R_{\rm obj}}{\partial q_i}\boldsymbol{n}^{\rm obj} \\ &+ \boldsymbol{v}^{\rm T}\frac{\partial\left[\boldsymbol{\omega}_{\rm obj}\times\right]}{\partial q_i}(\boldsymbol{x}_{\rm fin} - \boldsymbol{x}_{\rm obj} - a\boldsymbol{n}) \\ &+ \left(\frac{\partial R_{\rm obj}}{\partial q_i}\boldsymbol{v}^{\rm obj}\right)^{\rm T}\boldsymbol{\Delta}^{\rm tip}, \quad (i = 0, 1, 2, 3).\end{aligned} \tag{11.51}$$

11.4 Potential Energy of a Fingertip in Three-dimensional Grasping

Let us formulate the elastic potential energy stored in a hemispherical soft fingertip due to the contact with a planar surface of a rigid body. Recall that the surface can slide along $\boldsymbol{u}$ and $\boldsymbol{v}$ while maintaining constant $d_{\rm n}$. Let $d_{\rm t}$ be the displacement along the direction of inclination, and let $d_{\rm s}$ be the displacement perpendicular to the inclination. Displacement $d_{\rm n}$ is described in Eq. 11.24. Let us formulate $d_{\rm t}$, $d_{\rm s}$, and $\theta_{\rm p}$ in 3D grasping and manipulation.

Let us first derive the relative angle $\theta_{\rm p}$ between a fingertip and a planar surface. Let $\boldsymbol{b}^{\rm fin}$ be the unit normal vector of the plate behind a fingertip described in the finger coordinate system. The unit normal vector is then described in the spatial coordinate system as $\boldsymbol{b} = R_{\rm fin}\,\boldsymbol{b}^{\rm fin}$. The relative angle $\theta_{\rm p}$ is given by the angle between the two unit vectors $\boldsymbol{n}$ and $\boldsymbol{b}$. Thus, the relative angle is formulated as follows:

$$\cos\theta_{\rm p} = \boldsymbol{n}^{\rm T}\boldsymbol{b} = (R_{\rm obj}\boldsymbol{n}^{\rm obj})^{\rm T}(R_{\rm fin}\boldsymbol{b}^{\rm fin}), \tag{11.52}$$

$$\sin\theta_{\rm p} = \sqrt{1 - \cos^2\theta_{\rm p}}. \tag{11.53}$$

Any nonzero $\theta_{\rm p}$ shows that the object surface inclines to one direction with respect to the back plate of a fingertip. The direction of inclination is given by the projection of $\boldsymbol{b}$ on a planar surface composed of $\boldsymbol{u}$ and $\boldsymbol{v}$. The tangential deformation, introduced in Sect. 4.2, takes positive values in the negative direction of the projection. Consequently, the direction of tangential deformation is described as

$$(\boldsymbol{u}\boldsymbol{u}^{\rm T} + \boldsymbol{v}\boldsymbol{v}^{\rm T})(-\boldsymbol{b}) = (-\boldsymbol{u}^{\rm T}\boldsymbol{b})\boldsymbol{u} + (-\boldsymbol{v}^{\rm T}\boldsymbol{b})\boldsymbol{v},$$

implying that components of the direction vector along $\boldsymbol{u}$ and $\boldsymbol{v}$ are given by $(-\boldsymbol{u}^{\rm T}\boldsymbol{b})$ and $(-\boldsymbol{v}^{\rm T}\boldsymbol{b})$. Letting $\phi_{\rm t}$ be the angle between $\boldsymbol{u}$ and the direction vector, we have

$$\cos\phi_{\rm t} = \frac{-\boldsymbol{u}^{\rm T}\boldsymbol{b}}{\{(\boldsymbol{u}^{\rm T}\boldsymbol{b})^2 + (\boldsymbol{v}^{\rm T}\boldsymbol{b})^2\}^{1/2}}, \tag{11.54}$$

$$\sin\phi_{\rm t} = \frac{-\boldsymbol{v}^{\rm T}\boldsymbol{b}}{\{(\boldsymbol{u}^{\rm T}\boldsymbol{b})^2 + (\boldsymbol{v}^{\rm T}\boldsymbol{b})^2\}^{1/2}}. \tag{11.55}$$

Displacements $d_{\rm u}$ and $d_{\rm v}$ are converted into $d_{\rm t}$ and $d_{\rm s}$ as follows:

$$\begin{bmatrix} d_{\rm t} \\ d_{\rm s} \end{bmatrix} = \begin{bmatrix} \cos\phi_{\rm t} & \sin\phi_{\rm t} \\ -\sin\phi_{\rm t} & \cos\phi_{\rm t} \end{bmatrix} \begin{bmatrix} d_{\rm u} \\ d_{\rm v} \end{bmatrix}. \tag{11.56}$$

The potential energy given in Eq. 4.2 is then replaced by

$$P = \frac{1}{2}\int_{ell} k\left\{(PQ + d_{\rm t}\sin\theta_{\rm p})^2 + (d_{\rm t}\cos\theta_{\rm p})^2 + (d_{\rm s})^2\right\}. \tag{11.57}$$

Computing the above equation, we have

$$P = P_{\rm normal} + P_{\rm tangent} + P_{\rm side}, \tag{11.58}$$

where

$$P_{\rm normal}(d_{\rm n}, \theta_{\rm p}) = \frac{\pi E d_{\rm n}^3}{3\cos^2\theta_{\rm p}}, \tag{11.59}$$

$$P_{\rm tangent}(d_{\rm n}, d_{\rm t}, \theta_{\rm p}) = \pi E\{d_{\rm n}^2 d_{\rm t}\tan\theta_{\rm p} + d_{\rm n} d_{\rm t}^2\}, \tag{11.60}$$

$$P_{\rm side}(d_{\rm n}, d_{\rm s}) = \pi E d_{\rm n} d_{\rm s}^2. \tag{11.61}$$

The potential energy $P_{\rm normal}$ is caused by normal displacement $d_{\rm t}$. Potential energies $P_{\rm tangent}$ and $P_{\rm side}$ are caused by tangential displacements $d_{\rm t}$ and $d_{\rm s}$, respectively. Note that letting $d_{\rm s} = 0$, the above equation corresponds to the 2D model Eq. 4.4. Substituting Eq. 11.56 into the above equations, we finally have

$$P = \frac{\pi E d_{\rm n}^3}{3\cos^2\theta_{\rm p}} + \pi E d_{\rm n}^2 d_{\rm t}\tan\theta_{\rm p} + \pi E d_{\rm n}(d_{\rm u}^2 + d_{\rm v}^2). \tag{11.62}$$

Lagrange equations of motion require partial derivatives of potential energy with respect to generalized coordinates. Assume that the orientation of the fingertip, which is denoted by $R_{\rm fin}$, depends on a generalized coordinate $\theta_{\rm fin}$. Then, angle $\theta_{\rm p}$ depends on q_0 through q_3 and $\theta_{\rm fin}$. Since $\phi_{\rm t}$ depends on q_0 through q_3 and $\theta_{\rm fin}$, displacement $d_{\rm t}$ depends on $d_{\rm u}$, $d_{\rm v}$, q_0 through q_3, and $\theta_{\rm fin}$. Let us compute the partial derivatives of $\theta_{\rm p}$ and $d_{\rm t}$ with respect to such generalized coordinates. Recall that

$$\frac{\partial \boldsymbol{n}}{\partial q_i} = \left(\frac{\partial R_{\rm obj}}{\partial q_i}\right)^{\rm T} \boldsymbol{n}^{\rm obj}.$$

Let $\boldsymbol{\sigma}_{\text{obj},i}$ be the projection of the above vector onto a plane defined by $\boldsymbol{n}$ and $\boldsymbol{b}$. Note that $\boldsymbol{n} = R_{\text{obj}}\boldsymbol{n}^{\text{obj}}$ depends on q_0 through q_3 but $\boldsymbol{b} = R_{\text{fin}}\boldsymbol{b}^{\text{fin}}$ does not depend on these variables. Differentiating Eq. 11.52 with respect to q_0 through q_3, we then have

$$\frac{\partial \theta_{\text{p}}}{\partial q_i} = \begin{cases} \|\, \boldsymbol{\sigma}_{\text{obj},i} \,\| & \text{if } \boldsymbol{\sigma}_{\text{obj},i}^{\text{T}}\, \boldsymbol{b} \leq 0, \\ -\|\, \boldsymbol{\sigma}_{\text{obj},i} \,\| & \text{if } \boldsymbol{\sigma}_{\text{obj},i}^{\text{T}}\, \boldsymbol{b} \geq 0, \end{cases} \qquad (i = 0, 1, 2, 3). \tag{11.63}$$

(See Appendix B.3.) Since $\boldsymbol{b}$ depends on θ_{fin}, we have

$$\frac{\partial \boldsymbol{b}}{\partial \theta_{\text{fin}}} = \left(\frac{\partial R_{\text{fin}}}{\partial \theta_{\text{fin}}} \right)^{\text{T}} \boldsymbol{b}^{\text{fin}}.$$

Let $\boldsymbol{\sigma}_{\text{fin}}$ be the projection of the above vector onto a plane defined by $\boldsymbol{b}$ and $\boldsymbol{n}$. Note that $\boldsymbol{b} = R_{\text{fin}}\boldsymbol{b}^{\text{fin}}$ depends on angle θ_{fin} but $\boldsymbol{n} = R_{\text{obj}}\boldsymbol{n}^{\text{obj}}$ does not depend on this variable. Differentiating Eq. 11.52 with respect to θ_{fin}, we also have

$$\frac{\partial \theta_{\text{p}}}{\partial \theta_{\text{fin}}} = \begin{cases} \|\, \boldsymbol{\sigma}_{\text{fin}} \,\| & \text{if } \boldsymbol{\sigma}_{\text{fin}}^{\text{T}}\, \boldsymbol{n} \leq 0, \\ -\|\, \boldsymbol{\sigma}_{\text{fin}} \,\| & \text{if } \boldsymbol{\sigma}_{\text{fin}}^{\text{T}}\, \boldsymbol{n} \geq 0. \end{cases} \tag{11.64}$$

(See Appendix B.4.) Note that Eq. 11.56 directly yields

$$\frac{\partial d_{\text{t}}}{\partial d_{\text{u}}} = \cos\phi_{\text{t}} = \frac{-\boldsymbol{u}^{\text{T}}\boldsymbol{b}}{\{(\boldsymbol{u}^{\text{T}}\boldsymbol{b})^2 + (\boldsymbol{v}^{\text{T}}\boldsymbol{b})^2\}^{1/2}}, \tag{11.65}$$

$$\frac{\partial d_{\text{t}}}{\partial d_{\text{v}}} = \sin\phi_{\text{t}} = \frac{-\boldsymbol{v}^{\text{T}}\boldsymbol{b}}{\{(\boldsymbol{u}^{\text{T}}\boldsymbol{b})^2 + (\boldsymbol{v}^{\text{T}}\boldsymbol{b})^2\}^{1/2}}. \tag{11.66}$$

Differentiating Eq. 11.56 with respect to q_i and θ_{fin} yields

$$\frac{\partial d_{\text{t}}}{\partial q_i} = (-d_{\text{u}}\sin\phi_{\text{t}} + d_{\text{v}}\cos\phi_{\text{t}})\frac{\partial \phi_{\text{t}}}{\partial q_i}, \tag{11.67}$$

$$\frac{\partial d_{\text{t}}}{\partial \theta_{\text{fin}}} = (-d_{\text{u}}\sin\phi_{\text{t}} + d_{\text{v}}\cos\phi_{\text{t}})\frac{\partial \phi_{\text{t}}}{\partial \theta_{\text{fin}}}. \tag{11.68}$$

Differentiating $\phi_{\text{t}} = \operatorname{atan2}(-\boldsymbol{v}^{\text{T}}\boldsymbol{b},\, -\boldsymbol{u}^{\text{T}}\boldsymbol{b})$ with respect to q_i, we have

$$\frac{\partial \phi_{\text{t}}}{\partial q_i} = \frac{1}{(\boldsymbol{u}^{\text{T}}\boldsymbol{b})^2 + (\boldsymbol{v}^{\text{T}}\boldsymbol{b})^2} \left\{ \boldsymbol{u}^{\text{T}}\boldsymbol{b}\, \frac{\partial\, \boldsymbol{v}^{\text{T}}\boldsymbol{b}}{\partial q_i} - \frac{\partial\, \boldsymbol{u}^{\text{T}}\boldsymbol{b}}{\partial q_i}\, \boldsymbol{v}^{\text{T}}\boldsymbol{b} \right\}, \tag{11.69}$$

where

$$\frac{\partial\, \boldsymbol{u}^{\text{T}}\boldsymbol{b}}{\partial q_i} = \left(\frac{\partial R_{\text{obj}}}{\partial q_i} \boldsymbol{u}^{\text{obj}} \right)^{\text{T}} \boldsymbol{b}, \qquad \frac{\partial\, \boldsymbol{v}^{\text{T}}\boldsymbol{b}}{\partial q_i} = \left(\frac{\partial R_{\text{obj}}}{\partial q_i} \boldsymbol{v}^{\text{obj}} \right)^{\text{T}} \boldsymbol{b}.$$

(See Appendix B.5.) Differentiating $\phi_{\text{t}} = \operatorname{atan2}(-\boldsymbol{v}^{\text{T}}\boldsymbol{b},\, -\boldsymbol{u}^{\text{T}}\boldsymbol{b})$ with respect to θ_{fin}, we have

$$\frac{\partial \phi_{\rm t}}{\partial \theta_{\rm fin}} = \frac{1}{(\boldsymbol{u}^{\rm T}\boldsymbol{b})^2 + (\boldsymbol{v}^{\rm T}\boldsymbol{b})^2} \left\{ \boldsymbol{u}^{\rm T}\boldsymbol{b}\, \frac{\partial\, \boldsymbol{v}^{\rm T}\boldsymbol{b}}{\partial \theta_{\rm fin}} - \frac{\partial\, \boldsymbol{u}^{\rm T}\boldsymbol{b}}{\partial \theta_{\rm fin}}\, \boldsymbol{v}^{\rm T}\boldsymbol{b} \right\}, \qquad (11.70)$$

where

$$\frac{\partial\, \boldsymbol{u}^{\rm T}\boldsymbol{b}}{\partial \theta_{\rm fin}} = \left(\frac{\partial R_{\rm fin}}{\partial \theta_{\rm fin}} \boldsymbol{b}^{\rm fin} \right)^{\rm T} \boldsymbol{u}, \qquad \frac{\partial\, \boldsymbol{v}^{\rm T}\boldsymbol{b}}{\partial \theta_{\rm fin}} = \left(\frac{\partial R_{\rm fin}}{\partial \theta_{\rm fin}} \boldsymbol{b}^{\rm fin} \right)^{\rm T} \boldsymbol{v}.$$

Let us compute the partial derivatives of potential energy P with respect to the generalized coordinates. Since $d_{\rm n}$, $d_{\rm u}$, and $d_{\rm v}$ are independent state variables, while $\theta_{\rm p}$ and $d_{\rm t}$ depend on q_i, we have

$$\begin{aligned} \frac{\partial P}{\partial q_i} &= \frac{\partial P}{\partial \theta_{\rm p}} \frac{\partial \theta_{\rm p}}{\partial q_i} + \frac{\partial P}{\partial d_{\rm t}} \frac{\partial d_{\rm t}}{\partial q_i} \\ &= \left\{ \frac{2\pi E d_{\rm n}^3 \sin\theta_{\rm p}}{3\cos^3\theta_{\rm p}} + \frac{\pi E d_{\rm n}^2 d_{\rm t}}{\cos^2\theta_{\rm p}} \right\} \frac{\partial \theta_{\rm p}}{\partial q_i} + \pi E d_{\rm n}^2 \tan\theta_{\rm p} \frac{\partial d_{\rm t}}{\partial q_i}. \qquad (11.71) \end{aligned}$$

Partial derivatives $\partial\theta_{\rm p}/\partial q_i$ and $\partial d_{\rm t}/\partial q_i$ are given in Eqs. 11.63 and 11.67. Similarly,

$$\begin{aligned} \frac{\partial P}{\partial \theta_{\rm fin}} &= \frac{\partial P}{\partial \theta_{\rm p}} \frac{\partial \theta_{\rm p}}{\partial \theta_{\rm fin}} + \frac{\partial P}{\partial d_{\rm t}} \frac{\partial d_{\rm t}}{\partial \theta_{\rm fin}} \\ &= \left\{ \frac{2\pi E d_{\rm n}^3 \sin\theta_{\rm p}}{3\cos^3\theta_{\rm p}} + \frac{\pi E d_{\rm n}^2 d_{\rm t}}{\cos^2\theta_{\rm p}} \right\} \frac{\partial \theta_{\rm p}}{\partial \theta_{\rm fin}} + \pi E d_{\rm n}^2 \tan\theta_{\rm p} \frac{\partial d_{\rm t}}{\partial \theta_{\rm fin}}. \qquad (11.72) \end{aligned}$$

Partial derivatives $\partial\theta_{\rm p}/\partial\theta_{\rm fin}$ and $\partial d_{\rm t}/\partial\theta_{\rm fin}$ are given in Eqs. 11.64 and 11.68. In addition,

$$\frac{\partial P}{\partial d_{\rm n}} = \frac{\pi E d_{\rm n}^2}{\cos^2\theta_{\rm p}} + 2\pi E d_{\rm n} d_{\rm t} \tan\theta_{\rm p} + \pi E (d_{\rm u}^2 + d_{\rm v}^2), \qquad (11.73)$$

$$\frac{\partial P}{\partial d_{\rm u}} = (\pi E d_{\rm n}^2 \tan\theta_{\rm p}) \frac{\partial d_{\rm t}}{\partial d_{\rm u}} + 2\pi E d_{\rm n} d_{\rm u}, \qquad (11.74)$$

$$\frac{\partial P}{\partial d_{\rm v}} = (\pi E d_{\rm n}^2 \tan\theta_{\rm p}) \frac{\partial d_{\rm t}}{\partial d_{\rm v}} + 2\pi E d_{\rm n} d_{\rm v}. \qquad (11.75)$$

Partial derivatives $\partial d_{\rm t}/\partial d_{\rm u}$ and $\partial d_{\rm t}/\partial d_{\rm v}$ are given in Eqs. 11.65 and 11.66.

11.5 Grasping and Manipulation by Three 1-DOF Fingers

11.5.1 Observation

Let us observe the grasping and manipulation of a rigid object by three fingers with soft fingertips. Figure 11.2 shows the grasping of a rigid cylindrical object by three 1-DOF fingers with hemispherical soft fingertips. Figure 11.2a shows the initial location at which the joint angles at the base of three fingers are constant and equal to 71°. The object is perpendicular to the planar surface. Figure 11.2b shows one location at which two joint angles of the left finger and the middle finger are 71° and the joint angle of the right finger is 69°. The object inclines toward the right finger. Figure 11.2c shows one location at which the joint angles of the left, middle, and right fingers are 67°, 71°, and 74°, respectively. The cylindrical object grasped by the three fingers is

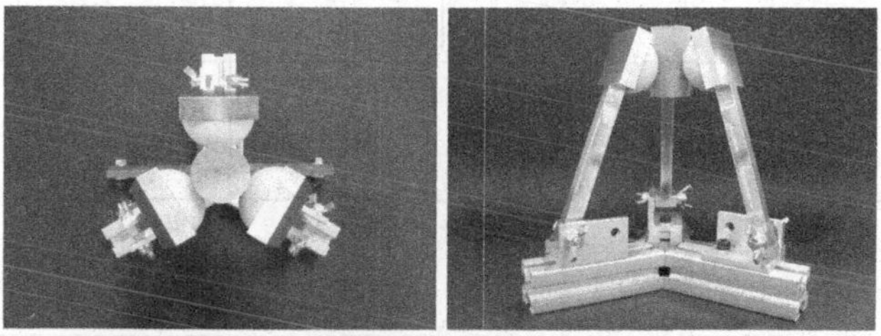

(a) 71°, 71°, 71°

(b) 71°, 71°, 69°

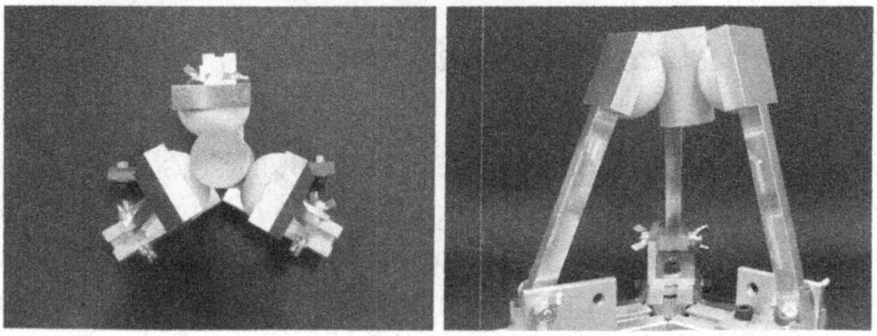

(c) 67°, 71°, 74°

Fig. 11.2 Grasping of a cylindrical object by three 1-DOF fingers with soft fingertips

Table 11.4 Geometric parameters of contacts between an object and fingers

	contact 1 ($k = 1$)	contact 2 ($k = 2$)	contact 3 ($k = 3$)
$\boldsymbol{n}_k^{\text{obj}}$	$\begin{bmatrix} 1 \\ 0 \\ 0 \end{bmatrix}$	$\begin{bmatrix} -1/2 \\ \sqrt{3}/2 \\ 0 \end{bmatrix}$	$\begin{bmatrix} -1/2 \\ -\sqrt{3}/2 \\ 0 \end{bmatrix}$
$\boldsymbol{u}_k^{\text{obj}}$	$\begin{bmatrix} 0 \\ 1 \\ 0 \end{bmatrix}$	$\begin{bmatrix} -\sqrt{3}/2 \\ -1/2 \\ 0 \end{bmatrix}$	$\begin{bmatrix} \sqrt{3}/2 \\ -1/2 \\ 0 \end{bmatrix}$
$\boldsymbol{v}_k^{\text{obj}}$	$\begin{bmatrix} 0 \\ 0 \\ 1 \end{bmatrix}$	$\begin{bmatrix} 0 \\ 0 \\ 1 \end{bmatrix}$	$\begin{bmatrix} 0 \\ 0 \\ 1 \end{bmatrix}$
$\boldsymbol{b}_k^{\text{fin}}$	$\begin{bmatrix} 1 \\ 0 \\ 0 \end{bmatrix}$	$\begin{bmatrix} 1 \\ 0 \\ 0 \end{bmatrix}$	$\begin{bmatrix} 1 \\ 0 \\ 0 \end{bmatrix}$
$\boldsymbol{x}_k^{\text{fin}}$	$\begin{bmatrix} -d_{\text{f}} \\ 0 \\ L \end{bmatrix}$	$\begin{bmatrix} -d_{\text{f}} \\ 0 \\ L \end{bmatrix}$	$\begin{bmatrix} -d_{\text{f}} \\ 0 \\ L \end{bmatrix}$
$\boldsymbol{r}_k$	$\begin{bmatrix} r \\ 0 \\ 0 \end{bmatrix}$	$\begin{bmatrix} (-1/2)r \\ (\sqrt{3}/2)r \\ 0 \end{bmatrix}$	$\begin{bmatrix} (-1/2)r \\ (-\sqrt{3}/2)r \\ 0 \end{bmatrix}$
$\boldsymbol{t}_k$	$\begin{bmatrix} 0 \\ -1 \\ 0 \end{bmatrix}$	$\begin{bmatrix} \sqrt{3}/2 \\ 1/2 \\ 0 \end{bmatrix}$	$\begin{bmatrix} -\sqrt{3}/2 \\ 1/2 \\ 0 \end{bmatrix}$

inclined to the left side. Through such experiments we found that both the grasping force and the direction of the cylindrical object could be regulated by the joint angles of the three fingers. Since the direction of the cylindrical object is described by two parameters, this implies that this spatial hand system consisting of three fingers can regulate three variables: grasping force and two variables representing the direction of a cylindrical object.

11.5.2 Mathematical Description

Assume that three 1-DOF fingers with hemispherical soft fingertips grasp a rigid hexagonal prism, as shown in Fig. 11.3. The hexagonal prism is assumed to be a cylinder of radius w. The radius of each fingertip is given by a. Fingertip 1 is in contact with side face S_1. Fingertip 2 is in contact with side face S_2. Fingertip 3 is in contact with side face S_3 of the hexagonal prism. Let $\mathrm{O} - xyz$ be the coordinate system attached to space. Let $\mathrm{G} - \xi\eta\zeta$ be the coordinate system attached to the prism. Let $\boldsymbol{t}_k$ be the unit vector specifying

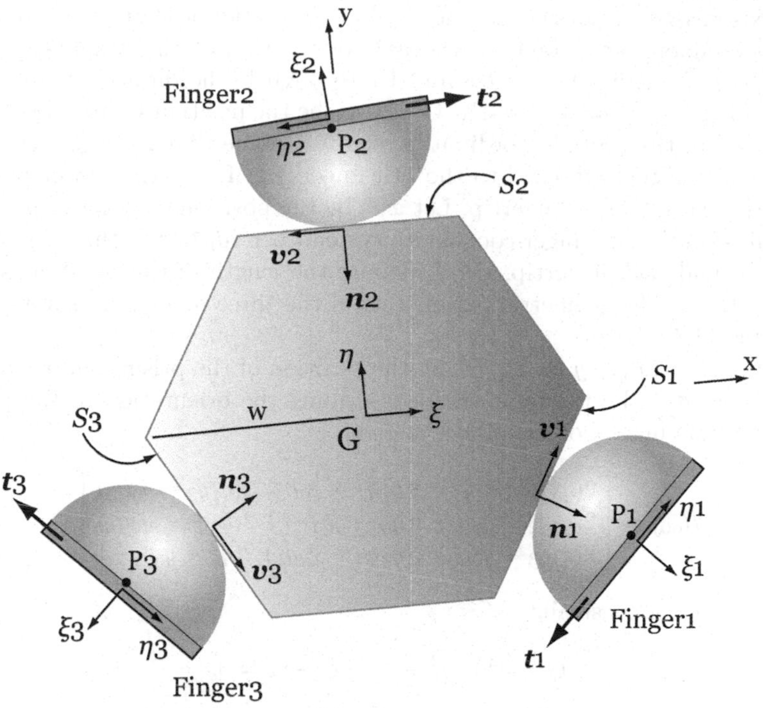

(**a**) Top view

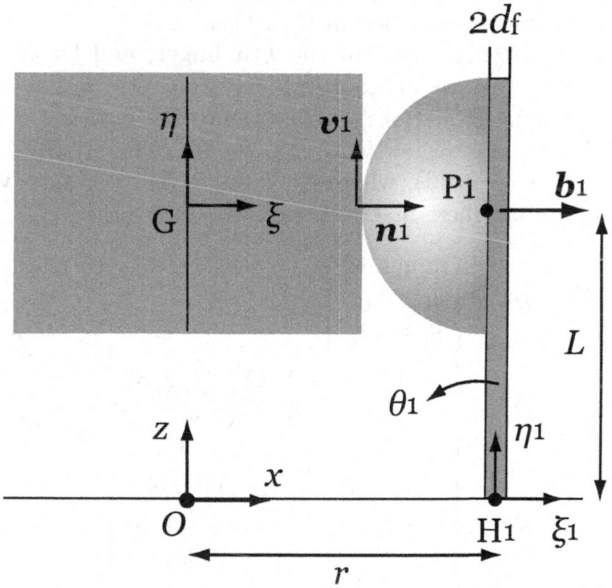

(**b**) Side view

Fig. 11.3 Three 1-DOF fingers grasping a rigid hexagonal prism

the axis of rotation of the kth finger. Let F_k be the nearest point from the origin O among any point on the axis of rotation of the kth finger. Note that point F_k is fixed in space and F_1 through F_3 lie on a circle of radius $r = (\sqrt{3}/2)w + a + d_{\mathrm{f}}$ centered at O. Let $\boldsymbol{r}_k$ be the position vector of point F_k described in the spatial coordinate system. Assume that $\mathrm{F}_k - \xi_k\eta_k\zeta_k$ is the coordinate system attached to the kth finger. Let P_k be the center point of the kth hemispherical fingertip. Let $\boldsymbol{x}_k^{\mathrm{fin}}$ be the positional vector of point P_k described in the kth finger coordinate system. Let $2d_{\mathrm{f}}$ be the thickness of the plate behind each fingertip. Let L denote the length of the plate, as shown in the figure. The geometric parameters of the three fingers are summarized in Table 11.4.

Let $\boldsymbol{x}_{\mathrm{obj}} = [\, x_{\mathrm{obj}},\, y_{\mathrm{obj}},\, z_{\mathrm{obj}} \,]^{\mathrm{T}}$ be the position of the prism, and let $\boldsymbol{q}_{\mathrm{obj}} = [\, q_0,\, q_1,\, q_2,\, q_3 \,]^{\mathrm{T}}$ be the quaternion describing the orientation of the prism. The rotation matrix of the object is given by

$$R_{\mathrm{obj}} = \begin{bmatrix} 2(q_0^2 + q_1^2) - 1 & 2(q_1q_2 + q_0q_3) & 2(q_1q_3 - q_0q_2) \\ 2(q_1q_2 - q_0q_3) & 2(q_0^2 + q_2^2) - 1 & 2(q_2q_3 + q_0q_1) \\ 2(q_1q_3 + q_0q_2) & 2(q_2q_3 - q_0q_1) & 2(q_0^2 + q_3^2) - 1 \end{bmatrix}$$

with geometric constraint

$$Q = q_0^2 + q_1^2 + q_2^2 + q_3^2 - 1 = 0.$$

Let $\boldsymbol{\omega}_{\mathrm{obj}} = [\, \omega_\xi,\, \omega_\eta,\, \omega_\zeta \,]^{\mathrm{T}}$ be the angular velocity of the prism. The angular velocity components are described in Eqs. 11.4–11.6.

Let θ_k be the rotation angle of the kth finger, and let R_k be the rotation matrix from the spatial coordinate system $\mathrm{O} - xyz$ to the kth finger coordinate system $\mathrm{F}_k - \xi_k\eta_k\zeta_k$. The orientation matrices corresponding to the three fingers are $R_1 = R(\boldsymbol{t}_1, \theta_1)$, $R_2 = R(\boldsymbol{t}_2, \theta_2)\, R(\boldsymbol{e}_z, 2\pi/3)$, and $R_3 = R(\boldsymbol{t}_3, \theta_3)\, R(\boldsymbol{e}_z, -2\pi/3)$, where $\boldsymbol{e}_z = [\, 0,\, 0,\, 1 \,]^{\mathrm{T}}$ is a unit vector along the z-axis. Computing these matrices, we have

$$R_1 = \begin{bmatrix} C_1 & 0 & -S_1 \\ 0 & 1 & 0 \\ S_1 & 0 & C_1 \end{bmatrix}, \tag{11.76}$$

$$R_2 = \begin{bmatrix} (-1/2)C_2 & -\sqrt{3}/2 & (1/2)S_2 \\ (\sqrt{3}/2)C_2 & -1/2 & (-\sqrt{3}/2)S_2 \\ S_2 & 0 & C_2 \end{bmatrix}, \tag{11.77}$$

$$R_3 = \begin{bmatrix} (-1/2)C_3 & \sqrt{3}/2 & (1/2)S_3 \\ (-\sqrt{3}/2)C_3 & -1/2 & (\sqrt{3}/2)S_3 \\ S_3 & 0 & C_3 \end{bmatrix}, \tag{11.78}$$

where C_k and S_k are abbreviations for $\cos\theta_k$ and $\sin\theta_k$. The partial derivatives are

Table 11.5 Vector $\boldsymbol{\kappa}^{\mathrm{n}}_{i,k}$ in three-fingered grasping

	$k=1$	$k=2$	$k=3$
$\boldsymbol{\kappa}^{\mathrm{n}}_{0,k}$	$\begin{bmatrix} 4q_0 \\ -2q_3 \\ 2q_2 \end{bmatrix}$	$\begin{bmatrix} -2q_0+\sqrt{3}q_3 \\ q_3+2\sqrt{3}q_0 \\ -q_2-\sqrt{3}q_1 \end{bmatrix}$	$\begin{bmatrix} -2q_0-\sqrt{3}q_3 \\ q_3-2\sqrt{3}q_0 \\ -q_2+\sqrt{3}q_1 \end{bmatrix}$
$\boldsymbol{\kappa}^{\mathrm{n}}_{1,k}$	$\begin{bmatrix} 4q_1 \\ 2q_2 \\ 2q_3 \end{bmatrix}$	$\begin{bmatrix} -2q_1+\sqrt{3}q_2 \\ -q_2 \\ -q_3-\sqrt{3}q_0 \end{bmatrix}$	$\begin{bmatrix} -2q_1-\sqrt{3}q_2 \\ -q_2 \\ -q_3+\sqrt{3}q_0 \end{bmatrix}$
$\boldsymbol{\kappa}^{\mathrm{n}}_{2,k}$	$\begin{bmatrix} 0 \\ 2q_1 \\ 2q_0 \end{bmatrix}$	$\begin{bmatrix} \sqrt{3}q_1 \\ -q_1+2\sqrt{3}q_2 \\ -q_0+\sqrt{3}q_3 \end{bmatrix}$	$\begin{bmatrix} -\sqrt{3}q_1 \\ -q_1-2\sqrt{3}q_2 \\ -q_0-\sqrt{3}q_3 \end{bmatrix}$
$\boldsymbol{\kappa}^{\mathrm{n}}_{2,k}$	$\begin{bmatrix} 0 \\ -2q_0 \\ 2q_1 \end{bmatrix}$	$\begin{bmatrix} \sqrt{3}q_0 \\ q_0 \\ -q_1+\sqrt{3}q_2 \end{bmatrix}$	$\begin{bmatrix} -\sqrt{3}q_0 \\ q_0 \\ -q_1-\sqrt{3}q_2 \end{bmatrix}$

$$\frac{\partial R_1}{\partial \theta_1} = \begin{bmatrix} -S_1 & 0 & -C_1 \\ 0 & 0 & 0 \\ C_1 & 0 & -S_1 \end{bmatrix}, \tag{11.79}$$

$$\frac{\partial R_2}{\partial \theta_2} = \begin{bmatrix} (1/2)S_2 & 0 & (1/2)C_2 \\ (-\sqrt{3}/2)S_2 & 0 & (-\sqrt{3}/2)C_2 \\ C_2 & 0 & -S_2 \end{bmatrix}, \tag{11.80}$$

$$\frac{\partial R_3}{\partial \theta_3} = \begin{bmatrix} (1/2)S_3 & 0 & (1/2)C_3 \\ (\sqrt{3}/2)S_3 & 0 & (\sqrt{3}/2)C_3 \\ C_3 & 0 & -S_3 \end{bmatrix}. \tag{11.81}$$

Let us introduce the following vectors:

$$\boldsymbol{\chi}_1 \triangleq \frac{\partial R_1}{\partial \theta_1}\boldsymbol{x}_1^{\mathrm{fin}} = \begin{bmatrix} d_{\mathrm{f}}S_1 - LC_1 \\ 0 \\ -d_{\mathrm{f}}C_1 - LS_1 \end{bmatrix}, \tag{11.82}$$

$$\boldsymbol{\chi}_2 \triangleq \frac{\partial R_2}{\partial \theta_2}\boldsymbol{x}_2^{\mathrm{fin}} = \begin{bmatrix} -(1/2)d_{\mathrm{f}}S_2 + (1/2)LC_2 \\ (\sqrt{3}/2)d_{\mathrm{f}}S_2 - (\sqrt{3}/2)LC_2 \\ -d_{\mathrm{f}}C_2 - LS_2 \end{bmatrix}, \tag{11.83}$$

$$\boldsymbol{\chi}_3 \triangleq \frac{\partial R_3}{\partial \theta_3}\boldsymbol{x}_3^{\mathrm{fin}} = \begin{bmatrix} -(1/2)d_{\mathrm{f}}S_3 + (1/2)LC_3 \\ -(\sqrt{3}/2)d_{\mathrm{f}}S_2 + (\sqrt{3}/2)LC_3 \\ -d_{\mathrm{f}}C_3 - LS_3 \end{bmatrix}. \tag{11.84}$$

Computing the angular velocities of the three fingers, we have

$$\boldsymbol{\omega}_k = \boldsymbol{t}_k\dot{\theta}_k, \quad (k=1,2,3). \tag{11.85}$$

Table 11.6 Vector $\boldsymbol{\kappa}^{\mathrm{u}}_{i,k}$ in three-fingered grasping

	$k=1$	$k=2$	$k=3$
$\boldsymbol{\kappa}^{\mathrm{u}}_{0,k}$	$\begin{bmatrix} 2q_3 \\ 4q_0 \\ -2q_1 \end{bmatrix}$	$\begin{bmatrix} -2\sqrt{3}q_0 - q_3 \\ \sqrt{3}q_3 - 2q_0 \\ -\sqrt{3}q_2 + q_1 \end{bmatrix}$	$\begin{bmatrix} 2\sqrt{3}q_0 - q_3 \\ -\sqrt{3}q_3 - 2q_0 \\ \sqrt{3}q_2 + q_1 \end{bmatrix}$
$\boldsymbol{\kappa}^{\mathrm{u}}_{1,k}$	$\begin{bmatrix} 2q_2 \\ 0 \\ -2q_0 \end{bmatrix}$	$\begin{bmatrix} -2\sqrt{3}q_1 - q_2 \\ -\sqrt{3}q_2 \\ -\sqrt{3}q_3 + q_0 \end{bmatrix}$	$\begin{bmatrix} 2\sqrt{3}q_1 - q_2 \\ \sqrt{3}q_2 \\ \sqrt{3}q_3 + q_0 \end{bmatrix}$
$\boldsymbol{\kappa}^{\mathrm{u}}_{2,k}$	$\begin{bmatrix} 2q_1 \\ 4q_2 \\ 2q_3 \end{bmatrix}$	$\begin{bmatrix} -q_1 \\ -\sqrt{3}q_1 - 2q_2 \\ -\sqrt{3}q_0 - q_3 \end{bmatrix}$	$\begin{bmatrix} -q_1 \\ \sqrt{3}q_1 - 2q_2 \\ \sqrt{3}q_0 - q_3 \end{bmatrix}$
$\boldsymbol{\kappa}^{\mathrm{u}}_{2,k}$	$\begin{bmatrix} 2q_0 \\ 0 \\ 2q_2 \end{bmatrix}$	$\begin{bmatrix} -q_0 \\ \sqrt{3}q_0 \\ -\sqrt{3}q_1 - q_2 \end{bmatrix}$	$\begin{bmatrix} -q_0 \\ -\sqrt{3}q_0 \\ \sqrt{3}q_1 - q_2 \end{bmatrix}$

Table 11.7 Vector $\boldsymbol{\kappa}^{\mathrm{v}}_{i,k}$ in three-fingered grasping

	$k=1$	$k=2$	$k=3$
$\boldsymbol{\kappa}^{\mathrm{v}}_{0,k}$	$\begin{bmatrix} -2q_2 \\ 2q_1 \\ 4q_0 \end{bmatrix}$	$\begin{bmatrix} -2q_2 \\ 2q_1 \\ 4q_0 \end{bmatrix}$	$\begin{bmatrix} -2q_2 \\ 2q_1 \\ 4q_0 \end{bmatrix}$
$\boldsymbol{\kappa}^{\mathrm{v}}_{1,k}$	$\begin{bmatrix} 2q_3 \\ 2q_0 \\ 0 \end{bmatrix}$	$\begin{bmatrix} 2q_3 \\ 2q_0 \\ 0 \end{bmatrix}$	$\begin{bmatrix} 2q_3 \\ 2q_0 \\ 0 \end{bmatrix}$
$\boldsymbol{\kappa}^{\mathrm{v}}_{2,k}$	$\begin{bmatrix} -2q_0 \\ 2q_3 \\ 0 \end{bmatrix}$	$\begin{bmatrix} -2q_0 \\ 2q_3 \\ 0 \end{bmatrix}$	$\begin{bmatrix} -2q_0 \\ 2q_3 \\ 0 \end{bmatrix}$
$\boldsymbol{\kappa}^{\mathrm{v}}_{2,k}$	$\begin{bmatrix} 2q_1 \\ 2q_2 \\ 4q_3 \end{bmatrix}$	$\begin{bmatrix} 2q_1 \\ 2q_2 \\ 4q_3 \end{bmatrix}$	$\begin{bmatrix} 2q_1 \\ 2q_2 \\ 4q_3 \end{bmatrix}$

The position vector of point P_k is formulated as follows:

$$\boldsymbol{x}_k = R_k\, \boldsymbol{x}_k^{\mathrm{fin}} + \boldsymbol{r}_k, \qquad (k = 1, 2, 3). \tag{11.86}$$

The velocity vector of point P_k is described as follows:

$$\dot{\boldsymbol{x}}_k = \dot{R}_k\, \boldsymbol{x}_k^{\mathrm{fin}} = \dot{\theta}_k \frac{\partial R_k}{\partial \theta_k} \boldsymbol{x}_k^{\mathrm{fin}} = \dot{\theta}_k \boldsymbol{\chi}_k, \qquad (k = 1, 2, 3). \tag{11.87}$$

Differentiating $\boldsymbol{x}_k$ with respect to θ_k, we have

$$\frac{\partial \boldsymbol{x}_k}{\partial \theta_k} = \boldsymbol{\chi}_k, \qquad (k = 1, 2, 3). \tag{11.88}$$

Differentiating $\dot{\boldsymbol{x}}_k$ with respect to θ_k and $\dot{\theta}_k$, we have

$$\frac{\partial \dot{\boldsymbol{x}}_k}{\partial \theta_k} = \left[\, \boldsymbol{\omega}_k \times \right] \boldsymbol{\chi}_k, \quad \frac{\partial \dot{\boldsymbol{x}}_k}{\partial \dot{\theta}_k} = \boldsymbol{\chi}_k, \qquad (k = 1, 2, 3). \tag{11.89}$$

Letting $\nu_k^{\rm n} = \boldsymbol{n}_k^{\rm T} \boldsymbol{b}_k$, $\nu_k^{\rm u} = \boldsymbol{u}_k^{\rm T} \boldsymbol{b}_k$, and $\nu_k^{\rm v} = \boldsymbol{v}_k^{\rm T} \boldsymbol{b}_k$, we have

$$\cos\theta_{{\rm p}k} = \nu_k^{\rm n}, \qquad \sin\theta_{{\rm p}k} = \sqrt{1 - (\nu_k^{\rm n})^2}, \tag{11.90}$$

$$\cos\phi_{{\rm t}k} = \frac{-\nu_k^{\rm u}}{\{(\nu_k^{\rm u})^2 + (\nu_k^{\rm v})^2\}^{1/2}}, \qquad \sin\phi_{{\rm t}k} = \frac{-\nu_k^{\rm v}}{\{(\nu_k^{\rm u})^2 + (\nu_k^{\rm v})^2\}^{1/2}}. \tag{11.91}$$

Let us introduce the following:

$$\boldsymbol{\kappa}_{i,k}^{\rm n} = \frac{\partial R_{\rm obj}}{\partial q_i} \boldsymbol{n}_k^{\rm obj}, \qquad \boldsymbol{\kappa}_{i,k}^{\rm u} = \frac{\partial R_{\rm obj}}{\partial q_i} \boldsymbol{u}_k^{\rm obj}, \qquad \boldsymbol{\kappa}_{i,k}^{\rm v} = \frac{\partial R_{\rm obj}}{\partial q_i} \boldsymbol{v}_k^{\rm obj}, \tag{11.92}$$
$$(i = 0, 1, 2, 3,\ k = 1, 2, 3).$$

Vectors $\boldsymbol{\kappa}_{0,1}^{\rm n}$ through $\boldsymbol{\kappa}_{3,3}^{\rm n}$, vectors $\boldsymbol{\kappa}_{0,1}^{\rm u}$ through $\boldsymbol{\kappa}_{3,3}^{\rm u}$, and vectors $\boldsymbol{\kappa}_{0,1}^{\rm v}$ through $\boldsymbol{\kappa}_{3,3}^{\rm v}$ are shown in Table 11.5, 11.6, and 11.7, respectively. Let us introduce the following quantities:

$$\boldsymbol{w}_{i,k} = \left\{ \left[\, \boldsymbol{\omega}_k \times \right] - \left[\, \boldsymbol{\omega}_{\rm obj} \times \right] \right\} a \boldsymbol{\kappa}_{i,k}^{\rm n}, \tag{11.93}$$

$$\boldsymbol{\sigma}_{i,k} = \frac{\partial \left[\, \boldsymbol{\omega}_{\rm obj} \times \right]}{\partial q_i} (\boldsymbol{x}_k - \boldsymbol{x}_{\rm obj} - a \boldsymbol{n}_k), \tag{11.94}$$

$$\hat{\boldsymbol{\sigma}}_{i,k} = \frac{\partial \left[\, \boldsymbol{\omega}_{\rm obj} \times \right]}{\partial \dot{q}_i} (\boldsymbol{x}_k - \boldsymbol{x}_{\rm obj} - a \boldsymbol{n}_k), \tag{11.95}$$
$$(i = 0, 1, 2, 3,\ k = 1, 2, 3).$$

Apply Table 11.3 for $\partial \left[\, \boldsymbol{\omega}_{\rm obj} \times \right] / \partial q_i$ and $\partial \left[\, \boldsymbol{\omega}_{\rm obj} \times \right] / \partial \dot{q}_i$.

11.5.3 Lagrange Equations of Motion

Let us derive the Lagrange equations of motion for manipulation by three 1-DOF fingers. The generalized coordinates are $\boldsymbol{x}_{\rm obj}$, $\boldsymbol{q}_{\rm obj}$, $\boldsymbol{\theta}_{\rm fin} = [\, \theta_1,\, \theta_2,\, \theta_3 \,]^{\rm T}$, $\boldsymbol{d}_{\rm n} = [\, d_{\rm n1},\, d_{\rm n2},\, d_{\rm n3} \,]^{\rm T}$, $\boldsymbol{d}_{\rm u} = [\, d_{\rm u1},\, d_{\rm u2},\, d_{\rm u3} \,]^{\rm T}$, and $\boldsymbol{d}_{\rm v} = [\, d_{\rm v1},\, d_{\rm v2},\, d_{\rm v3} \,]^{\rm T}$. The kinetic energy is described as follows:

$$K = \frac{1}{2} m_{\rm obj}\, \dot{\boldsymbol{x}}_{\rm obj}^{\rm T} \dot{\boldsymbol{x}}_{\rm obj} + K_{\rm rot} + \sum_{k=1}^{3} \frac{1}{2} I_{\rm fin} \dot{\theta}_k^2$$
$$+ \sum_{k=1}^{3} \frac{1}{2} m_{\rm n} \dot{d}_{{\rm n}k}^2 + \sum_{k=1}^{3} \frac{1}{2} m_{\rm u} \dot{d}_{{\rm u}k}^2 + \sum_{k=1}^{3} \frac{1}{2} m_{\rm v} \dot{d}_{{\rm v}k}^2. \tag{11.96}$$

Potential energy is formulated as

$$P = P_1 + P_2 + P_3, \tag{11.97}$$

where

$$P_k = \frac{\pi E d_{{\rm n}k}^3}{3\cos^2\theta_{{\rm p}k}} + \pi E d_{{\rm n}k}^2 d_{{\rm t}k} \tan\theta_{{\rm p}k} + \pi E d_{{\rm n}k} (d_{{\rm u}k}^2 + d_{{\rm v}k}^2), \qquad (k = 1, 2, 3).$$

The work done by external torques is given by

$$W = \tau_1 \theta_1 + \tau_2 \theta_2 + \tau_3 \theta_3. \tag{11.98}$$

Since the distance between the origin G and the three side faces is constant and equal to $(\sqrt{3}/2)w$, the normal constraint at the kth fingertip is described as follows:

$$C_{{\rm n}k} = (R_{\rm obj}\, \boldsymbol{n}_k^{\rm obj})^{\rm T} (\boldsymbol{x}_{\rm obj} - R_k\, \boldsymbol{x}_k^{\rm fin} - \boldsymbol{r}_k) - d_{{\rm n}k} + a + (\sqrt{3}/2)w = 0.$$

Two tangential constraints at the kth fingertip are described as follows:

$$\dot{C}_{{\rm u}k} = (R_{\rm obj} \boldsymbol{u}_k^{\rm obj})^{\rm T} \boldsymbol{\Delta}_k^{\rm tip} - d_{{\rm u}k} = 0,$$
$$\dot{C}_{{\rm v}k} = (R_{\rm obj} \boldsymbol{v}_k^{\rm obj})^{\rm T} \boldsymbol{\Delta}_k^{\rm tip} - d_{{\rm v}k} = 0,$$

where

$$\boldsymbol{\Delta}_k^{\rm tip} = \left[\boldsymbol{\omega}_{\rm obj} \times\right] (\boldsymbol{x}_k - \boldsymbol{x}_{\rm obj} - a\boldsymbol{n}_k)$$
$$+ \left[\boldsymbol{\omega}_k \times\right] a\boldsymbol{n}_k + \dot{\boldsymbol{x}}_{\rm obj} - \dot{\boldsymbol{x}}_k.$$

Summing the holonomic constraints with Lagrange multipliers yields

$$C_{\rm H} = \lambda_{\rm Q} Q + \lambda_{\rm n1} C_{\rm n1} + \lambda_{\rm n2} C_{\rm n2} + \lambda_{\rm n3} C_{\rm n3}. \tag{11.99}$$

Summing Pfaffian constraints with Lagrange multipliers yields

$$\dot{C}_{\rm P} = \mu_{\rm u1} \dot{C}_{\rm u1} + \mu_{\rm u2} \dot{C}_{\rm u2} + \mu_{\rm u3} \dot{C}_{\rm u3}$$
$$+ \mu_{\rm v1} \dot{C}_{\rm v1} + \mu_{\rm v2} \dot{C}_{\rm v2} + \mu_{\rm v3} \dot{C}_{\rm v3}.$$

Let us introduce collective vectors consisting of Lagrange multipliers: $\boldsymbol{\lambda}_{\rm n} = [\,\lambda_{\rm n1},\ \lambda_{\rm n2},\ \lambda_{\rm n3},\,]^{\rm T}$, $\boldsymbol{\mu}_{\rm u} = [\,\mu_{\rm u1},\ \mu_{\rm u2},\ \mu_{\rm u3},\,]^{\rm T}$, and $\boldsymbol{\mu}_{\rm v} = [\,\mu_{\rm v1},\ \mu_{\rm v2},\ \mu_{\rm v3},\,]^{\rm T}$. The Lagrangian of the system is then given by recalling Eqs. 11.96 through 11.99

as follows:

$$\mathcal{L} = K - P + W + C_{\mathrm{H}}. \tag{11.100}$$

Lagrange equation of motion with respect to $\boldsymbol{x}_{\mathrm{obj}}$

Let $M_{\mathrm{obj}} = \mathrm{diag}\{\, m_{\mathrm{obj}},\, m_{\mathrm{obj}},\, m_{\mathrm{obj}} \,\}$ and introduce the following 3×3 matrices:

$$N_{\mathrm{obj}} = \left[\boldsymbol{n}_1 \; \boldsymbol{n}_2 \; \boldsymbol{n}_3 \right],$$
$$U_{\mathrm{obj}} = \left[\boldsymbol{u}_1 \; \boldsymbol{u}_2 \; \boldsymbol{u}_3 \right],$$
$$V_{\mathrm{obj}} = \left[\boldsymbol{v}_1 \; \boldsymbol{v}_2 \; \boldsymbol{v}_3 \right].$$

Equations 11.31, 11.38, and 11.42 then yield the Lagrange equation of motion with respect to $\boldsymbol{x}_{\mathrm{obj}}$ as follows:

$$M_{\mathrm{obj}} \ddot{\boldsymbol{x}}_{\mathrm{obj}} - N_{\mathrm{obj}} \boldsymbol{\lambda}_{\mathrm{n}} - U_{\mathrm{obj}} \boldsymbol{\mu}_{\mathrm{u}} - V_{\mathrm{obj}} \boldsymbol{\mu}_{\mathrm{v}} = \boldsymbol{0}_3, \tag{11.101}$$

where $\boldsymbol{0}_3$ is the 3D zero vector.

Lagrange equations of motion with respect to $\boldsymbol{q}_{\mathrm{obj}}$

Let N_{quat} be a 4×3 matrix of which the (i,k)th element is given by

$$(N_{\mathrm{quat}})_{i,k} = \left(\boldsymbol{\kappa}^{\mathrm{n}}_{i,k}\right)^{\mathrm{T}} (\boldsymbol{x}_{\mathrm{obj}} - \boldsymbol{x}_k), \qquad (i = 0, 1, 2, 3, \;\; k = 1, 2, 3).$$

Let U_{quat} and V_{quat} be 4×3 matrices of which the (i,k)th elements are given by

$$(U_{\mathrm{quat}})_{i,k} = \boldsymbol{u}_k^{\mathrm{T}} \hat{\boldsymbol{\sigma}}_{i,k}, \qquad (i = 0, 1, 2, 3, \;\; k = 1, 2, 3),$$
$$(V_{\mathrm{quat}})_{i,k} = \boldsymbol{v}_k^{\mathrm{T}} \hat{\boldsymbol{\sigma}}_{i,k}, \qquad (i = 0, 1, 2, 3, \;\; k = 1, 2, 3).$$

Vectors $\hat{\boldsymbol{\sigma}}_{i,k}$ are given in Eq. 11.95. Let $\boldsymbol{f}_{\mathrm{quat}} = [\, f_{\mathrm{quat},0},\, f_{\mathrm{quat},1},\, f_{\mathrm{quat},2},\, f_{\mathrm{quat},3} \,]^{\mathrm{T}}$, where

$$f_{\mathrm{quat},i} = -\sum_{k=1}^{3} \frac{\partial P_k}{\partial q_i}, \qquad (i = 0, 1, 2, 3),$$

and

$$\frac{\partial P_k}{\partial q_i} = \left\{ \frac{2\pi E d_{\mathrm{n}k}^3 \sin\theta_{\mathrm{p}k}}{3\cos^3\theta_{\mathrm{p}k}} + \frac{\pi E d_{\mathrm{n}k}^2 d_{\mathrm{t}k}}{\cos^2\theta_{\mathrm{p}k}} \right\} \frac{\partial \theta_{\mathrm{p}k}}{\partial q_i} + \pi E d_{\mathrm{n}k}^2 \tan\theta_{\mathrm{p}k} \frac{\partial d_{\mathrm{t}k}}{\partial q_i}.$$

Equations 11.10–11.12, 11.32, 11.39, 11.43, and 11.71 yield the Lagrange equation of motion with respect to $\boldsymbol{q}_{\rm obj}$ as follows:

$$\begin{aligned}&J\ddot{\boldsymbol{q}} - \boldsymbol{h}_0\lambda_{\rm Q} - N_{\rm quat}\boldsymbol{\lambda}_{\rm n} - U_{\rm quat}\boldsymbol{\mu}_{\rm u} - V_{\rm quat}\boldsymbol{\mu}_{\rm v} = \\ &W_{\rm quat}\dot{\boldsymbol{q}} + Y_{\rm quat}\dot{\boldsymbol{q}} + \boldsymbol{f}_{\rm quat}.\end{aligned} \tag{11.102}$$

Apply Eqs. 11.20 through 11.23 to compute J, $W_{\rm quat}$, and $Y_{\rm quat}$.

Lagrange equations of motion with respect to $\boldsymbol{\theta}_{\rm fin}$

Differentiating $C_{{\rm n}k}$ with respect to θ_k, we have

$$\frac{\partial C_{{\rm n}k}}{\partial \theta_k} = \left(\frac{\partial C_{{\rm n}k}}{\partial \boldsymbol{x}_k}\right)^{\rm T} \frac{\partial \boldsymbol{x}_k}{\partial \theta_k} = -\boldsymbol{n}_k^{\rm T}\boldsymbol{\chi}_k.$$

Differentiating $\dot{C}_{{\rm u}k}$ and $\dot{C}_{{\rm v}k}$ with respect to θ_k and $\dot{\theta}_k$, we have

$$\frac{\partial \dot{C}_{{\rm u}k}}{\partial \dot{\theta}_k} = \left(\frac{\partial \dot{C}_{{\rm u}k}}{\partial \boldsymbol{\omega}_k}\right)^{\rm T} \frac{\partial \boldsymbol{\omega}_k}{\partial \dot{\theta}_k} + \left(\frac{\partial \dot{C}_{{\rm u}k}}{\partial \dot{\boldsymbol{x}}_k}\right)^{\rm T} \frac{\partial \dot{\boldsymbol{x}}_k}{\partial \dot{\theta}_k} = a\boldsymbol{v}_k^{\rm T}\boldsymbol{t}_k - \boldsymbol{u}_k^{\rm T}\boldsymbol{\chi}_k,$$

$$\frac{\partial \dot{C}_{{\rm v}k}}{\partial \dot{\theta}_k} = \left(\frac{\partial \dot{C}_{{\rm v}k}}{\partial \boldsymbol{\omega}_k}\right)^{\rm T} \frac{\partial \boldsymbol{\omega}_k}{\partial \dot{\theta}_k} + \left(\frac{\partial \dot{C}_{{\rm v}k}}{\partial \dot{\boldsymbol{x}}_k}\right)^{\rm T} \frac{\partial \dot{\boldsymbol{x}}_k}{\partial \dot{\theta}_k} = -a\boldsymbol{u}_k^{\rm T}\boldsymbol{t}_k - \boldsymbol{v}_k^{\rm T}\boldsymbol{\chi}_k.$$

Let $J_{\rm fin} = {\rm diag}\{\, I_{\rm fin},\, I_{\rm fin},\, I_{\rm fin}\,\}$ and introduce the following 3×3 matrices:

$$\begin{aligned}N_{\rm fin} &= {\rm diag}\{\, -\boldsymbol{n}_1^{\rm T}\boldsymbol{\chi}_1,\, -\boldsymbol{n}_2^{\rm T}\boldsymbol{\chi}_2,\, -\boldsymbol{n}_3^{\rm T}\boldsymbol{\chi}_3\,\},\\ U_{\rm fin} &= {\rm diag}\{\, a\boldsymbol{v}_1^{\rm T}\boldsymbol{t}_1 - \boldsymbol{u}_1\boldsymbol{\chi}_1,\, a\boldsymbol{v}_2^{\rm T}\boldsymbol{t}_2 - \boldsymbol{u}_2\boldsymbol{\chi}_2,\, a\boldsymbol{v}_3^{\rm T}\boldsymbol{t}_3 - \boldsymbol{u}_3\boldsymbol{\chi}_3\,\},\\ V_{\rm fin} &= {\rm diag}\{\, -a\boldsymbol{u}_1^{\rm T}\boldsymbol{t}_1 - \boldsymbol{v}_1\boldsymbol{\chi}_1,\, -a\boldsymbol{u}_2^{\rm T}\boldsymbol{t}_2 - \boldsymbol{v}_2\boldsymbol{\chi}_2,\, -a\boldsymbol{u}_3^{\rm T}\boldsymbol{t}_3 - \boldsymbol{v}_3\boldsymbol{\chi}_3\,\}.\end{aligned}$$

Let $\boldsymbol{f}_{\rm fin} = [\, f_{{\rm fin},1},\, f_{{\rm fin},2},\, f_{{\rm fin},3},\,]^{\rm T}$, where

$$\begin{aligned}f_{{\rm fin},k} &= -\frac{\partial P_k}{\partial \theta_k}\\ &= -\left\{\frac{2\pi E d_{{\rm n}k}^3 \sin\theta_{{\rm p}k}}{3\cos^3\theta_{{\rm p}k}} - \frac{\pi E d_{{\rm n}k}^2 d_{{\rm t}k}}{\cos^2\theta_{{\rm p}k}}\right\}\frac{\partial \theta_{{\rm p}k}}{\partial \theta_k} - \pi E d_{{\rm n}k}^2 \tan\theta_{{\rm p}k}\,\frac{\partial d_{{\rm t}k}}{\partial \theta_k}.\end{aligned}$$

Equations 11.33, 11.40, 11.44, and 11.72 yield the Lagrange equation of motion with respect to $\boldsymbol{\theta}_{\rm fin}$ as follows:

$$J_{\rm fin}\ddot{\boldsymbol{\theta}}_{\rm fin} - N_{\rm fin}\boldsymbol{\lambda}_{\rm n} - U_{\rm fin}\boldsymbol{\mu}_{\rm u} - V_{\rm fin}\boldsymbol{\mu}_{\rm v} = \boldsymbol{f}_{\rm fin} + \boldsymbol{\tau}, \tag{11.103}$$

where $\boldsymbol{\tau} = [\, \tau_1,\, \tau_2,\, \tau_3\,]^{\rm T}$.

Lagrange equations of motion with respect to $\boldsymbol{d}_\mathrm{n}$

Equations 11.34 and 11.73 yield the Lagrange equation of motion with respect to $\boldsymbol{d}_\mathrm{n}$ as follows:

$$M_\mathrm{n}\ddot{\boldsymbol{d}}_\mathrm{n} - (-E_{3\times 3})\boldsymbol{\lambda}_\mathrm{n} = \boldsymbol{f}_\mathrm{n}, \tag{11.104}$$

where $M_\mathrm{n} = \mathrm{diag}\{\, m_\mathrm{n},\, m_\mathrm{n},\, m_\mathrm{n}\,\}$, $\boldsymbol{f}_\mathrm{n} = [\, f_{\mathrm{n}1},\, f_{\mathrm{n}1},\, f_{\mathrm{n}1}\,]^\mathrm{T}$, and

$$f_{\mathrm{n}k} = -\frac{\partial P_k}{\partial d_{\mathrm{n}k}} = -\frac{\pi E d_{\mathrm{n}k}^2}{\cos^2\theta_{\mathrm{p}k}} - 2\pi E d_{\mathrm{n}k} d_{\mathrm{t}k} \tan\theta_{\mathrm{p}k} - \pi E(d_{\mathrm{u}k}^2 + d_{\mathrm{v}k}^2).$$

Lagrange equations of motion with respect to $\boldsymbol{d}_\mathrm{u}$ and $\boldsymbol{d}_\mathrm{v}$

Equations 11.41 and 11.74 yield the Lagrange equation of motion with respect to $\boldsymbol{d}_\mathrm{u}$ as follows:

$$M_\mathrm{u}\ddot{\boldsymbol{d}}_\mathrm{u} - (-E_{3\times 3})\boldsymbol{\mu}_\mathrm{u} = \boldsymbol{f}_\mathrm{u}, \tag{11.105}$$

where $M_\mathrm{u} = \mathrm{diag}\{\, m_\mathrm{u},\, m_\mathrm{u},\, m_\mathrm{u}\,\}$, $\boldsymbol{f}_\mathrm{u} = [\, f_{\mathrm{u}1},\, f_{\mathrm{u}2},\, f_{\mathrm{u}3}\,]^\mathrm{T}$, and

$$f_{\mathrm{u}k} = -\frac{\partial P}{\partial d_{\mathrm{u}k}} = -(\pi E d_{\mathrm{n}k}^2 \tan\theta_{\mathrm{p}k})\frac{\partial d_{\mathrm{t}k}}{\partial d_{\mathrm{u}k}} - 2\pi E d_{\mathrm{n}k} d_{\mathrm{u}k}.$$

Equations 11.45 and 11.75 yield the Lagrange equation of motion with respect to $\boldsymbol{d}_\mathrm{v}$ as follows:

$$M_\mathrm{v}\ddot{\boldsymbol{d}}_\mathrm{v} - (-E_{3\times 3})\boldsymbol{\mu}_\mathrm{v} = \boldsymbol{f}_\mathrm{v}, \tag{11.106}$$

where $M_\mathrm{v} = \mathrm{diag}\{\, m_\mathrm{v},\, m_\mathrm{v},\, m_\mathrm{v}\,\}$, $\boldsymbol{f}_\mathrm{v} = [\, f_{\mathrm{v}1},\, f_{\mathrm{v}2},\, f_{\mathrm{v}3}\,]^\mathrm{T}$, and

$$f_{\mathrm{v}k} = -\frac{\partial P}{\partial d_{\mathrm{v}k}} = -(\pi E d_{\mathrm{n}k}^2 \tan\theta_{\mathrm{p}k})\frac{\partial d_{\mathrm{t}k}}{\partial d_{\mathrm{v}k}} - 2\pi E d_{\mathrm{n}k} d_{\mathrm{v}k}.$$

Stabilization of holonomic constraint Q

Holonomic constraint Q can be stabilized by

$$-\boldsymbol{h}_0^\mathrm{T}\ddot{\boldsymbol{q}}_\mathrm{obj} = 2\dot{\boldsymbol{q}}_\mathrm{obj}^\mathrm{T}\dot{\boldsymbol{q}}_\mathrm{obj} + 2\alpha\boldsymbol{h}_0^\mathrm{T}\dot{\boldsymbol{q}}_\mathrm{obj} + \alpha^2 Q. \tag{11.107}$$

Stabilization of holonomic constraint C_n

Let us introduce a 3×4 matrix $\Gamma_k = \left[\, \boldsymbol{\kappa}_{0,k}^\mathrm{n}\ \boldsymbol{\kappa}_{1,k}^\mathrm{n}\ \boldsymbol{\kappa}_{2,k}^\mathrm{n}\ \boldsymbol{\kappa}_{3,k}^\mathrm{n}\,\right]$. Let $\hat{Q}_k$ be a 4×4 matrix, the (i,j)th element of which is given by

Table 11.8 Vector $(\partial^2 R_{\text{obj}}/\partial q_i \partial q_j)\, \boldsymbol{n}_1^{\text{obj}}$ in three-fingered grasping

	$j=0$	$j=1$	$j=2$	$j=3$
$i=0$	$\begin{bmatrix}4\\0\\0\end{bmatrix}$	$\begin{bmatrix}0\\0\\0\end{bmatrix}$	$\begin{bmatrix}0\\0\\2\end{bmatrix}$	$\begin{bmatrix}0\\-2\\0\end{bmatrix}$
$i=1$	$\begin{bmatrix}0\\0\\0\end{bmatrix}$	$\begin{bmatrix}4\\0\\0\end{bmatrix}$	$\begin{bmatrix}0\\2\\0\end{bmatrix}$	$\begin{bmatrix}0\\0\\2\end{bmatrix}$
$i=2$	$\begin{bmatrix}0\\0\\2\end{bmatrix}$	$\begin{bmatrix}0\\2\\0\end{bmatrix}$	$\begin{bmatrix}0\\0\\0\end{bmatrix}$	$\begin{bmatrix}0\\0\\0\end{bmatrix}$
$i=3$	$\begin{bmatrix}0\\-2\\0\end{bmatrix}$	$\begin{bmatrix}0\\0\\2\end{bmatrix}$	$\begin{bmatrix}0\\0\\0\end{bmatrix}$	$\begin{bmatrix}0\\0\\0\end{bmatrix}$

$$\left(\hat{Q}_k\right)_{i,j} = \left(\frac{\partial^2 R_{\text{obj}}}{\partial q_i \partial q_j}\boldsymbol{n}_k^{\text{obj}}\right)^{\text{T}} (\boldsymbol{x}_{\text{obj}} - \boldsymbol{x}_k), \qquad (i,j = 0,1,2,3).$$

Vectors $(\partial^2 R_{\text{obj}}/\partial q_i \partial q_j)\boldsymbol{n}_k^{\text{obj}}$, $(i = 0,1,2,3,\ k = 1,2,3)$ are listed in Tables 11.8–11.10. Equations 11.35–11.37 yield an ordinary differential equation that stabilizes a set of holonomic constraints $\boldsymbol{C}_{\text{n}}$:

$$-N_{\text{obj}}^{\text{T}}\ddot{\boldsymbol{x}}_{\text{obj}} - N_{\text{quat}}^{\text{T}}\ddot{\boldsymbol{q}}_{\text{obj}} - N_{\text{fin}}^{\text{T}}\ddot{\boldsymbol{\theta}}_{\text{fin}} - (-E_{3\times 3})\ddot{\boldsymbol{d}}_{\text{n}} =$$
$$\boldsymbol{g}_{\text{n}} + 2\alpha\left\{N_{\text{obj}}^{\text{T}}\dot{\boldsymbol{x}}_{\text{obj}} + N_{\text{quat}}^{\text{T}}\dot{\boldsymbol{q}}_{\text{obj}} + N_{\text{fin}}^{\text{T}}\dot{\boldsymbol{\theta}}_{\text{fin}} + E_{3\times 3}\dot{\boldsymbol{d}}_{\text{n}}\right\} + \alpha^2\boldsymbol{C}_{\text{n}}, \qquad (11.108)$$

where $\boldsymbol{g}_{\text{n}} = [\, g_{\text{n}1},\ g_{\text{n}2},\ g_{\text{n}3}\,]^{\text{T}}$ and

$$g_{\text{n}k} = \left[\, \dot{\boldsymbol{x}}_{\text{obj}}^{\text{T}}\ \dot{\boldsymbol{q}}_{\text{obj}}^{\text{T}}\ \dot{\theta}_k^{\text{T}}\,\right] \begin{bmatrix} & \Gamma_k & \\ \Gamma_k^{\text{T}} & \hat{Q}_k & \Gamma_k^{\text{T}}\chi_k \\ & (\Gamma_k^{\text{T}}\chi_k)^{\text{T}} & \end{bmatrix} \begin{bmatrix} \dot{\boldsymbol{x}}_{\text{obj}} \\ \dot{\boldsymbol{q}}_{\text{obj}} \\ \dot{\theta}_k \end{bmatrix}, \qquad (k = 1,2,3).$$

Stabilization of Pfaffian constraints $\dot{\boldsymbol{C}}_{\text{u}}$ and $\dot{\boldsymbol{C}}_{\text{v}}$

From Eqs. 11.94 and 11.93 we have

$$\frac{\partial \dot{C}_{\text{u}k}}{\partial q_i} = \left(\boldsymbol{\kappa}_{i,k}^{\text{u}}\right)^{\text{T}} \boldsymbol{\Delta}_k^{\text{tip}} + \boldsymbol{u}_k^{\text{T}}(\boldsymbol{\sigma}_{i,k} + \boldsymbol{w}_{i,k}), \qquad (11.109)$$

$$\frac{\partial \dot{C}_{\text{v}k}}{\partial q_i} = \left(\boldsymbol{\kappa}_{i,k}^{\text{v}}\right)^{\text{T}} \boldsymbol{\Delta}_k^{\text{tip}} + \boldsymbol{v}_k^{\text{T}}(\boldsymbol{\sigma}_{i,k} + \boldsymbol{w}_{i,k}). \qquad (11.110)$$

Table 11.9 Vector $(\partial^2 R_{\text{obj}}/\partial q_i \partial q_j)\,\boldsymbol{n}_2^{\text{obj}}$ in three-fingered grasping

	$j=0$	$j=1$	$j=2$	$j=3$
$i=0$	$\begin{bmatrix} -2\sqrt{3} \\ -2 \\ 0 \end{bmatrix}$	$\begin{bmatrix} 0 \\ 0 \\ 1 \end{bmatrix}$	$\begin{bmatrix} 0 \\ 0 \\ -\sqrt{3} \end{bmatrix}$	$\begin{bmatrix} -1 \\ \sqrt{3} \\ 0 \end{bmatrix}$
$i=1$	$\begin{bmatrix} 0 \\ 0 \\ 1 \end{bmatrix}$	$\begin{bmatrix} -2\sqrt{3} \\ 0 \\ 0 \end{bmatrix}$	$\begin{bmatrix} -1 \\ -\sqrt{3} \\ 0 \end{bmatrix}$	$\begin{bmatrix} 0 \\ 0 \\ -\sqrt{3} \end{bmatrix}$
$i=2$	$\begin{bmatrix} 0 \\ 0 \\ -\sqrt{3} \end{bmatrix}$	$\begin{bmatrix} -1 \\ -\sqrt{3} \\ 0 \end{bmatrix}$	$\begin{bmatrix} 0 \\ -2 \\ 0 \end{bmatrix}$	$\begin{bmatrix} 0 \\ 0 \\ -1 \end{bmatrix}$
$i=3$	$\begin{bmatrix} -1 \\ \sqrt{3} \\ 0 \end{bmatrix}$	$\begin{bmatrix} 0 \\ 0 \\ -\sqrt{3} \end{bmatrix}$	$\begin{bmatrix} 0 \\ 0 \\ -1 \end{bmatrix}$	$\begin{bmatrix} 0 \\ 0 \\ 0 \end{bmatrix}$

Table 11.10 Vector $(\partial^2 R_{\text{obj}}/\partial q_i \partial q_j)\,\boldsymbol{n}_3^{\text{obj}}$ in three-fingered grasping

	$j=0$	$j=1$	$j=2$	$j=3$
$i=0$	$\begin{bmatrix} 2\sqrt{3} \\ -2 \\ 0 \end{bmatrix}$	$\begin{bmatrix} 0 \\ 0 \\ 1 \end{bmatrix}$	$\begin{bmatrix} 0 \\ 0 \\ \sqrt{3} \end{bmatrix}$	$\begin{bmatrix} -1 \\ -\sqrt{3} \\ 0 \end{bmatrix}$
$i=1$	$\begin{bmatrix} 0 \\ 0 \\ 1 \end{bmatrix}$	$\begin{bmatrix} 2\sqrt{3} \\ 0 \\ 0 \end{bmatrix}$	$\begin{bmatrix} -1 \\ \sqrt{3} \\ 0 \end{bmatrix}$	$\begin{bmatrix} 0 \\ 0 \\ \sqrt{3} \end{bmatrix}$
$i=2$	$\begin{bmatrix} 0 \\ 0 \\ \sqrt{3} \end{bmatrix}$	$\begin{bmatrix} -1 \\ \sqrt{3} \\ 0 \end{bmatrix}$	$\begin{bmatrix} 0 \\ -2 \\ 0 \end{bmatrix}$	$\begin{bmatrix} 0 \\ 0 \\ -1 \end{bmatrix}$
$i=3$	$\begin{bmatrix} -1 \\ -\sqrt{3} \\ 0 \end{bmatrix}$	$\begin{bmatrix} 0 \\ 0 \\ \sqrt{3} \end{bmatrix}$	$\begin{bmatrix} 0 \\ 0 \\ -1 \end{bmatrix}$	$\begin{bmatrix} 0 \\ 0 \\ 0 \end{bmatrix}$

Equations 11.46–11.48 yield an ordinary differential equation that stabilizes a set of Pfaffian constraints $\dot{\boldsymbol{C}}_{\text{u}}$:

$$-U_{\text{obj}}^{\text{T}} \ddot{\boldsymbol{x}}_{\text{obj}} - U_{\text{quad}}^{\text{T}} \ddot{\boldsymbol{q}}_{\text{obj}} - U_{\text{fin}}^{\text{T}} \ddot{\theta}_{\text{fin}} - (-E_{3\times 3})^{\text{T}} \ddot{\boldsymbol{d}}_{\text{u}} = \boldsymbol{g}_{\text{u}} + \beta \dot{\boldsymbol{C}}_{\text{u}}, \quad (11.111)$$

where $\boldsymbol{g}_{\text{u}} = [\, g_{\text{u}1},\, g_{\text{u}2},\, g_{\text{u}3} \,]^{\text{T}}$ and

$$g_{\text{u}k} = -\left([\, \boldsymbol{\omega}_{\text{obj}} \times] \, \boldsymbol{u}_k \right)^{\text{T}} \dot{\boldsymbol{x}}_{\text{obj}} + \sum_{i=0}^{3} \frac{\partial \dot{C}_{\text{u}k}}{\partial q_i} \dot{q}_i + \boldsymbol{u}_k^{\text{T}} \left[\boldsymbol{\omega}_{\text{obj}} \times \right] \chi_k \dot{\theta}_k .$$

Apply Eq. 11.109 for $\partial \dot{C}_{\text{u}k}/\partial q_i$. Equations 11.49–11.51 yield an ordinary differential equation that stabilizes a set of Pfaffian constraints $\dot{\boldsymbol{C}}_{\text{v}}$:

$$-V_{\rm obj}^{\rm T}\ddot{\boldsymbol{x}}_{\rm obj} - V_{\rm quad}^{\rm T}\ddot{\boldsymbol{q}}_{\rm obj} - V_{\rm fin}^{\rm T}\ddot{\boldsymbol{\theta}}_{\rm fin} - (-E_{3\times3})^{\rm T}\ddot{\boldsymbol{d}}_{\rm v} = \boldsymbol{g}_{\rm v} + \beta\dot{\boldsymbol{C}}_{\rm v}, \tag{11.112}$$

where $\boldsymbol{g}_{\rm v} = [\,g_{\rm v1},\, g_{\rm v2},\, g_{\rm v3}\,]^{\rm T}$ and

$$g_{{\rm v}k} = -\left(\left[\,\boldsymbol{\omega}_{\rm obj}\times\,\right]\boldsymbol{v}_k\right)^{\rm T}\dot{\boldsymbol{x}}_{\rm obj} + \sum_{i=0}^{3}\frac{\partial\dot{C}_{{\rm v}k}}{\partial q_i}\dot{q}_i + \boldsymbol{v}_k^{\rm T}\left[\,\boldsymbol{\omega}_{\rm obj}\times\,\right]\boldsymbol{\chi}_k\dot{\theta}_k.$$

Apply Eq. 11.110 for $\partial\dot{C}_{{\rm v}k}/\partial q_i$.

11.5.4 Simulation

Combining Lagrange equations of motion Eqs. 11.101–11.106 with constraint stabilizing Eqs. 11.107, 11.108, 11.111, and 11.112, we have a set of system equations.

Let $\boldsymbol{\Theta} = [\,\boldsymbol{x}_{\rm obj}^{\rm T},\, \boldsymbol{\theta}_{\rm fin}^{\rm T},\, \boldsymbol{d}_{\rm n}^{\rm T},\, \boldsymbol{d}_{\rm u}^{\rm T},\, \boldsymbol{d}_{\rm v}^{\rm T}\,]^{\rm T}$ and $\boldsymbol{\Psi} = [\,\ddot{\boldsymbol{q}}_{\rm obj}^{\rm T},\, \lambda_{\rm Q}\,]^{\rm T}$. Let $\boldsymbol{\Lambda} = [\,\boldsymbol{\lambda}_{\rm n}^{\rm T},\, \boldsymbol{\mu}_{\rm u}^{\rm T},\, \boldsymbol{\mu}_{\rm v}^{\rm T}\,]^{\rm T}$ be a set of Lagrange multipliers. Let us define

$$M = {\rm diag}\{M_{\rm obj},\, J_{\rm fin},\, M_{\rm n},\, M_{\rm u},\, M_{\rm v}\}, \qquad M_{\rm Q} = \begin{bmatrix} J & -\boldsymbol{h}_0 \\ -\boldsymbol{h}_0^{\rm T} & 0 \end{bmatrix},$$

$$\Phi = \begin{bmatrix} N_{\rm obj} & U_{\rm obj} & V_{\rm obj} \\ N_{\rm fin} & U_{\rm fin} & V_{\rm fin} \\ -E_{3\times3} & O_{3\times3} & O_{3\times3} \\ O_{3\times3} & -E_{3\times3} & O_{3\times3} \\ O_{3\times3} & O_{3\times3} & -E_{3\times3} \end{bmatrix}, \qquad \Phi_{\rm Q} = \left[\,N_{\rm quat}\; U_{\rm quat}\; V_{\rm quat}\,\right],$$

where M is a 15×15 diagonal matrix, $M_{\rm Q}$ is a 5×5 matrix, Φ is a 15×9 matrix, and $\Phi_{\rm Q}$ is a 5×9 matrix. Note that M, $M_{\rm Q}$, and Φ depend on generalized coordinates and generalized velocities but do not include second-order derivatives of the coordinates or Lagrange multipliers. Note that matrix $M_{\rm Q}$ is regular even though the 4×4 matrix J is not regular. A set of Lagrange equations of motion and equations for stabilization is then collectively described as follows:

$$\begin{bmatrix} M & & -\Phi \\ & M_{\rm Q} & -\Phi_{\rm Q} \\ -\Phi^{\rm T} & -\Phi_{\rm Q}^{\rm T} & \end{bmatrix}\begin{bmatrix} \ddot{\boldsymbol{\Theta}} \\ \boldsymbol{\Psi} \\ \boldsymbol{\Lambda} \end{bmatrix} = \begin{bmatrix} \boldsymbol{F} \\ \boldsymbol{F}_{\rm Q} \\ \boldsymbol{G} \end{bmatrix}, \tag{11.113}$$

where

$$\boldsymbol{F} = \begin{bmatrix} \mathbf{0}_3 \\ \boldsymbol{f}_{\text{fin}} + \boldsymbol{\tau} \\ \boldsymbol{f}_{\text{n}} \\ \boldsymbol{f}_{\text{u}} \\ \boldsymbol{f}_{\text{v}} \end{bmatrix}, \quad \boldsymbol{F}_{\text{Q}} = \begin{bmatrix} W_{\text{quat}}\dot{\boldsymbol{q}} + Y_{\text{quat}}\dot{\boldsymbol{q}} + \boldsymbol{f}_{\text{quat}} \\ 2\dot{\boldsymbol{q}}_{\text{obj}}^{\text{T}}\dot{\boldsymbol{q}}_{\text{obj}} + 2\alpha\boldsymbol{h}_0^{\text{T}}\dot{\boldsymbol{q}}_{\text{obj}} + \alpha^2 Q \end{bmatrix},$$

$$\boldsymbol{G} = \begin{bmatrix} \boldsymbol{g}_{\text{n}} + 2\alpha\left\{N_{\text{obj}}^{\text{T}}\dot{\boldsymbol{x}}_{\text{obj}} + N_{\text{quat}}^{\text{T}}\dot{\boldsymbol{q}}_{\text{obj}} + N_{\text{fin}}^{\text{T}}\dot{\boldsymbol{\theta}}_{\text{fin}} + E_{3\times 3}\dot{\boldsymbol{d}}_{\text{n}}\right\} + \alpha^2\boldsymbol{C}_{\text{n}} \\ \boldsymbol{g}_{\text{u}} + \beta\dot{\boldsymbol{C}}_{\text{u}} \\ \boldsymbol{g}_{\text{v}} + \beta\dot{\boldsymbol{C}}_{\text{v}} \end{bmatrix}.$$

Note that $\boldsymbol{F}$, $\boldsymbol{F}_{\text{Q}}$, and $\boldsymbol{G}$ depend on generalized coordinates and generalized velocities but do not include second-order derivatives of the coordinates or Lagrange multipliers. Solving Eq. 11.113, we obtain $\ddot{\boldsymbol{\Theta}}$, $\boldsymbol{\Psi}$, and $\boldsymbol{\Lambda}$. Applying numerical integration of ODEs, we can sketch the behavior of the three-fingered hand with soft fingertips.

Let us simulate the spatial grasping and manipulation of a hexagonal prism via three 1-DOF fingers with soft fingertips. The size of the prism is given by $W = 50\,\text{mm}$ and its inertial properties are $m_{\text{obj}} = 300\,\text{g}$ and $I_{\text{obj}} = 417\,\text{kg}\cdot\text{mm}^2$. The inertial properties of each finger are given by $m_{\text{fin}} = 100\,\text{g}$ and $I_{\text{fin}} = 582\,\text{kg}\cdot\text{mm}^2$. The other parameters employ the same values shown in the previous chapters. In this simulation, we have assumed that the gravitational effect is negligible during grasping and manipulation. We have applied a simple PID law to control individual finger joints to observe the motion of the grasped object during the control of three fingers. Control input $\boldsymbol{\tau}_k$ for the kth finger joint is expressed as

$$\tau_k = -K_{\text{P}}(\theta_k - \theta_k^{\text{d}}) - K_{\text{D}}\dot{\theta}_k - K_{\text{I}}\int_0^t \{\theta_k(\tau) - \theta_k^{\text{d}}\}\,\text{d}\tau, \qquad (11.114)$$

where K_{P}, K_{D}, and K_{I} denote the proportional, differential, and integral gains, respectively. Let us apply the same gain values of $K_{\text{P}} = 300\,\text{Nm}$, $K_{\text{D}} = 1\,\text{Nm}\cdot\text{s}$, and $K_{\text{I}} = 0.2\,\text{Nm/s}$ to the three fingers.

Figures 11.4–11.6 show the simulation results when the desired angles of the three finger joints are given at time 0. The desired angles are $\theta_1^{\text{d}} = 0°$, $\theta_2^{\text{d}} = 5°$, and $\theta_3^{\text{d}} = 5°$ in Fig. 11.4. As shown in Figs. 11.4a–c, the joint angles converge to their desired values within 0.2 s. Let us investigate the direction of the central axis of the hexagonal prism. The unit vector along the ζ-axis, which specifies the central axis, is described as $[\,2(q_1q_3 - q_0q_2),\, 2(q_2q_3 + q_0q_1),\, 2(q_0^2 + q_3^2) - 1\,]^{\text{T}}$. Note that this vector coincides with the third column of the rotation matrix. Figure 11.4d denotes the angle between the ζ-axis and the z-axis, which is given by $\cos^{-1}(2(q_0^2 + q_3^2) - 1)$. This angle represents the inclination of the prism. Figure 11.4e denotes the direction of the projection of the ζ-axis on the $x - y$ plane, which is given by $\text{atan2}\,(2(q_2q_3 + q_0q_1), 2(q_1q_3 - q_0q_2))$. This angle represents the direction of the inclined prism. Quaternion elements during grasping and manipulation are plotted in Figs. 11.4f–i. Figures 11.4j–l denote the coordinates of the

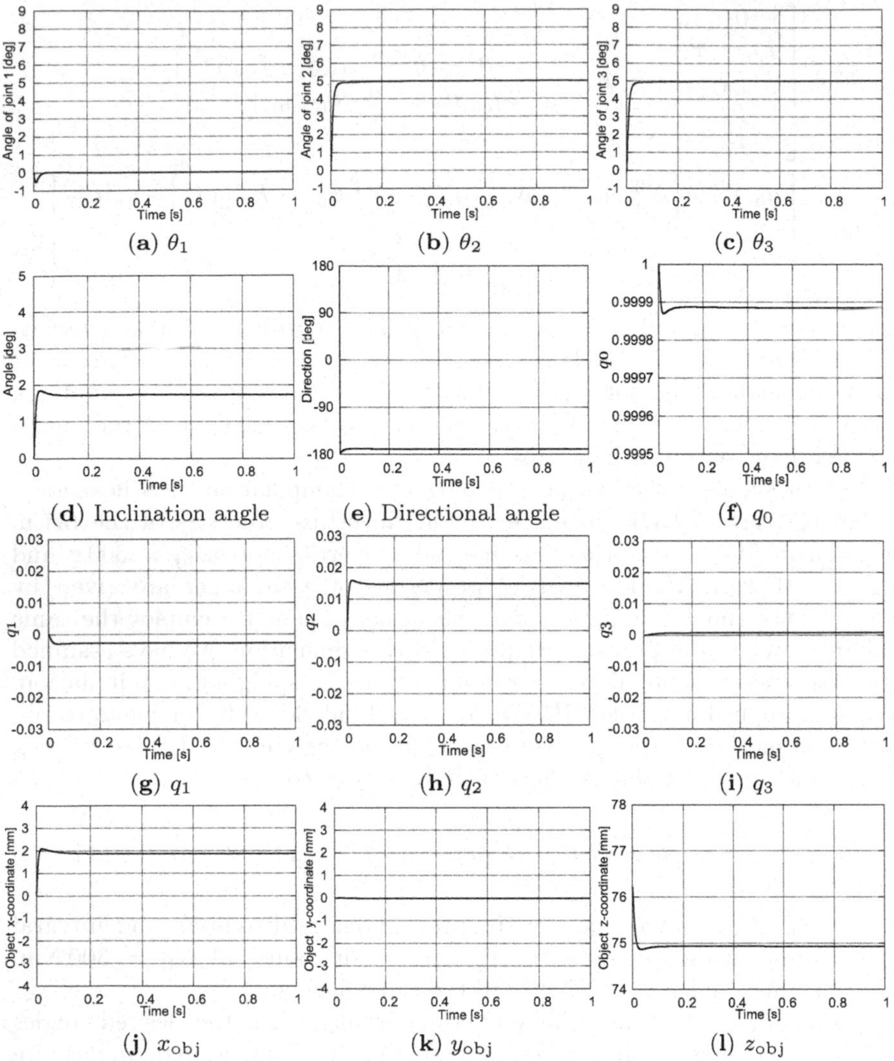

Fig. 11.4 Simulation result of three-fingered grasping ($\theta_1^{\mathrm{d}} = 0°$, $\theta_2^{\mathrm{d}} = 5°$, $\theta_3^{\mathrm{d}} = 5°$)

grasped object. The coordinates converge, implying that the object is stable with respect to the motion of three fingers. The desired angles are $\theta_1^{\mathrm{d}} = 5°$, $\theta_2^{\mathrm{d}} = 0°$, and $\theta_3^{\mathrm{d}} = 5°$ in Fig. 11.5. Thc desired angles are $\theta_1^{\mathrm{d}} = 5°$, $\theta_2^{\mathrm{d}} = 5°$, and $\theta_3^{\mathrm{d}} = 0°$ in Fig. 11.4. Figures 11.4d, 11.5d, and 11.6d indicate that the three fingers can incline the prism toward any direction, suggesting that a set of three 1-DOF fingers with soft tips can regulate the inclination and the direction of the grasped object.

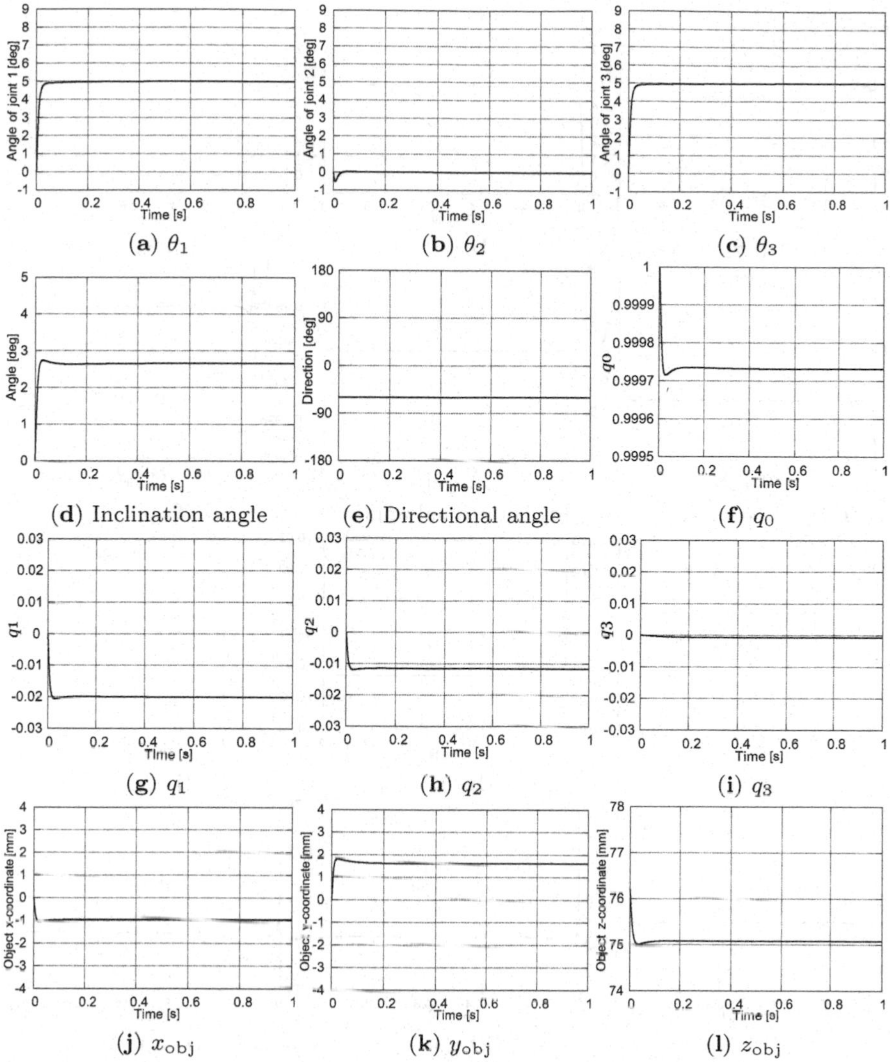

Fig. 11.5 Simulation result of three-fingered grasping ($\theta_1^{\mathrm{d}} = 5°$, $\theta_2^{\mathrm{d}} = 0°$, $\theta_3^{\mathrm{d}} = 5°$)

11.6 Concluding Remarks

This chapter focused on spatial grasping and manipulation via soft fingertips. Normal and tangential geometric constraints between a rigid object surface and hemispherical soft fingertips were formulated. Here, we have two tangential constraints, Pfaffian and nonholonomic, implying that we cannot integrate these constraints analytically. Note that static analysis cannot incorporate nonholonomic Pfaffian constraints, which include the time deriva-

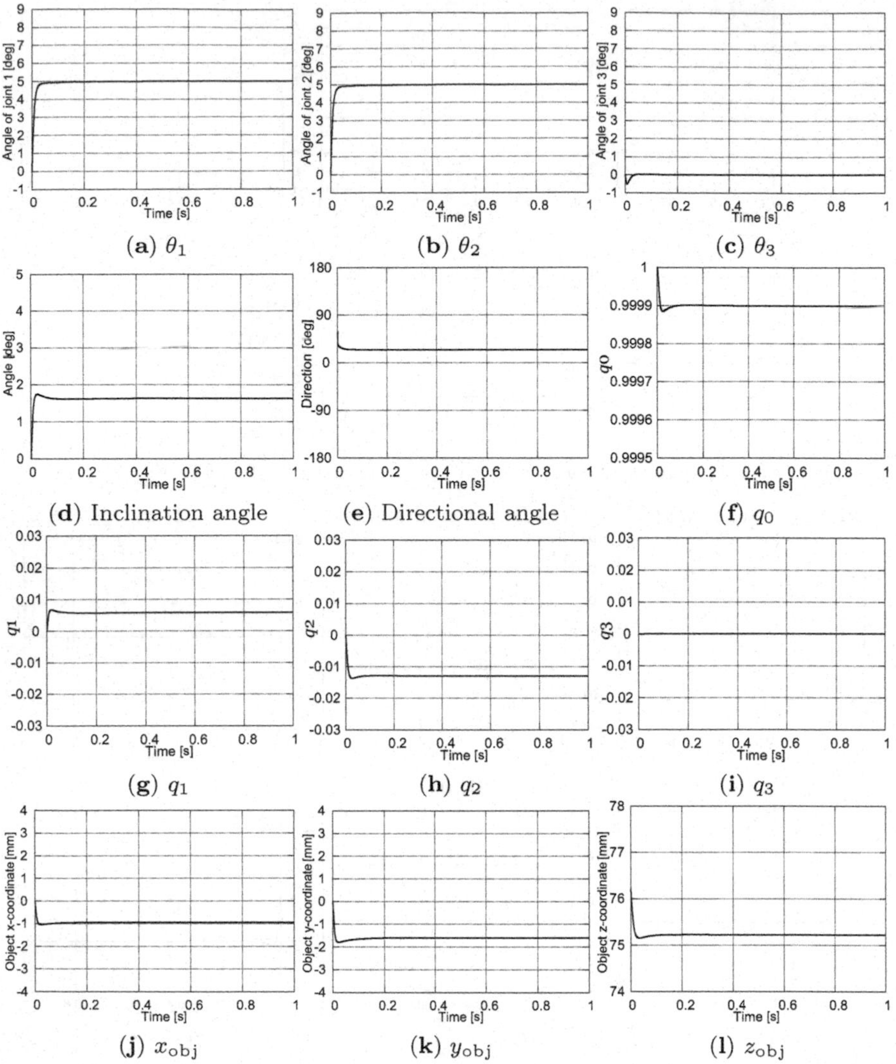

Fig. 11.6 Simulation result of three-fingered grasping ($\theta_1^{\mathrm{d}} = 5°$, $\theta_2^{\mathrm{d}} = 5°$, $\theta_3^{\mathrm{d}} = 0°$)

tives of system variables. As a result, we cannot formulate spatial grasping in a static manner. In the formulation of the potential energy of a fingertip, we introduced one normal term and two tangential terms. A simple example of three 1-DOF fingers with hemispherical soft fingertips grasping a rigid hexagonal prism was described in detail in order to illustrate the formulation of spatial grasping and manipulation.

This chapter focused on the formulation of spatial grasping and manipulation. Detailed analysis and experimental verification remain to be performed.

Chapter 12
Conclusions

12.1 Main Contribution

The present study has observed the static and dynamic behavior of a rigid and rectangular object grasped by a simple two-fingered robotic hand that consists of a pair of 1-DOF revolute joint and hemispherical soft fingertips. We found that the target object could easily be manipulated from side to side by means of a simple robot constructed with stepping motors. Generally, since stepping motors are used for angular control based not on feedback control but rather on open-loop control, the control law for the elementary manipulation does not contain information of the finger joints or the target object during the observation experiment. In case of verification experiments using DC motors, we also recognized that object information during manipulation is not necessary for stable grasping and pinching motions. However, the object can be securely pinched and freely manipulated by a minimal-DOF robotic hand. We defined this intrinsic manipulation performance of soft fingertips with no object information as *conformable manipulation*.

First, this manipulation ability stems from a mechanical phenomenon that produces restoring forces and moments needed for pulling back a grasped object to a stable position and orientation using soft fingers. That is, the soft fingers have a minimal condition in terms of their elastic energy that varies according to the magnitude of deformation of the fingers. The unique or multiple minimum value of the energy is referred to as the *local minimum of elastic potential energy (LMEE)*. This mechanical idea of the elastic energy had been observed even in fundamental compression tests of soft fingertips. The discriminative characteristics of the LMEE provide a consistent stability that enables successful pinching movements and dynamic manipulation motions. In particular, through actual manipulating motions, the LMEE was extended to the *LMEE with constraints (LMEEwC)*, in which usual motions occur accompanying the contact between fingers and the object.

The present study concentrated on explicit descriptions and formulation procedures of the dynamics of soft-fingered manipulation in the presence of holonomic and Pfaffian constraints. A unique technique for analyzing a dynamic system having these constraints is the constraint stabilization method of which the computations of the system dynamics were performed on the basis. We described, in a step-by-step manner, the computation procedure required for the application of this method.

These analyses led us to consider what types of tasks could be performed using only a 2-DOF mechanism that consists of a combination of rotational and translational movements for revolute and prismatic joints. Through several simulations and verifications, we could categorize skillfully performed tasks of the minimal-DOF robotic hand from the viewpoint of the relationship between the number of mechanical degrees of freedom and the number of feasible manipulation tasks. In this analysis, we found a special manipulation capability that ensures the control of many more tasks than the number of mechanical degrees of freedom by employing different types of independent joint mechanisms. Furthermore, we revealed that the rotational structure of a robotic finger is superior to that of the prismatic joint because the former mechanism is able to control an individual object variable by converting the target task, which is determined preliminarily. This control method, which can be fulfilled only through soft-fingered manipulation, was defined as *task-selection control.* In addition, we indicated that *Jacobian-free control design* could be achieved using the considerably straightforward control laws of the present study. These control laws do not require grasping forces that appear on soft fingertips due to their deformations or compensators for the gravitational effect on the entire system. This is because the LMEEwC corresponds to a unique equilibrium point and a minimum value of potential energy that involves even the gravitational potential of the system.

In this study, we formulated a 2D elastic model of the soft fingertip, which is a function of three independent variables including even a tangential deformation caused by object slide motions on the fingertip. All of the above observations and relevant definitions were derived on the basis of the extended 2D model. Some qualitative and analytical explanations of the LMEEwC were also added based on the model in terms of how a unique equilibrium state of elastic energy is generated during manipulation. In addition, we extended the 2D model and newly derived a 3D fingertip model required for the dynamic analysis of the entire system of robotic hands in 3D space. In this process, quaternion descriptions were used in an attempt to avoid trigonometric functions that can be complicated in a series of dynamic analyses, resulting in a reduction in computation time. The 3D fingertip model newly contains an additional tangential displacement normal to another displacement already used in the 2D model. The robot configuration in the 3D space was constructed as a set of three fingers that are placed so as to form an equilateral triangular arrangement. Even in this case, *conformable manipulation* could be achieved in the sense that the dynamic information of a target object pinched by the three fingers was not required for designing control strategies for free manipulation movements.

12.2 Future Work

To clarify the dynamics of manipulation movements of soft-fingered robotic hands in 3D space, in the last chapter we formulated three-fingered handling and executed a numerical simulation for free manipulation motions. This simulation focused on neither the direct control of the pinched object nor the achievement of sophisticated tasks, but rather only on the stable pinch performances accomplished by simple joint angle control. Therefore, we must realize the position and orientation control of the pinched object by making use of a three-fingered robotic hand and should indicate an optimal control strategy for fine and dexterous manipulations in 3D space. In addition, in both cases of 2D and 3D handling problems, we must demonstrate the control law in actual manipulation experiments such as peg-in-hole tasks in contact with real environments.

Appendix A
Static Modeling of Fingertips

A.1 Contact Plane Formula

As illustrated in Fig. 3.1, point C with respect to the origin O is described in vector form as

$$\overrightarrow{\mathrm{OC}} = \begin{bmatrix} (a-d)\sin\theta_{\mathrm{p}} \\ 0 \\ (a-d)\cos\theta_{\mathrm{p}} \end{bmatrix}. \tag{A.1}$$

In addition, a normal unit vector with respect to the contact surface is represented as

$$\mathrm{n} = \begin{bmatrix} \sin\theta_{\mathrm{p}} \\ 0 \\ \cos\theta_{\mathrm{p}} \end{bmatrix}. \tag{A.2}$$

Since the contact plane can be written in an inner product form as

$$\left\{[x, y, z]^T - \overrightarrow{\mathrm{OC}}\right\} \cdot \mathrm{n} = 0, \tag{A.3}$$

the plane equation is described as follows:

$$x\sin\theta_{\mathrm{p}} + z\cos\theta_{\mathrm{p}} = a - d. \tag{A.4}$$

A.2 Spring Constant Formulation

Linear elasticity relates stress σ and strain ϵ as follows:

$$\sigma = E\epsilon, \tag{A.5}$$

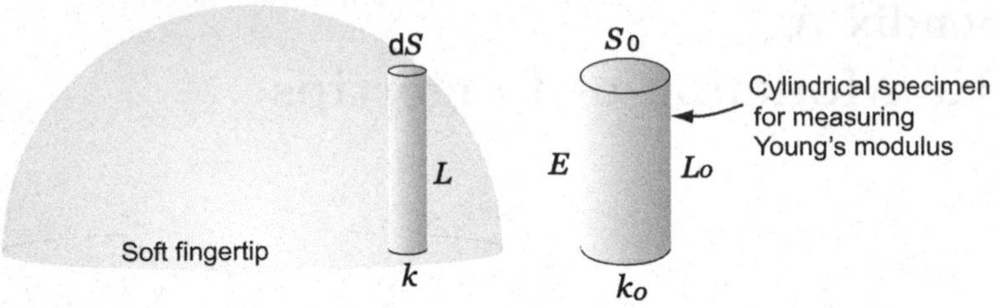

Fig. A.1 Spring constant inside the soft fingertip, where the Young modulus E is identified by the basic compression test of specimens shown in Fig. 3.6a

where E denotes Young's modulus. As shown in Fig. A.1, let k_0, $\mathrm{d}S_0$, and L_0 be the spring constant, the sectional area, and the natural length of a specimen, respectively, for measuring Young's modulus. Letting F denote a force applied to the specimen and δx a displacement in the identification test, we have

$$\sigma = \frac{F}{S_0}, \quad \epsilon = \frac{\delta x}{L_0}, \quad F = k_0 \delta x. \tag{A.6}$$

Then, Young's modulus is described as

$$E = \frac{\sigma}{\epsilon} = \frac{F}{S_0}\frac{L_0}{\delta x} = k_0 \frac{L_0}{S_0}. \tag{A.7}$$

Since in Chaps. 3–8 it is assumed that Young's modulus is an invariant physical value for an individual material, the following equation is satisfied, as illustrated in Fig. A.1:

$$k \frac{L}{\mathrm{d}S} = k_0 \frac{L_0}{S_0} = E, \tag{A.8}$$

which directly yields

$$k = E \frac{\mathrm{d}S}{L} = \frac{E \mathrm{d}S}{\sqrt{a^2 - (x^2 + y^2)}}. \tag{A.9}$$

A.3 Coordinate Conversion to Derive Fingertip Stiffness

As illustrated in Fig. A.2, let $\sum'$ be the coordinate system obtained by displacing O of the $\sum$ frame to O$'$, and let $\sum''$ be the cylindrical coordinate system inclined by θ_p from the z'-axis. Let r be the arbitrary radius on the contact circle of origin C, and let ϕ be the common rotational angle around the z-, z'-, and z''-axes. The relationship between (x', y') on the $\sum'$ frame and (r, ϕ) on the $\sum''$ frame is then expressed as

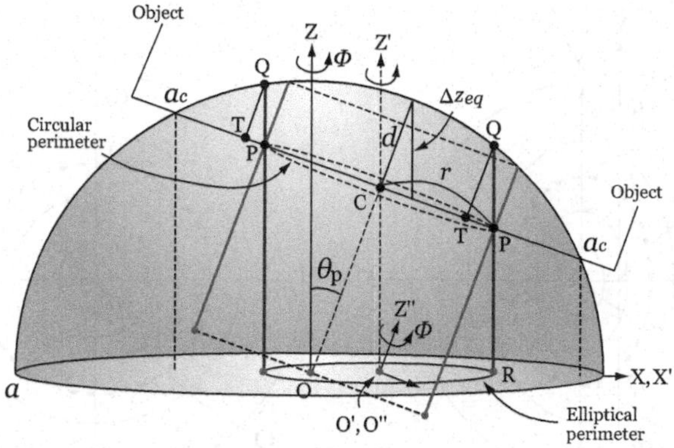

Fig. A.2 Equivalent fingertip stiffness with respect to the $\sum''$-coordinate system

$$x' = r\cos\phi\cos\theta_{\mathrm{p}}, \tag{A.10}$$

$$y' = r\sin\phi. \tag{A.11}$$

Since the relationship between (x, y) and (x', y') is described as $x = x' + (a - d)\sin\theta_{\mathrm{p}}$ and $y = y'$, the variable transformation through the coordinate systems $\sum$ and $\sum''$ can be expressed as

$$x = r\cos\phi\cos\theta_{\mathrm{p}} + (a - d)\sin\theta_{\mathrm{p}}, \tag{A.12}$$

$$y = r\sin\phi. \tag{A.13}$$

Simultaneously, the elliptical region at the bottom surface of the fingertip shown in Fig. A.2 can be converted into a circular region according to the above transformation rule, that is, the integration area of (r, ϕ) varies at $[0, a_c]$ and $[0, 2\pi]$, respectively.

Next, let us consider the physical meaning of the single integration of $B(r, \phi)$ used in Eq. 3.7 as:

$$\int_0^{2\pi} B(r,\phi)\,\mathrm{d}\phi = \int_0^{2\pi} \frac{\cos\theta_{\mathrm{p}}\,\mathrm{d}\phi}{\sqrt{a^2 - \{x^2(r,\phi) + y^2(r,\phi)\}}}. \tag{A.14}$$

Equation A.14 corresponds to the stiffness of an elliptical perimeter, the longitudinal radius of which is r, as shown in Fig. A.3a. In addition, substituting $\theta_{\mathrm{p}} = 0$ into Eq. A.14 enables an equivalent stiffness to be obtained on a circular perimeter of radius r, as shown in Fig. A.3b.

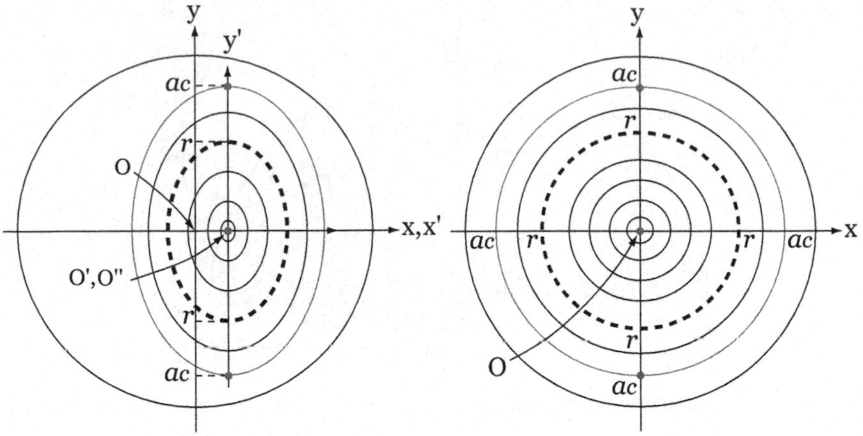

(a) Elliptical perimeter (b) Circular perimenter

Fig. A.3 Fingertip stiffness on a certain perimeter. **a** Elliptical perimeter. **b** Circular perimenter

A.4 Approximation Method for a Nonlinear Curve

As shown in Fig. 9.2b, the stress-strain curve obtained by averaging six raw data depicted in Fig. 9.2a passes through an original point on the figure. We therefore introduce a stress formula that does not include constant terms as follows:

$$\sigma(\epsilon) = \sum_{i=1}^{n} a_i \epsilon^i. \tag{A.15}$$

In what follows, we determine the maximum order n and the constant a_i used in Eq. A.15. For comparison, we apply several model functions of $n = 3, 4, 5, 6, 7$ to Eq. A.15 and plot each approximated result of the functions in Table A.1. Note that [%] is the asymptotical standard error (ASE) and corresponds to the standard error of each constant a_i. Furthermore, the root-mean-square value of the error (RSME) is used in the least-squares method as a dimensionless index to evaluate the level of approximation. For example, as the RSME of a result obtained in a test approaches zero, the approximation becomes better.

Table A.1 Asymptotical standard error

	3	4	5	6	7
ASE(a_1)	2.94	4.132	4.223e-7	6.753e-7	1.012e-6
ASE(a_2)	2.307	1.977	5.48e-7	1.232e-6	2.463e-6
ASE(a_3)	1.501	1.449	3.236e-7	1.078e-6	2.963e-6
ASE(a_4)		0.9937	2.187e-7	1.202e-6	4.806e-6
ASE(a_5)			1.396e-7	1.674e-6	1.088e-5
ASE(a_6)				71.43	67.42
ASE(a_7)					72.11
RMSE	27.1249	7.32683	2.88568e-7	2.88394e-7	2.88227e-7

Appendix B
Three-dimensional Modeling of Fingertips

B.1 Derivatives of Angular Velocity Matrix

Since

$$\frac{\partial\left[\boldsymbol{\omega}\times\right]}{\partial q_i} = \frac{\partial \omega_\xi}{\partial q_i}\frac{\partial\left[\boldsymbol{\omega}\times\right]}{\partial \omega_\xi} + \frac{\partial \omega_\eta}{\partial q_i}\frac{\partial\left[\boldsymbol{\omega}\times\right]}{\partial \omega_\eta} + \frac{\partial \omega_\zeta}{\partial q_i}\frac{\partial\left[\boldsymbol{\omega}\times\right]}{\partial \omega_\zeta},$$

$$\frac{\partial\left[\boldsymbol{\omega}\times\right]}{\partial \dot{q}_i} = \frac{\partial \omega_\xi}{\partial \dot{q}_i}\frac{\partial\left[\boldsymbol{\omega}\times\right]}{\partial \omega_\xi} + \frac{\partial \omega_\eta}{\partial \dot{q}_i}\frac{\partial\left[\boldsymbol{\omega}\times\right]}{\partial \omega_\eta} + \frac{\partial \omega_\zeta}{\partial \dot{q}_i}\frac{\partial\left[\boldsymbol{\omega}\times\right]}{\partial \omega_\zeta},$$

$$(i = 0, 1, 2, 3)$$

and

$$S_\xi \triangleq \frac{\partial\left[\boldsymbol{\omega}\times\right]}{\partial \omega_\xi} = \begin{bmatrix} 0 & 0 & 0 \\ 0 & 0 & -1 \\ 0 & 1 & 0 \end{bmatrix},$$

$$S_\eta \triangleq \frac{\partial\left[\boldsymbol{\omega}\times\right]}{\partial \omega_\eta} = \begin{bmatrix} 0 & 0 & 1 \\ 0 & 0 & 0 \\ -1 & 0 & 0 \end{bmatrix},$$

$$S_\zeta \triangleq \frac{\partial\left[\boldsymbol{\omega}\times\right]}{\partial \omega_\zeta} = \begin{bmatrix} 0 & -1 & 0 \\ 1 & 0 & 0 \\ 0 & 0 & 0 \end{bmatrix},$$

we obtain

$$\frac{\partial\left[\boldsymbol{\omega}\times\right]}{\partial q_0}=(-2\dot{q}_1)S_\xi+(-2\dot{q}_2)S_\eta+(-2\dot{q}_3)S_\zeta=\begin{bmatrix}0 & 2\dot{q}_3 & -2\dot{q}_2\\ -2\dot{q}_3 & 0 & 2\dot{q}_1\\ 2\dot{q}_2 & -2\dot{q}_1 & 0\end{bmatrix},$$

$$\frac{\partial\left[\boldsymbol{\omega}\times\right]}{\partial q_1}=(2\dot{q}_0)S_\xi+(2\dot{q}_3)S_\eta+(-2\dot{q}_2)S_\zeta=\begin{bmatrix}0 & 2\dot{q}_2 & 2\dot{q}_3\\ -2\dot{q}_2 & 0 & -2\dot{q}_0\\ -2\dot{q}_3 & 2\dot{q}_0 & 0\end{bmatrix},$$

$$\frac{\partial\left[\boldsymbol{\omega}\times\right]}{\partial q_2}=(-2\dot{q}_3)S_\xi+(2\dot{q}_0)S_\eta+(2\dot{q}_1)S_\zeta=\begin{bmatrix}0 & -2\dot{q}_1 & 2\dot{q}_0\\ 2\dot{q}_1 & 0 & 2\dot{q}_3\\ -2\dot{q}_0 & -2\dot{q}_3 & 0\end{bmatrix},$$

$$\frac{\partial\left[\boldsymbol{\omega}\times\right]}{\partial q_3}=(2\dot{q}_2)S_\xi+(-2\dot{q}_1)S_\eta+(2\dot{q}_0)S_\zeta=\begin{bmatrix}0 & -2\dot{q}_0 & -2\dot{q}_1\\ 2\dot{q}_0 & 0 & -2\dot{q}_2\\ 2\dot{q}_1 & 2\dot{q}_2 & 0\end{bmatrix},$$

and

$$\frac{\partial\left[\boldsymbol{\omega}\times\right]}{\partial \dot{q}_0}=(2q_1)S_\xi+(2q_2)S_\eta+(2q_3)S_\zeta=\begin{bmatrix}0 & -2q_3 & 2q_2\\ 2q_3 & 0 & -2q_1\\ -2q_2 & 2q_1 & 0\end{bmatrix},$$

$$\frac{\partial\left[\boldsymbol{\omega}\times\right]}{\partial \dot{q}_1}=(-2q_0)S_\xi+(-2q_3)S_\eta+(2q_2)S_\zeta=\begin{bmatrix}0 & -2q_2 & -2q_3\\ 2q_2 & 0 & 2q_0\\ 2q_3 & -2q_0 & 0\end{bmatrix},$$

$$\frac{\partial\left[\boldsymbol{\omega}\times\right]}{\partial \dot{q}_2}=(2q_3)S_\xi+(-2q_0)S_\eta+(-2q_1)S_\zeta=\begin{bmatrix}0 & 2q_1 & -2q_0\\ -2q_1 & 0 & -2q_3\\ 2q_0 & 2q_3 & 0\end{bmatrix},$$

$$\frac{\partial\left[\boldsymbol{\omega}\times\right]}{\partial \dot{q}_3}=(-2q_2)S_\xi+(2q_1)S_\eta+(-2q_0)S_\zeta=\begin{bmatrix}0 & 2q_0 & 2q_1\\ -2q_0 & 0 & 2q_2\\ -2q_1 & -2q_2 & 0\end{bmatrix}.$$

B.2 Bilinear Form of the Outer Product Matrix

Let $\left[\boldsymbol{\omega}\times\right]$ be a skew-symmetric matrix describing the outer product with vector $\boldsymbol{\omega}=\left[\omega_x,\,\omega_y,\,\omega_z\right]^{\mathrm{T}}$, which is described as

$$\left[\boldsymbol{\omega}\times\right]=\begin{bmatrix}0 & -\omega_z & \omega_y\\ \omega_z & 0 & -\omega_x\\ -\omega_y & \omega_x & 0\end{bmatrix}.$$

Recall that

$$\frac{\partial\left[\boldsymbol{\omega}\times\right]}{\partial\omega_x}=S_\xi,\qquad \frac{\partial\left[\boldsymbol{\omega}\times\right]}{\partial\omega_y}=S_\eta,\qquad \frac{\partial\left[\boldsymbol{\omega}\times\right]}{\partial\omega_z}=S_\zeta.$$

Let f be a bilinear form of matrix $\left[\boldsymbol{\omega}\times\right]$:

$$f = \boldsymbol{a}^{\mathrm{T}}\left[\boldsymbol{\omega}\times\right]\boldsymbol{b},$$

where $\boldsymbol{a}$ and $\boldsymbol{b}$ are 3D constant vectors. Differentiating the above equation with respect to ω_x, ω_y, and ω_z, we have

$$\frac{\partial f}{\partial \omega_x} = \boldsymbol{a}^{\mathrm{T}} S_\xi\, \boldsymbol{b} = a_z b_y - a_y b_z,$$
$$\frac{\partial f}{\partial \omega_x} = \boldsymbol{a}^{\mathrm{T}} S_\eta\, \boldsymbol{b} = a_x b_z - a_z b_x,$$
$$\frac{\partial f}{\partial \omega_x} = \boldsymbol{a}^{\mathrm{T}} S_\zeta\, \boldsymbol{b} = a_y b_x - a_x b_y,$$

which directly yields

$$\frac{\partial\, \boldsymbol{a}^{\mathrm{T}}\left[\boldsymbol{\omega}\times\right]\boldsymbol{b}}{\partial\, \boldsymbol{\omega}} = \boldsymbol{b}\times\boldsymbol{a}.$$

B.3 Derivatives of Relative Angle with Respect to Quaternion Elements

Recalling that vectors $\boldsymbol{n}$, $\boldsymbol{b}$, and $\boldsymbol{n}\times\boldsymbol{b}$ are unit vectors that span 3D space, the derivative $\partial\boldsymbol{n}/\partial q_i$ is described by a linear combination of these vectors as follows:

$$\frac{\partial \boldsymbol{n}}{\partial q_i} = \alpha\boldsymbol{n} + \beta\boldsymbol{b} + \gamma(\boldsymbol{n}\times\boldsymbol{b}), \tag{B.1}$$

where α, β, and γ are coefficients. Differentiating $\boldsymbol{n}^{\mathrm{T}}\boldsymbol{n} = 1$ with respect to q_i, we have

$$\left(\frac{\partial \boldsymbol{n}}{\partial q_i}\right)^{\mathrm{T}} \boldsymbol{n} = 0. \tag{B.2}$$

Substituting Eq. B.1 into the above equation with $\boldsymbol{n}^{\mathrm{T}}\boldsymbol{b} = \cos\theta_{\mathrm{p}}$ yields the following equation:

$$\alpha + \beta\cos\theta_{\mathrm{p}} = 0.$$

Differentiating $\boldsymbol{n}^{\mathrm{T}}\boldsymbol{b} = \cos\theta_{\mathrm{p}}$ with respect to q_i, we have

$$\left(\frac{\partial \boldsymbol{n}}{\partial q_i}\right)^{\mathrm{T}} \boldsymbol{b} = -\sin\theta_{\mathrm{p}}\frac{\partial\theta_{\mathrm{p}}}{\partial q_i}. \tag{B.3}$$

Substituting Eq. B.1 into the above equation yields

$$\alpha \cos\theta_{\mathrm{p}} + \beta = -\sin\theta_{\mathrm{p}} \frac{\partial \theta_{\mathrm{p}}}{\partial q_i}.$$

Consequently, we have

$$\alpha = \frac{\cos\theta_{\mathrm{p}}}{\sin\theta_{\mathrm{p}}} \frac{\partial \theta_{\mathrm{p}}}{\partial q_i}, \qquad \beta = -\frac{1}{\sin\theta_{\mathrm{p}}} \frac{\partial \theta_{\mathrm{p}}}{\partial q_i}. \tag{B.4}$$

Since $\boldsymbol{\sigma}_{\mathrm{obj},i}$ is the projection of vector $\partial \boldsymbol{n}/\partial q_i$ onto a plane defined by $\boldsymbol{n}$ and $\boldsymbol{b}$, we have

$$\boldsymbol{\sigma}_{\mathrm{obj},i} = \alpha \boldsymbol{n} + \beta \boldsymbol{b}. \tag{B.5}$$

Then,

$$\| \boldsymbol{\sigma}_{\mathrm{obj},i} \|^2 = \alpha^2 + \beta^2 + 2\alpha\beta \cos\theta_{\mathrm{p}} = \left(\frac{\partial \theta_{\mathrm{p}}}{\partial q_i} \right)^2.$$

Since $(\partial \boldsymbol{n}/\partial q_i)^{\mathrm{T}} \boldsymbol{b} = \boldsymbol{\sigma}_{\mathrm{obj},i}^{\mathrm{T}} \boldsymbol{b}$, we have

$$\boldsymbol{\sigma}_{\mathrm{obj},i}^{\mathrm{T}} \boldsymbol{b} = -\sin\theta_{\mathrm{p}} \frac{\partial \theta_{\mathrm{p}}}{\partial q_i}. \tag{B.6}$$

Note that $\sin\theta_{\mathrm{p}}$ is always nonnegative. Thus, the above equation suggests that $\partial\theta_{\mathrm{p}}/\partial q_i \geq 0$ when $\boldsymbol{\sigma}_{\mathrm{obj},i}^{\mathrm{T}} \boldsymbol{b} \leq 0$, while $\partial\theta_{\mathrm{p}}/\partial q_i \leq 0$ when $\boldsymbol{\sigma}_{\mathrm{obj},i}^{\mathrm{T}} \boldsymbol{b} \geq 0$. As a result,

$$\frac{\partial \theta_{\mathrm{p}}}{\partial q_i} = \begin{cases} \| \boldsymbol{\sigma}_{\mathrm{obj},i} \| & \text{if } \boldsymbol{\sigma}_{\mathrm{obj},i}^{\mathrm{T}} \boldsymbol{b} \leq 0, \\ -\| \boldsymbol{\sigma}_{\mathrm{obj},i} \| & \text{if } \boldsymbol{\sigma}_{\mathrm{obj},i}^{\mathrm{T}} \boldsymbol{b} \geq 0. \end{cases} \tag{B.7}$$

B.4 Derivatives of Relative Angle with Respect to Finger Angle

Similar to Appendix B.3, we have

$$\frac{\partial \boldsymbol{b}}{\partial \theta_{\mathrm{fin}}} = \alpha' \boldsymbol{b} + \beta' \boldsymbol{n} + \gamma' (\boldsymbol{b} \times \boldsymbol{n}), \tag{B.8}$$

where α', β', and γ' are coefficients. Differentiating $\boldsymbol{b}^{\mathrm{T}}\boldsymbol{b} = 1$ with respect to q_i, we have

$$\left(\frac{\partial \boldsymbol{b}}{\partial \theta_{\mathrm{fin}}} \right)^{\mathrm{T}} \boldsymbol{b} = 0. \tag{B.9}$$

Substituting Eq. B.8 into the above equation with $\boldsymbol{b}^{\mathrm{T}}\boldsymbol{n} = \cos\theta_{\mathrm{p}}$ yields the following equation:

$$\alpha' + \beta' \cos\theta_{\mathrm{p}} = 0.$$

Differentiating $\boldsymbol{b}^{\mathrm{T}}\boldsymbol{n} = \cos\theta_{\mathrm{p}}$ with respect to θ_{fin}, we have

$$\left(\frac{\partial \boldsymbol{b}}{\partial \theta_{\mathrm{fin}}}\right)^{\mathrm{T}} \boldsymbol{n} = -\sin\theta_{\mathrm{p}} \frac{\partial \theta_{\mathrm{p}}}{\partial \theta_{\mathrm{fin}}}. \tag{B.10}$$

Substituting Eq. B.8 into the above equation yields

$$\alpha' \cos\theta_{\mathrm{p}} + \beta' = -\sin\theta_{\mathrm{p}} \frac{\partial \theta_{\mathrm{p}}}{\partial \theta_{\mathrm{fin}}}.$$

Consequently, we have

$$\alpha = \frac{\cos\theta_{\mathrm{p}}}{\sin\theta_{\mathrm{p}}} \frac{\partial \theta_{\mathrm{p}}}{\partial \theta_{\mathrm{fin}}}, \qquad \beta = -\frac{1}{\sin\theta_{\mathrm{p}}} \frac{\partial \theta_{\mathrm{p}}}{\partial \theta_{\mathrm{fin}}}. \tag{B.11}$$

Since $\boldsymbol{\sigma}_{\mathrm{fin}}$ is the projection of vector $\partial \boldsymbol{b}/\partial \theta_{\mathrm{fin}}$ onto a plane defined by $\boldsymbol{b}$ and $\boldsymbol{n}$, we have

$$\boldsymbol{\sigma}_{\mathrm{fin}} = \alpha' \boldsymbol{b} + \beta' \boldsymbol{n}. \tag{B.12}$$

Then,

$$\| \boldsymbol{\sigma}_{\mathrm{fin}} \|^2 = (\alpha')^2 + (\beta')^2 + 2\alpha'\beta' \cos\theta_{\mathrm{p}} = \left(\frac{\partial \theta_{\mathrm{p}}}{\partial \theta_{\mathrm{fin}}}\right)^2.$$

Since $(\partial \boldsymbol{b}/\partial \theta_{\mathrm{fin}})^{\mathrm{T}}\boldsymbol{n} = \boldsymbol{\sigma}_{\mathrm{fin}}^{\mathrm{T}}\,\boldsymbol{n}$, we have

$$\boldsymbol{\sigma}_{\mathrm{fin}}^{\mathrm{T}}\,\boldsymbol{n} = -\sin\theta_{\mathrm{p}} \frac{\partial \theta_{\mathrm{p}}}{\partial \theta_{\mathrm{fin}}}. \tag{B.13}$$

Note that $\sin\theta_{\mathrm{p}}$ is always nonnegative. Thus, the above equation suggests that $\partial\theta_{\mathrm{p}}/\partial\theta_{\mathrm{fin}} \geq 0$ when $\boldsymbol{\sigma}_{\mathrm{fin}}^{\mathrm{T}}\,\boldsymbol{n} \leq 0$, while $\partial\theta_{\mathrm{p}}/\partial\theta_{\mathrm{fin}} \leq 0$ when $\boldsymbol{\sigma}_{\mathrm{fin}}^{\mathrm{T}}\,\boldsymbol{n} \geq 0$. As a result,

$$\frac{\partial \theta_{\mathrm{p}}}{\partial \theta_{\mathrm{fin}}} = \begin{cases} \| \boldsymbol{\sigma}_{\mathrm{fin}} \| & \text{if } \boldsymbol{\sigma}_{\mathrm{fin}}^{\mathrm{T}}\,\boldsymbol{n} \leq 0, \\ -\| \boldsymbol{\sigma}_{\mathrm{fin}} \| & \text{if } \boldsymbol{\sigma}_{\mathrm{fin}}^{\mathrm{T}}\,\boldsymbol{n} \geq 0. \end{cases} \tag{B.14}$$

B.5 Derivative of the Arctangent Function

Assume that $\theta = \mathrm{atan2}\,(y, x)$ and the variables x and y depend on an independent variable q. Let us compute the derivative of θ with respect to q.

From the definition of function atan2 (y, x) we have

$$
\begin{aligned}
x &= (x^2 + y^2)^{1/2} \cos\theta, \\
y &= (x^2 + y^2)^{1/2} \sin\theta.
\end{aligned} \tag{B.15}
$$

Differentiating the above equations with respect to q yields

$$
\begin{aligned}
\frac{\partial x}{\partial q} &= (x^2 + y^2)^{-1/2} \left(x\frac{\partial x}{\partial q} + y\frac{\partial y}{\partial q} \right) \cos\theta - (x^2 + y^2)^{1/2} \frac{\partial \theta}{\partial q} \sin\theta, \\
\frac{\partial y}{\partial q} &= (x^2 + y^2)^{-1/2} \left(x\frac{\partial x}{\partial q} + y\frac{\partial y}{\partial q} \right) \sin\theta + (x^2 + y^2)^{1/2} \frac{\partial \theta}{\partial q} \cos\theta.
\end{aligned} \tag{B.16}
$$

Substituting Eq. B.15 into the above, we have

$$
\begin{aligned}
\frac{\partial x}{\partial q} &= x \left(x\frac{\partial x}{\partial q} + y\frac{\partial y}{\partial q} \right) (x^2 + y^2)^{-1} - y\frac{\partial \theta}{\partial q}, \\
\frac{\partial y}{\partial q} &= y \left(x\frac{\partial x}{\partial q} + y\frac{\partial y}{\partial q} \right) (x^2 + y^2)^{-1} + x\frac{\partial \theta}{\partial q}.
\end{aligned} \tag{B.17}
$$

Consequently, we have

$$
-y\frac{\partial x}{\partial q} + x\frac{\partial y}{\partial q} = (y^2 + x^2)\frac{\partial \theta}{\partial q},
$$

which directly yields

$$
\frac{\partial \theta}{\partial q} = \frac{1}{x^2 + y^2} \left\{ x\frac{\partial y}{\partial q} - \frac{\partial x}{\partial q} y \right\}. \tag{B.18}
$$

References

[AA88] H. Asada and Y. Asari. The Direct Teaching of Tool Manipulation Skills Via the Impedance Identification of Human Motions. In *Proc. IEEE Int. Conf. on Robotics and Automation*, pages 1269–1274, 1988.

[AAH88] C. An, C. Atkeson, and J. Hollerbach. Model-Based Control of a Robot Manipulator. *MIT Press*, 1988.

[ADN+02] S. Arimoto, Z. Doulgeri, P. Nguyen, and J. Fasoulas. Stable pinching by a pair of robot fingers with soft tips under the effect of gravity. *Robotica*, 20:241–249, 2002.

[AM85] S. Arimoto and F. Miyazaki. Asymptotic Stability of Feedback Control Laws for Robot Manipulator. In *Proc. the 1th Int. IFAC Symposium on Robot Control*, pages 221–226, 1985.

[ANH+00] S. Arimoto, P. Nguyen, H. Y. Han, and Z. Doulgeri. Dynamics and control of a set of dual fingers with soft tips. *Robotica*, 18:71–80, 2000.

[Ari96] S. Arimoto. Control Theory of Non-Linear Mechanical Systems: A Passivity-Based and Circuit-Theoretic Approach. *Oxford Univ Pr on Demand*, 1996.

[AS88] R.J. Anderson and M.W. Spong. Hybrid Impedance Control of Robotic Manipulators. *IEEE Trans. Robotics and Automation*, 4(5):549–556, 1988.

[Asa83] H. Asada. A Geometric Representation of Manipulator Dynamics and its Application to Arm Design. *ASME Journal of Dynamics Systems, Measurement, and Control*, 105:131–142, 1983.

[Asa84] H. Asada. Dynamic analysis and design of robot manipulators using inertia ellipsoids. In *Proc. IEEE Int. Conf. on Robotics and Automation*, pages 94–102, 1984.

[ATÖ+06] H. Atilla, O. Tekeli, K. Örnek, F. Batioglu, A.H. Elhan, and T. Eryilmaz. Pattern electroretinography and visual evoked potentials in optic nerve diseases. *Journal of Clinical Neuroscience*, 13(1):55–59, 2006.

[ATY+01] S. Arimoto, K. Tahara, M. Yamaguchi, P. Nguyen, and H. Y. Han. Principle of superposition for controlling pinch motions by means of robot fingers with soft tips. *Robotica*, 19:21–28, 2001.

[BAC+84] E. Bizzi, N. Accornero, W. Chapple, and N. Hogan. Posture Control and Trajectory Formation During Arm Movement. *Journal of Neuroscience*, 4(11):2738–2744, 1984.

[Bau72] J. Baumgarte. Stabilization of constraints and integrals of motion in dynamical systems. *Computer Methods in Applied Mechanics and Engineering*, 1:1–16, 1972.

[BHJ+82] M. Brady, J. Hollerbach, T. Johnson, T. Lozano-Perez, and M. Mason. Robot Motion: Planning and Control. *MIT Press*, Reprinted papers, 1982.

[BPM76] E. Bizzi, A. Polit, and P. Morasso. Mechanisms underlying achievement of final head position. *Journal of Neurophysiology*, 39:435–444, 1976.

[Bur81] R.E. Burke. Motor units: anatomy, physiology, and functional organization. In Handbook of Physiology, *American Physiological Society*, Vol.2, pages 345–422, 1981.

[CB04] J. Cho and P. Benkeser. Elastically deformable model-based motion-tracking of left ventricle. In *Proc. IEEE Int. Conf. Engineering in Medicine and Biology Society*, pages 1925–1928, 2004.

[CC95] D. Chang and M. Cutkosky. Rolling with deformable fingertips. In *Proc. IEEE/RSJ Int. Conf. Intelligent Robots and Systems*, pages 194–199, 1995.

[CKK+68] S. H. Crandall, D. C. Karnopp, E. F. Kurts, and D. C. Pridmore-Brown. Dynamics of Mechanical and Electromechanical Systems. *McGraw-Hill*, 1968.

[CGV08] R.H. Clewley, J.M. Guckenheimer, and F.J. Valero-Cuevas. Estimating Effective Degrees of Freedom in Motor Systems. *IEEE Trans. on Biomedical Engineering*, 55(2):430–442, 2008.

[CHS89] A.B.A. Cole, J.E. Hauser, and S.S. Sastry. Kinematics and Control of Multifingered Hands with Rolling Contact. *IEEE Trans on Automatic Control*, 34(4):398–404, 1989.

[CK89] M. Cutkosky and I. Kao. Computing and controlling the compliance of a robotic hand. *IEEE Trans. Robotics and Automation*, 5(2):151–165, 1989.

[CLC+81] W.P. Cooney and M.J. Lucca and E.Y.S. Chao and R.L. Linscheid. The Kinesiology of the Thumb Trapeziometacarpal Joint. *Journal of Bone and Joint Surgery*, 63(9):1371–1381, 1981.

[CM06] L.Y. Chang and Y. Matsuoka. A Kinematic Thumb Model for the ACT Hand. In *Proc. IEEE Int. Conf. Robotics and Automation*, pages 1000–1005, 2006.

[Coo79] J.D. Cooke. Dependence of human arm movements on limb mechanical properties. *Brain Research*, 165:366–369, 1979.

[CR79] J. Craig and M. Raibert. A systematic method of hybrid position/force control of a manipulator. In *Proc. IEEE Int. Conf. on Computer Software and Applications*, pages 446–451, 1979.

[CW86] M.R. Cutkosky and P.K. Wright. Friction, Stability and the Design of Robotic Fingers. *Int. Journal of Robotics Research*, 5(4):20–37, 1986.

[Dev00] S.R. Devasahayam. Signals and Systems in Biomedical Engineering. *Plenum Pub Corp*, 2000.

[DF03] Z. Doulgeri and J. Fasoulas. Grasping control of rolling manipulation with deformable fingertips. *IEEE/ASME Trans. Mechatronics*, 8(2):283–286, 2003.

[DFA02] Z. Doulgeri, J. Fasoulas, and S. Arimoto. Feedback control for object manipulation by a pair of soft tip fingers. *Robotica*, 20:1–11, 2002.

[Feh68] E. Fehlberg. Classical Fifth-, Sixth-, Seventh, and Eighth-Order Runge-Kutta Formulas with Stepsize Control. *NASA Technical Report*, NASA TR R-287, October, 1968.

[Feh69] E. Fehlberg. Low-order classical Runge-Kutta formulas with stepsize control and their application to some heat transfer problems. *NASA Technical Report*, NASA TR R-315, July, 1969.

[Fel66] A.G. Feldman. Functional tuning of the nervous system during control of movement or maintenance of a steady posture III: Mechanographic analysis of the execution by man of the simplest motor tasks. *Biophysics*, 11:766–775, 1966.

[GHB+95] D.J. Giurintano, A.M. Hollister, W.L. Bufold, D.E. Thompson, and L.M. Myers. A virtual five-link model of the thumb. *Medical Engineering and Physics*, 17(4):297–303, 1995.

[GK96] H. Gomi and M. Kawato. Equilibrium-point control hypothesis examined by measured arm-stiffness during multi-joint movement. *Science*, 272(5258):117–120, 1996.

[Gol88] A.A. Goldenberg. Implementation of force and impedance control in robot manipulators. In *Proc. IEEE Int. Conf. on Robotics and Automation*, pages 1626–1632, 1988.

[GPS02] H. Goldstein, C. P. Poole, and J. L. Safko. Classical Mechanics, Third Edition. Chapter 2 Variational Principles and Lagrange's Equations, *Addison–Wesley*, 2002.

[HA77a] H. Hanafusa and H. Asada. A robot hand with elastic fingers and its application to assembly process. In *IFAC Symp. on Information and Control Problems in Manufacturing Technology*, pages 127–138, 1977.

[HA77b] H. Hanafusa and H. Asada. Stable prehension by a robot hand with elastic fingers. In *Proc. 7th Int. Symp. on Industrial Robots*, pages 361–368,1977.

[HAK+01] H. Y. Han, S. Arimoto, M. Yamaguchi, K. Tahara, and P. Nguyen. Robotic pinching by means of a pair of soft fingers with sensory feedback. In *Proc. IEEE Int. Conf. Robotics and Automation*, pages 97–102, 2001.

[HBM+92] A. Hollister, W.L. Bufold, L.M. Myers, D.J. Giurintano, and A. Novick. The Axes of Rotation of the Thumb Carpometacarpal Joint. *Journal of Orthopaedic Research*, 10(3):454–460, 1992.

[Hes69] M. R. Hestenes. Multiplier and Gradient Methods. *Journal of Optimization Theory and Applications*, Vol. 4, No. 5, pp.303–320, Nov., 1969.

[HH00] A.J. Hodgson and N. Hogan. A model-independent definition of attractor behavior applicable tointeractive tasks. *IEEE Trans. Systems, Man, and Cybernetics, Part C: Applications and Reviews*, 30(1):105–118, 2000.

[HK03] J. Hamill and K.M. Knutzen. Biomechanical Basis of Human Movement 2nd Ed.. *Lippincott Williams & Wilkins*, Chap. 4, 2003.

[Hog80] N. Hogan. Mechanical impedance control in assistive devices and manipulators. In *Proc. on Joint Automatic Control Conference*, pages TA10–B, 1980.

[Hog84a] N. Hogan. Impedance Control of Industrial Robots. *Robotics & Computer-Integrated Manufacturing*, 1(1):97–113, 1984.

[Hog84b] N. Hogan. Adaptive Control of Mechanical Impedance by Coactivation of Antagonist Muscles. *IEEE Trans. Automatic Control*, 29(8):681–690, 1984.

[Hog85a] N. Hogan. Impedance control: An approach to manipulation: Part 1 - theory. *ASME Journal of Dynamics Systems, Measurement, and Control*, 107(1):1–7, 1985.

[Hog85b] N. Hogan. Impedance control: An approach to manipulation: Part 2 - implementation. *ASME Journal of Dynamics Systems, Measurement, and Control*, 107(1):8–16, 1985.

[Hog85c] N. Hogan. Impedance control: An approach to manipulation: Part 3 - application. *ASME Journal of Dynamics Systems, Measurement, and Control*, 107(1):17–24, 1985.

[Hog87] N. Hogan. Stable Execution of Contact Tasks Using Impedance Control. In *Proc. IEEE Int. Conf. on Robotics and Automation*, pages 1047–1054, 1987.

[Hog88] N. Hogan. On the Stability of Manipulators Performing Contact Tasks. *IEEE Trans. Robotics and Automation*, 4(6):677–686, 1988.

[Hol90] O. Holmes. Human Neurophysiology. *Unwin Hyman*, 1990.

[IH06] T. Inoue and S. Hirai. Elastic model of deformable fingertip for soft-fingered manipulation. *IEEE Trans. Robotics*, 22(6):1273–1279, 2006.

[IH07a] T. Inoue and S. Hirai. A Two-phased Object Orientation Controller on Soft Finger Operations. In *Proc. IEEE Int. Conf. Intelligent Robots and Systems*, pages 2528–2533, 2007.

[IH07b] T. Inoue and S. Hirai. Dynamic Stable Manipulation via Soft-fingered Hand. In *Proc. IEEE Int. Conf. Robotics and Automation*, pages 586–591, 2007.

[Ino71] H. Inoue. Computer controlled bilateral manipulator. *Bulletin of the Japan Society of Mechanical Engineers*, 14(69):199–207, 1971.

[JE93] J. Jalón and E.Bayo. *Kinematic and Dynamic Simulation of Multibody Systems*. Springer Verlag, 1993. Chaper 5.

[JR88] Z. Ji and B. Roth. Direct Computation of Grasping Force for Three-Finger Tip-Prehension Grasps. *ASME Trans. on Mechanisms, Transmissions, and Automation in Design*, 110:405–413, 1988.

[JJ82] J.Salisbury and J.Craig. Articulated hands: Force control and kinematic issues. *Int. Journal of Robotics Research*, 1(1):4–17, 1982.

[Joh85] J. L. Johnson. Contact Mechanics. *Cambridge University Press*, 1985.

[Kap07] A.I. Kapandji. The Physiology of the Joints, Volume 1: Upper Limb. *Churchill Livingstone*, 6th edition, 2007.

[Kaz88] H. Kazerooni. Direct-Drive Active Compliant End Effector (Active RCC). *IEEE Trans. Robotics and Automation*, 4(3):324–333, 1988.

[Kel77] J.A.S. Kelso. Motor control mechanisms underlying human movement reproduction. *Journal of Experimental Psychology*, 3:529–543, 1977.

[KH80] J.A.S. Kelso and K.G. Holt. Exploring a vibratory system analysis of human movement production. *Journal of Neurophysiology*, 43:1183–1196, 1980.

[KHS86a] H. Kazerooni, P. Houpt, and T. Sheridan. The fundamental concepts of robust compliant motion for robot manipulators. In *Proc. IEEE Int. Conf. on Robotics and Automation*, pages 418–427, 1986.

[KHS86b] H. Kazerooni, P. Houpt, and T. Sheridan. Robust compliant motion for manipulators, part 2: Design method. *IEEE Trans on Robotics and Automation*, 2(2):93–105, 1986.

[Kob85] H. Kobayashi. Control and Geometrical Considerations for an Articulated Robot Hand. *Int. Journal of Robotics Research*, 4(1):3–12, 1985.

[KR86] J. Kerr and B. Roth. Analysis of Multifingered Hands. *Int. Journal of Robotics Research*, 4(4):3–17, 1986.

[KSH86] H. Kazerooni, T. Sheridan, and P. Houpt. Robust compliant motion for manipulators, part 1: The fundamental concepts of compliant motion. *IEEE Trans on Robotics and Automation*, 2(2):83–92, 1986.

[Kui02] J.B. Kuipers. Quaternions and Rotation Sequences: A Primer with Applications to Orbits, Aerospace, and Virtual Reality. *Princeton University Press*, 2002.

[KY04] I. Kao and F. Yang. Stiffness and contact mechanics for soft fingers in grasping and manipulation. *IEEE Trans. Robotics and Automation*, 20(1):132–135, 2004.

[Lak78] K. Lakshminarayana. Mechanics of Form Closure. *ASME Technical Paper*, 78-DET-32, 1978.

[LFP83] J. Luh, W. Fisher, and R. Paul. Joint torque control by a direct feedback for industrial robots. *IEEE Trans on Automatic Control*, 28(2):153–161, 1983.

[LHS89] Z. Li, P. Hsu, and S. Sastry. Grasping and coordinated manipulation by a multifingered robot hand. *Int. Journal of Robotics Research*, 8(4):33–50, 1989.

[LS88] Z. Li and S.S. Sastry. Taks-Oriented Optimal Grasping by Multifingered Robot Hands. *IEEE Trans. Robotics and Automation*, 4(1):32–44, 1988.

[LWP80] J. Luh, M. Walker, and R. Paul. Resolved-acceleration control of mechanical manipulators. *IEEE Trans on Automatic Control*, 25(3):468–474, 1980.

[MA85] F. Miyazaki and S. Arimoto. Sensory Feedback for Robot Manipulators. *Journal of Robotic Systems*, 2(1):53–71, 1985.

[Mas81] M. Mason. Compliance and force control for computer-controlled manipulators. *IEEE Trans on Systems, Man, and Cybernetics*, 11(6):418–432, 1981.

[MKT92] H. Maekawa, K. Komoriya, and K. Tanie. Manipulation of an unknown object by multifingered hands with rolling contact using tactile feedback. In *Proc. IEEE Int. Conf. Intelligent Robots and Systems*, pages 1877–1882, 1992.

[MLS94] R. Murray, Z. Li, and S. Sastry. *A Mathematical Introduction to Robotic Manipulation*. CRC Press, 1994.

[MS85] M. Mason and J. Salisbury. *Robot Hands and the Mechanics of Manipulation*. MIT Press, 1985.

[MT96] T. McInerney and D. Terzopoulos. Deformable models in medical image analysis: A survey. *Medical Image Analysis*, 1(2):91–108, 1996.

[MTK+92] H. Maekawa, K. Tanie, K. Komoriya, M. Kaneko, C. Horiguchi, and T. Sugawara. Development of a finger-shaped tactile sensor and its evaluation by active touch. In *Proc. IEEE Int. Conf. Robotics and Automation*, pages 1327–1334, 1992.

[NA01] P. Nguyen and S. Arimoto. Performance of pinching motions of two multi-dof robotic fingers with soft-tips. In *Proc. IEEE Int. Conf. on Robotics and Automation*, pages 2344–2349, 2001.

[Ngu86] V. Nguyen. Constructing Force-Closure Graspes. In *Proc. IEEE Int. Conf. on Robotics and Automation*, pages 1368–1373, 1986.

[Ngu87] V. Nguyen. Constructing Force-Closure Graspes in 3D. In *Proc. IEEE Int. Conf. on Robotics and Automation*, pages 240–245, 1987.

[Ngu88] V. Nguyen. Constructing force-closure grasps. *Int. Journal of Robotics Research*, 7(3):3–16, 1988.

[NH76] T.R. Nichols and J.C. Houk. Improvement in linearity and regulation of stiffness that results from actions of stretch reflex. *Journal of Neurophysiology*, 39:119–142, 1976.

[NM65] J. A. Nelder and R. Mead. A simplex method for function minimization. *Computer Journal*, Vol. 7, pp.308–313, 1965.

[NNY89] Y. Nakamura, K. Nagai, and T. Yoshikawa. Dynamics and Stability in Coordination of Multiple Robotic Mechanisms. *Int. Journal of Robotics Research*, 8(2):44–61, 1989.

[NW73] J. Nevins and D. Whitney. The force vector assembler concept. In *Proc. 1st Int. Conf. on Robots and Manipulator Systems*, pages 273–288, 1973.

[NW75] J. Nevins and D. Whitney. *Adaptable-Programmable Assembly Systems: An Information and Control Problem.* The Charles Stark Draper Laboratory, Inc., 1975.

[Pau72] R. Paul. *Modelling, Trajectory Calculation and Servoing of a Computer Controlled Arm.* Stanford Univ. Calif Dept. of Computer Science, thesis edition, 1972.

[Pau79a] R. Paul. Manipulator cartesian path control. *IEEE Trans on Systems, Man, and Cybernetics*, 9(11):702–711, 1979.

[Pau79b] R. Paul. Robots, models, and automation. *IEEE Magazine Computer*, 12(7):19–27, 1979.

[PB79] A. Polit and E. Bizzi. Characteristics of the motor programs underlying arm movements in monkeys. *Journal of Neurophysiology*, 42:183–194, 1979.

[PH86] T. Park and E. Haug. A hybrid numerical integration method for machine dynamic simulation. *ASME Trans. on Mechanisms, Transmissions, and Automation in Design*, 108:211–216, 1986.

[PS76] R. Paul and B. Shimano. Compliance and control. In *Proc. on Joint Automatic Control Conference*, pages 694–699, 1976.

[RC81] M. Raibert and J. Craig. Hybrid position/force control of manipulators. *ASME Journal of Dynamics Systems, Measurement, and Control*, 102:126–133, 1981.

[RH78] M. Raibert and B. Horn. Manipulator control using the configuration space method. *Industrial Robot*, 5:69–73, 1978.

[RK76] F. Reuleaux and A.B.W. Kennedy (ed. & trans.). Kinematics of Machinery. *London: Macmillan and Co.*, access online, 1876.

[Sal80] J. Salisbury. Active stiffness control of a manipulator in cartesian coordinates. In *Proc. IEEE Int. Conf. on Decision and Control*, pages 95–100, 1980.

[Sch19] G. Schlesinger. *Der Mechanische Aufbau der künstlichen Glieder, Part II of Ersatzglieder und Arbeitshilfen.* Springer Berlin, 1919.

[SHV05] M. Spong, S. Hutchinson, and M. Vidyasagar. *Robot Modeling and Control.* John Wiley & Sons Inc, 2005.

[Ski75] F. Skinner. Designing a multiple prehension manipulator. *Mechanical Engineering*, 97(9):30–37, 1975.

[SL85] K. Shin and C. Lee. Compliant control of robotic manipulators with resolved acceleration. In *Proc. IEEE Int. Conf. on Decision and Control*, pages 350–357, 1985.

[SL89] S. Sastry and Z. Li. Robot Motion Planning with Nonholonomic Constraints. In *Proc. IEEE Int. Conf. on Decision and Control*, pages 211–216, 1989.

[SR83] J.K. Salisbury and B. Roth. Kinematic and Force Analysis of Articulated Mechanical Hands. *ASME Trans. on Mechanisms, Transmissions, and Automation in Design*, 105:35–41, 1983.

[TA81] M. Takegaki and S. Arimoto. A New Feedback Method for Dynamic Control of Manipulators. *ASME Journal of Dynamics Systems, Measurement, and Control*, 102:119–125, 1981.

[Tak74] K. Takase. The design of an articulated manipulator with torque control ability. In *Proc. 4th Int. Symp. on Industrial Robots*, pages 261–270, 1974.

[Tak77] K. Takase. Task-oriented variable control of manipulator and its software servoing system. In *Proc. IFAC Symp. on Information Control in Manufacturing*, pages 139–145, 1977.

[Tat91] Y. Tatara. On compression of rubber elastic sphere over a large range of displacements-part i:theoretical study. *ASME J. Engineering Materials and Technology*, 113:285–291, 1991.

[TGI76] K. Takeyasu, T. Goto, and T. Inoyama. Precision insertion control robot and its application. *ASME Journal of Engineering for Industry*, 98(4):1313–1317, 1976.

[TS55] C. Taylor and R. Schwarz. The anatomy and mechanics of the human hand. *Artificial Limbs*, 2(2):22–35, 1955.

[TVF+93] S. Teukolsky, W. Vetterling, B. Flannery, and W. Press. *Numerical Recipes in C.* Cambridge University Press, 1993. Chapter 10.

[VJT03] F.J. Valero-Cuevas, M.E. Johanson, and J.D. Towles. Towards a realistic biomechanical model of the thumb: the choice of kinematic description may be more critical than the solution method or the variability/uncertainly of musculoskeletal parameters. *Journal of Biomechanics*, 36:1019–1030, 2003.

[Whi76] D. Whitney. Force feedback control of manipulator fine motions. In *Proc. on Joint Automatic Control Conference*, pages 687–693, 1976.

[Whi77] D. Whitney. Force feedback control of manipulator fine motions. *ASME Journal of Dynamics Systems, Measurement, and Control*, 99(2):91–97, 1977.

[WP80] C. Wu and R. Paul. Manipulator compliance based on joint torque control. In *Proc. IEEE Int. Conf. on Decision and Control*, pages 88–94, 1980.

[XBK00] N. Xydas, M. Bhagavat, and I. Kao. Study of soft-finger contact mechanics using finite elements analysis and experiments. In *Proc. IEEE Int. Conf. Robotics and Automatation*, pages 2179–2184, 2000.

[XK99] N. Xydas and I. Kao. Modeling of contact mechanics and friction limit surfaces for soft fingers in robotics, with experimental results. *Int. Journal of Robotics Research*, 18(8):941–950, 1999.

[YAB+06] M. Yoshida, S. Arimoto, J.-H. Bae, and Y. Kishi. Modeling and Computer Simulation of 3D Object Grasping and Manipulation by Dual Fingers under Nonholonomic Constraints. in *2006 IEEE/RSJ Int. Conf. on Intelligent Robots and Systems*, pages 5675–5681, 2006.

[YAB08] M. Yoshida, S. Arimoto, and Z.-W. Luo. Three-Dimensional Object Manipulation by Two Robot Fingers with Soft Tips and Minimum D.O.F.. in *2008 IEEE Int. Conf. on Robotics and Automation*, pages 4707–4714, 2008.

[YN87] T. Yoshikawa and K. Nagai. Manipulating and Grasping Forces in Manipulation by Multi-fingered Hands. In *Proc. IEEE Int. Conf. on Robotics and Automation*, pages 1998–2004, 1987.

[YN88] T. Yoshikawa and K. Nagai. Evaluation and Determination of Grasping Forces for Multi-fingered Hands. In *Proc. IEEE Int. Conf. on Robotics and Automation*, pages 245–248, 1988.

[Yos84] T. Yoshikawa. Analysis and Control of Robot Manipulators with Redundancy. In *Proc. First Int. Symposium of Robotics Research*, pages 735–747, 1984.

[Yos85] T. Yoshikawa. Dynamic Manipulability of Robot Manipulators. *Journal of Robotic Systems*, 2(1):113–124, 1985.

[YSY99] Y. Yokokohji, M. Sakamoto, and T. Yoshikawa. Vision-Aided Object Manipulation by a Multifingered Hand with Soft Fingertips. In *Proc. IEEE Int. Conf. Robotics and Automation*, pages 3201–3208, 1999.

[YSY00] Y. Yokokohji, M. Sakamoto, and T. Yoshikawa. Object manipulation by soft fingers and vision. In *Proc. 9th Int. Symposium of Robotics Research*, pages 365–374, 2000.

Index